AF388976

Encyclopedia of ZEMCH Research and Development

Encyclopedia of ZEMCH Research and Development

Editors

Masa Noguchi
Antonio Frattari
Carlos Torres Formoso
Haşim Altan
John Odhiambo Onyango
Jun-Tae Kim
Kheira Anissa Tabet Aoul
Mehdi Amirkhani
Sara Jane Wilkinson
Shaila Bantanur

MDPI • Basel • Beijing • Wuhan • Barcelona • Belgrade • Manchester • Tokyo • Cluj • Tianjin

Editors

Masa Noguchi
The University of Melbourne
Melbourne, VIC, Australia

Antonio Frattari
University of Trento
Trento, Italy

Carlos Torres Formoso
Federal University of Rio
Grande do Sul
Porto Alegre, Brazil

Haşim Altan
Prince Mohammad bin Fahd
University
Dhahran, Saudi Arabia

John Odhiambo Onyango
University of Notre Dame
Notre Dame, IN, USA

Jun-Tae Kim
Kongju National University
Cheonan, Republic of Korea

Kheira Anissa Tabet Aoul
United Arab Emirates
University
Al Ain, United Arab Emirates

Mehdi Amirkhani
University of South Australia
Adelaide, SA, Australia

Sara Jane Wilkinson
University of Technology
Sydney
Sydney, NSW, Australia

Shaila Bantanur
BMS School of Architecture
Bengaluru, India

Editorial Office
MDPI
St. Alban-Anlage 66
4052 Basel, Switzerland

This is a reprint of articles from the Topical Collection published online in the open access journal *Encyclopedia* (ISSN 2673-8392) (available at: https://www.mdpi.com/journal/encyclopedia/topical_collections/ZEMCH).

For citation purposes, cite each article independently as indicated on the article page online and as indicated below:

LastName, A.A.; LastName, B.B.; LastName, C.C. Article Title. *Journal Name* **Year**, *Volume Number*, Page Range.

ISBN 978-3-0365-8144-6 (Hbk)
ISBN 978-3-0365-8145-3 (PDF)

Contents

About the Editors

Masa Noguchi

Dr. Masa Noguchi is an Associate Professor in Environmental Design at the Faculty of Architecture, Building and Planning, University of Melbourne. He specialises in environmental experience design (EXD) decision-making analysis based on a mass customisation framework that embraces machine learning and value engineering techniques for the improvement of operational energy efficiency, affordability, and occupants' wellbeing in built environments. In parallel to EXD studies, he also initiated the global movement of the 'Zero Energy Mass Custom home' (ZEMCH), and researches and develops vertical village/subdivision plug-in housing systems for the evolution of future-proof cities.

Antonio Frattari

Prof. Dr. Antonio Frattari was an Associate Professor at the University of Roma "La Sapienza" from 1975 to 1986. From 1986, he was a Professor at University of Trento-Italy (UNITN) until retiring in 2017. From 2017, he has been a teaching fellow of wooden architecture at UNITN. He was head of the UNITN Laboratory of Building Design (LBD) from 1986 to 2017, and head of the UNITN University Centre for Smart Buildings (CUNEDI) from 2004 to 2017. He is a member of the 'Zero Energy Mass Custom Home' (ZEMCH) group, a member of the International Association for Housing Science (IAHS), and a member of the International Council for Monument and Sites (ICOMOS). His current research focuses on sustainable building; on buildings for emergencies; on building automation for enhancing quality of living for senior citizens and people living with impairments; and on the conservation and reclamation of wooden buildings. His main research topics are: emergency buildings, design for all, sustainable architecture, active building design, wooden buildings, conservation, and reclamation of wooden buildings. He is the author of more than 250 publications, including monographs, articles in national and international journals, and papers on conference proceedings.

Carlos Torres Formoso

Prof. Dr. Carlos Formoso is a Professor of Construction Management at the Federal University of Rio Grande do Sul (UFRGS), Brazil. He has a B.Sc in Civil Engineering (UFRGS) and an M.Sc in Construction Engineering (UFRGS), and was awarded a PhD by the University of Salford, UK. His main research interests are lean construction, mass customisation, production planning and control, safety management, social housing, and value management. For more than thirty years, over his academic career, he has developed several research projects in partnership with industry actors, including contractors, real estate developers, housing funding agencies, and companies that deliver prefabricated systems. Regarding international academic collaborations, he has been a visiting scholar in several universities and research institutions outside Brazil, including the University of California, Berkeley (USA), NBRI and University of Agder (Norway), Strathclyde University, University of Huddersfield, and University of Salford (UK), and ESPE (Ecuador). Also, he is currently a member of the editorial boards of six international scientific journals. He has been a senior advisor of the Brazilian Ministry of the Cities for the National Program for Quality and Productivity in the Habitat (PBQP-H) for eleven years. Prof. Formoso has published more than 110 journal papers and around 40 books and book chapters. He has also supervised 113 MSc dissertations and 25 PhD theses.

Haşim Altan

Prof. Dr. Haşim Altan is a Professor of Sustainable Design in the College of Architecture and Design (COAD) at Prince Mohammad bin Fahd University (PMU) in Dhahran, Saudi Arabia. He is a chartered architect (RIBA) and a chartered engineer (CIBSE) with over 20 years of academic and practice experience in the field of architecture, engineering, and construction (AEC) in built environments in the UK, Europe, the Middle East, and North Africa (MENA) regions. Prof.Dr. Altan sits on a number of editorial boards and reviews project proposals for the European Commission, UK Research Councils, and Qatar National Research Fund (QNRF) by the Qatar Foundation. Since 2004, he has, singly or jointly, secured and directed 32 research grants worth over GBP 21 million. He is a founding member of the International Network on 'Zero Energy Mass Custom Home' (ZEMCH), which has so far organised nine international conferences, several design workshops, and numerous technical visits. As well as having supervised 17 successful PhD theses (11 as first supervisor), and over 80 Masters (MSc, March, and MEng) dissertations, Prof. Dr. Altan has published more than 300 publications in refereed journals and international conference proceedings, in addition to 14 book chapters and 11 books (edited/authored) published by Taylor & Francis, Springer, and others. These include titles such as "Handbook of Retrofitting High Density Residential Buildings Building Retrofitting", "Advances in Architecture, Engineering & Technology", "Smart Techniques in Urban Planning & Technology", "Transgressive Design Strategies for Utopian Cities", and "Innovating Strategies and Solutions for Urban Performance & Regeneration".

John Odhiambo Onyango

Dr. John Odhiambo Onyango is an academic, a scholar and is currently an Associate Professor of Architecture at the School of Architecture, University of Notre Dame (USA). He has a Doctorate of Philosophy in Architecture from the University of Glasgow (Scotland, UK); a Masters in Architecture degree from the University of Notre Dame (USA); a Certificate in Professional Practice & Management in Architecture (RIBA Part 3—Architect's License in the United Kingdom) from the Bartlett School of Architecture, University College London (England, UK), and a Bachelor's degree with honours in Architecture from the University of Nairobi (Kenya). John's research primarily focuses on sustainability. In a broad sense, his research takes holistic approaches to creative practices at the levels of building and urban design, that are both interdisciplinary as well as multidisciplinary in collaboration with other colleagues in allied fields such as engineering, sociology, healthcare, and IT communications. The CIB working paper for "W108: Climate Change and the Built Environment" suggested future research should focus on two main hazards: floods and heat waves. These two areas continue to be active research areas. He has participated in over 13 research projects worldwide and, since 2012, has secured over USD 730,000 in competitive research grants.

Jun-Tae Kim

Prof. Dr. Jun-Tae Kim is a Professor in the Department of Architectural Engineering and in the Postgraduate Program of Energy Systems Engineering at Kongju National University, Republic of Korea. He earned a PhD from the University of New South Wales, Sydney, Australia, we he studied solar energy architecture. He started his career at the Korea Institute of Energy Research before taking up his position at the university in 1998. He was the president of the Korea Solar Energy Society from 2016 to 2017. His main research interests are solar energy applications in buildings and affordable zero-energy buildings. He has participated in various IEA projects on zero energy solar buildings, such as SHC Task 41 'Solar Energy and Architecture', SHC/EBC Task 40/Annex 52 'Net Zero Energy Buildings', and EBC Annex 65 'Long-Term Performance of Super-Insulation in Building Components & Systems'. His current activities include working for the IEA PVPS programme, Task 15 'Enabling Framework for BIPV acceleration', and as part of the project team of the IEC/ISO/IEA for BIPV international standards. As a part of his research activities, he has also engaged in the ZEMCH ('Zero Energy Mass Custom Home') Network. He chaired the ZEMCH 2019 international conference in Seoul, Korea.

Kheira Anissa Tabet Aoul

Prof. Dr. Kheira Anissa Tabet Aoul is a Professor of Sustainable Architecture and Chair of the Architectural Engineering Department at the United Arab Emirates University (UAE). Prof. Kheira draws from over 30 years of international academic and professional experience in the USA, North Africa, and the UAE. She is engaged in education- and practice-based research, focusing on advancing the design and execution of sustainable, socio-culturally sensitive built environments, particularly under hot climate. High-performance buildings and human factors in the built environment are her major areas of expertise. In 2019, she was appointed Director of the ZEMCH Network UAE regional center. She also chaired the 2021 ZEMCH international Conference in Dubai. Kheira holds a PhD in building science from Sheffield University (UK) and she is also a LEED Accredited Professional.

Mehdi Amirkhani

Dr. Mehdi Amirkhani is an academic at the University of South Australia (UniSA). Before joining UniSA, he worked as a Research Fellow and educator at the University of Melbourne, Deakin University, Victoria University, Queensland University of Technology (QUT), Griffith University, and some universities in Iran. Mehdi has significant experience teaching environmental technology, including units centred around lighting, energy estimation, thermal comfort, acoustics (speech privacy and speech intelligibility), and computational simulation programs. In particular, his expertise includes strategies for applying environmental technologies in sustainable design and building retrofitting that facilitates occupants' health and well-being. Dr. Amirkhani is an active researcher in building environmental technologies, building systems and services, building retrofitting, sustainable design, net-zero-energy buildings, and indoor environmental quality. His PhD thesis established an innovative approach to lighting design in office buildings that optimised the visual comfort (electric and daylight) and energy consumption performance of commercial facades. Dr. Amirkhani is a Nationwide House Energy Rating Scheme (NatHERS) Accredited Assessor and has more than twelve years of practical experience as an Architect in Australia and overseas.

Sara Jane Wilkinson

Prof. Dr. Sara Jane Wilkinson is a chartered building surveyor and Australia's first female professor of property. Her transdisciplinary research program sits at the intersection of sustainability, urban development and transformation, with a focus on green cities and preparing our urban environments for the challenges of climate change. Sara is Australia's most cited property academic and holds the highest ORCID, H, and i10 scores, overall, within the discipline. Her recent projects include the development of a STAR (Sustainable Temporary Adaptive Reuse) Toolkit, the use of virtual reality to assess customer willingness to invest in residential green infrastructure, a Bushfire Retrofit toolkit for housing, the development of a prototype wallbot to inspect and monitor high-rise green walls, and the performance evaluation of hempcrete wall panels. Sara engages in trans-disciplinary research with colleagues from engineering, science, health, and business as well as built environment disciplines.

Shaila Bantanur

Prof. Dr. Shaila Bantanur assumed the role of Director at the BMS School of Architecture in August 2020. She obtained her doctorate in 2015 from The Indian Institute of Technology, Roorkee (IIT Roorkee) India, in the Department of Architecture and Planning, and was the recipient of a GATE Scholarship from Ministry of Human Resource Development (MHRD) during this time. She holds two Master's Degrees: a Master's in Architecture from Rizvi College of Architecture, Mumbai (2009) and a Master's from the Institute of Town Planners India (ITPI), New Delhi (2008). With more than 20 years of experience as a researcher, academic, and practitioner, she is a life member of the Council of Architecture (CoA, New Delhi), INTACH, ITPI, and also a member of the Indian Institute of Architects (IIA). Regarding her own research, she has made enormous contributions, connecting with colleagues nationally and internationally. Her research works have been published in and presented at in various national and international peer-reviewed journals and conferences. She is currently the project investigator for two international collaborations: one with the University of Notre Dame, Indiana, United States and the other with Kathmandu University, Nepal. Prof. Bantanur is the coordinator for Unnat Baharat Abhayan, a flagship program of the MHRD. She has been part of the Scientific Program Committee for International Conferences and has been a member of many editorial boards for national and international journals. Throughout her career, she has received several commendations and awards. To name few significant ones, she was the recipient of the INTACH UK Scholarship Award (2017) from INTACH Heritage Academy, New Delhi, received an India National Trust for Art and Cultural Heritage award of merit (South Zone) for two consecutive years (2016 and 2017) for documenting unlisted heritage structures, and was the recipient of a Technical University Dresden (TU Dresden) travel fellowship.

 *encyclopedia*

Entry

Sustainable Architecture—What's Next?

Harald N. Røstvik

City- and Regional Planning, Institute of Safety, Economics and Planning (ISØP),
Faculty of Science and Technology, University of Stavanger, 4021 Stavanger, Norway; harald.n.rostvik@uis.no

Definition: Sustainable architecture encompasses more than energy efficiency, zero carbon dioxide (CO_2) emission or renewable energy use in the built environment. It also needs to alleviate overall impacts on the natural environment or ecosystem that surrounds it. It may be argued that primitive vernacular architecture (architecture without architects) built and operated using local techniques and resources alone can be considered to be sustainable. Yet later, after the 1992 Rio Conference and its declarations, more specific definitions emerged putting weight on the rational use of land area, materials and energy, preferably local, as well as area efficient planning, economy and recyclability. The advantage of this is to reduce the ecological footprint of buildings and the climate gas emissions from a sector that represents 35–50 percent of global climate gas emissions, depending on how one counts. This paper clarifies concepts, questions cemented truths and points a way forward by asking; what's next?

Keywords: sustainable; architecture; Zero Energy Mass Custom Homes (ZEMCH); energy; efficiency; forecasting; ethics; climate change; log burning

Citation: Røstvik, H.N. Sustainable Architecture—What's Next? *Encyclopedia* **2021**, *1*, 293–313. https://doi.org/10.3390/encyclopedia1010025

Academic Editor: Haşim Altan

Received: 1 February 2021
Accepted: 15 March 2021
Published: 23 March 2021

Publisher's Note: MDPI stays neutral with regard to jurisdictional claims in published maps and institutional affiliations.

1. Background

A few months ago, the author of this article had the pleasure to be part of a zoom-union, a 50 years celebration with classmates from first year at the School of Architecture, University of Manchester, England in 1970. The group was supposed to meet in that great city, but the reunion was cancelled, like so many other trips during 2020, due to the Covid-19 pandemic. The way all had adapted digitally since the lockdown started mid-March was impressive. Everybody has been intensively digitalized and the impact of that change was remarkable since it happened in a very short period of time and its long-term positive effects are not yet clear. This paper will not only deal with shaping the notion of sustainable architecture but also speculate how today's COVID-19 pandemic could change the way we think and act from viewpoints of the author's learning experiences as a professor, teacher, researcher, and an architect over the last 50 years. The intention is to communicate and share them in a structured way.

2. Climate Change

The consensus among the majority of the climate change experts in the UN's Intergovernmental Panel on Climate Change (IPCC) is that the climate is changed and it is manmade. While we continue to do our utmost as professionals and private citizens to avoid further climate change, EU has set the goal of reducing CO_2 emissions by minimum 55% by 2030 and China claims it will be climate neutral by 2060. The question is now how much the climate will change, in spite of these new goals, and how much will we have to adapt to the new situation or climate to deal with it [1].

According to one of the lead authors of IPCC's report now working on the next report not yet published, global average temperatures can rise as much as somewhere between 3.2 and 3.6 degrees C by 2100 compared to pre-industrial time, the period 1850–1880 [2]. We have already passed 1.25 degrees C. Several sources have earlier forecasted temperature rise by between 2- and 6-degrees C. Thus, 3.6 degrees C is hence moderate [3]. But what does it mean for buildings and cities, is it a lot or not?

The answer is simple; it is a lot.

If we look at Europe, the continent is basically split into roughly three climate zones: the northern (Stockholm), the middle (Zurich) and the southern (Milan). The annual average temperature between the northern and the middle is 2.4 degrees C and the same is the case between the middle and the southern, 2.4 degrees C. In other words, the design manuals for buildings in Stockholm in 2100 will have to show designs capable of managing the temperatures 3.6 degrees higher than now, like in climates in the regions halfway between Zurich and Milan [4] (pp. 95–101). That is hot and needs a total rethink.

In addition, and complicating the matter further, the "heat island effect" will mean that temperatures in cities will occasionally be much higher than the above. The above were averages only, but in cities, due to heat escaping from buildings, sheltered spaces, sunlight reflection from neighboring buildings "heat island effects" can mean temperatures 5–12 degrees above averages. Heat is going to become a huge challenge and the need for cooling through natural ventilation and/or air conditioning will increase dramatically. We, therefore, need a total rethink of how we design highly insulated buildings. We need to rethink all we have known until now and go back to the drawing board. That is the scale of the challenge.

3. Sustainable Architecture

Defining sustainability is almost impossible due to all the preconceptions connected to the word. It must be one of the most extensively used word of our time trying to describe something but not succeeding. Before the word was introduced words like ecology or environmental were often used and. Rachel Carson's *Silent Spring* in 1962 was a wake-up call about the use of pesticides and the damage to the ecological balance in nature, to the environment [5]. She wrote about the loss of insects leading to the possible extinction of birds, hence the title, no birds—no bird song—just silence. In Alvin Toffler's book *Future Shock*, ten years later in 1971, scientific trajectories were sketched. In the last chapter, number 20, Toffler writes about the strategy of social futurism, the death of technology and the humanization of the planner [6]. It is a signal of what is later to come. In architecture a "going back to nature" school emerged. Some architects designed buildings with turf on the roof, built from logs or local brick. Not much different from the vernacular architecture in earlier periods, yet the new awakening was coupled with modest attempts at using technology. Solar thermal systems for domestic hot water heating emerged, either home built or by local plumbers. They were very simple systems and many of them clogged up due to high levels of chalk in the local water. Some remote housing schemes using primitive windmills to pump water or try to produce some electricity emerged. Others used large log fires to produce direct heat and to heat water for storage. Finally, small solar Photovoltaic (PV) units delivering some electricity became available after the spacecraft industry had developed them for use to power communication links, satellites and space rockets. The advanced PV technology with no moving parts rubbed off into the building industry. Some of the early architectural experiments even used a combination of all of the above technologies. With all this gear inside, on or around the house it looked more like an engineer's dreamy Christmas tree than architecture.

However, and luckily, more and more architects engaged to make solar and other technologies as integrated parts of buildings. The early examples from 1988, as in the Chanelle house Figure 1, was one attempt at creating acceptable solar architecture and at the same time integrate as many technologies as possible.

(a)

(b)

Figure 1. (**a**) The Chanelle house, Stavanger, Norway was one of the first modern energy autonomous buildings in Europe trying to integrate renewables like solar PV, solar thermal, windmill and battery and thermal storage in a more architecturally acceptable way through building integration. The house designed by the author was an attempt at showing alternatives to the messy roofscape developing throughout Europe at the time, due to lack of engagement among architects, as this picture from the Eastern Europe apartment block and the Greek house show the roof chaos is total (**b**). All photos: the author.

From this period, the UN's work with the Commission on Environment and Development led by the Norwegian Prime Minister Gro Harlem Brundtland, hence the report's nickname "the Brundtland Commission Report" began to have an impact [7]. This was partly because it actually tried to define the concept sustainability. However, it struggled to define it because the report wanted to show some hope and combat critics that had argued the report will end with a lot of words trying to stop a natural technological development and global growth. After two world wars, people wanted growth and prosperity and not more obstacles. Hence, the concept of sustainable growth was coined, and it still stands as a viable surviving concept, mainly because it is round and not very concrete. It has been applied to all kinds of situations and professions. In architecture, the concept of sustainable architecture became dominating for a long time. The other slogan that came out of the report was our common future. This too lives on.

The Brundtland Commission documented that the world in many ways was on the way towards ecological disaster. We, especially us in the Western world, were not living sustainable lifestyles. Here the definitions emerged, and they basically stated that we need to make sure that whatever we do does not make life harder for coming generations. We must treat the natural environment as if we just borrowed it, to pass it on to the next generations. This, almost religious statement, is not sensational but very simple and self-explanatory in its general simplicity; sustainable development is development that meets the needs of the present without compromising the ability of future generations to meet their own need.

If there is anything people in the Western world have actually done during the last couple of generations, it is to do the contrary of what the Brundtland Commission preached. Furthermore, when you get down to the nitty-gritty of life and have to make actual decisions as an architect, for example, how do you do it in a sustainable way? While the Commission argued for the generational perspective in general terms it was very clear, especially in debates after the launch, that the use of solar energy is the only way to avoid the dangerous use of nuclear energy in the future. The commission also made it clear that buildings account for a huge proportion of energy use globally. Later, we have come to realize that buildings as a rule of thumb represent approximately 40% of global energy use and the figure for buildings' part of climate gas emissions including production and transport of building materials, represent between 40 and 50% of global annual emissions.

Then in 1992 followed the UN Rio Conference. The first of a range of international conferences on sustainability, although climate had become the overarching theme by now. Avoiding climate change was the main objective. Still, and mysteriously, at Rio it was not planned to deal with energy as a separate issue. Instead it was hidden in between all kinds of other headlines like agriculture, transportation and housing. It was dealt with as a minor issue. Through the work of Eurosolar, World Council Renewable Energy (WCRE), International Solar Energy Society (ISES) and World Watch Institute, through a process initiated by this author and German Hermann Scheer of WCRE, the issue of energy was lifted higher up on the agenda and today is one of the top priorities, along with energy efficiency and behavior change [8].

Further into the 1990s it appeared that to get the mixed solution to work in a built context you needed to be an engineer in order to live in the houses, but slowly but surely the simpler solutions won. At their base was always energy efficiency in order to reduce the need for energy. By doing that, the energy supply system could be smaller and cheaper to buy as well as better integrated. The emergence of zero energy buildings and later mass production removed the problems related to self-build mistakes and improved quality control. The concept of zero energy mass custom homes (ZEMCH) was born and its organization developed over time. Two events were crucial to this development; first the ZEMCH technical tour in 2006, later called ZEMCH Mission to Japan and in 2010 the establishment of the ZEMCH Network. Secondly, the international ZEMCH conference in Glasgow in 2012, which spiraled work in its intended direction, hence contributing further to the development of the fields of zero energy mass construction housing [9] (pp. 83–95). In order to see such changes some kind of forecasting skills are useful.

3.1. Insulation

The need for thick insulation will disappear with the kind of climate change indicated above. New insulation methods like the ones used in refrigerators and caravans that are more compact than glass wool and rockwool based ones, will probably be taking bigger market shares. Thus, making housing construction slimmer and cheaper. Designing buildings with in total 35–50 cm thick walls can become useless as overheating the inside of the building will become a problem, instead, 10 cm wall thicknesses could do the job most of the days over a year, even in the colder climates of the world. We already know the efficiency of increasing the thickness of wall- and roof insulation is minor over a certain insulation thickness, as shown in Figure 2. Increasing the insulation thickness from 10 to 15 cm increases the insulation value (U-value) by 0.06 W/m^{2}K. Increasing insulation thickness from 25 to 30 cm increases the U-value with only half, 0.03 W/m^{2}K. This is illustrated with the curve getting flatter and flatter as the insulation thickness increases. Increase from 45 to 50 cm insulation hence only increases the U-value by less than 0.015 W/m^{2}K.

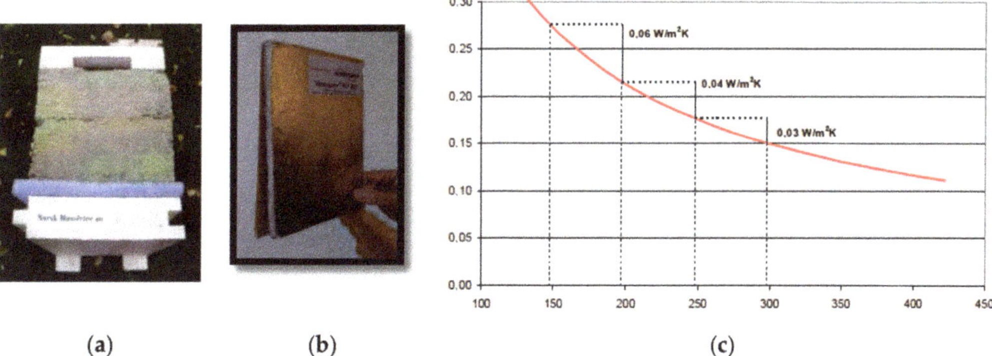

(a)   (b)   (c)

Figure 2. Housing using 35–50 cm thick walls, whereof most is insulation is very material-, transport- and workmanship demanding (**a**). The search has started for slimmer insulation methods inspired by the caravan and refrigeration industry, fridges that people use in kitchens is one example (**b**). The science of insulation thicknesses and insulation efficiency is clear. As illustrated for a Norwegian climate, the improvement of insulation is smaller and smaller the thicker the wall is (**c**). The *X*-axis shows insulation thickness in cm (rockwool or glass wool). The *Y*-axis shows u-value. All photos and illustrations: the author.

This was tried out in the Kolnes house in Randaberg, Norway where a 10-kW solar façade made the house net energy autonomous over the year by grid connection producing between 6500 and 9000 kWh a year. See Figure 3 The house's annual energy need is 6000 kWh. The cost of the solar modules, inverter and mounting is reduced by the saving in cladding materials that it is replacing. If that cladding was planned to be of an expensive type like marble or stone sheets, bricks, high quality metals or expensive timbers, the solar PV cladding would be cheaper than that. The cost of the solar system module part, normally at least 60 percent of the total solar system cost, is hence paid back instantly. The house is area efficient, a compact house of only 60 m^2 total floor area. In addition to the annual surplus production of solar electricity from the south east and south west facades, the designer has questioned the following:

Is the use of extremely thick insulated walls, nowadays ending up with up to 50 cm overall to design Net Zero Energy Buildings (NZEBs) in the Norwegian climate, wasteful? Instead, a refrigerator-type insulation was tried, leading to total wall thickness of 15 cm only. This led to savings in materials needed, their transport and workmanship hours as well as floor area. The area saving, footprint on site, was as high as 23% for a 60 m^2 house with 30 m^2 on two floors, if the difference between 50 cm and 15 cm walls are used as a basis for the calculation—Figure 3c.

Is the use of centralized balanced ventilation and heat recovery being essential in order to achieve NZEB standard in a cold climate like the Norwegian one? Ducting for transporting air to and from a central heat recovery unit is demanding necessary floor height and vertical ducts adding to material use and costs. Instead, a point-based, local heat recovery/fresh air unit is used, one for each room. It is a small 20 cm diameter duct like a prefabricated unit. There is one per room inserted through the external wall. The length of ducts saved is for this compact house hence 60 running duct meters and a room of 3 m^2 that a central air handling unit would have needed—see Figure 3b.

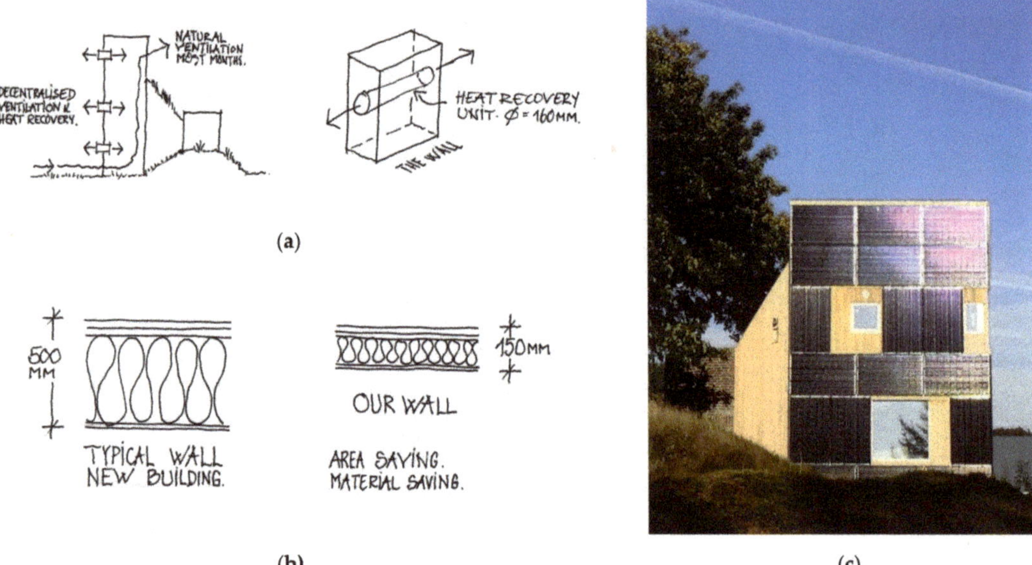

Figure 3. Kolnes compact house, Randaberg, Norway: 60 m^2 total floor area (**a**,**c**). New refrigerator insulation materials for envelope applied. Instead of the typical 50 cm total wall thickness in new NZEBs in Norway, the total wall thickness here was 15 cm, leading to huge material and area savings (**b**). Instead of a centralized ventilation and heat recovery system, demanding floor height and ducts a point fresh air and heat recovery system was used in each room. This saved 60 running meters of ventilation distribution ducts and a 3 m^2 room that a centralized system would have needed (**c**). All drawings and photos: the author, which is also the architect.

The massive use of insulation as one way of reducing heating need often seems like a one eyed and very short-sighted strategy because it is designed to reduce building envelope heat losses to almost zero during the coldest periods when outdoor temperatures are minus 1 to minus 10 degrees C. In many climates this happens very rarely. An alternative strategy would be to design the envelope's thickness based on the annual average temperatures. In the Norwegian cold climate example above this means that a traditionally insulated wall could be 20 cm instead of 50 cm total wall thickness and heating systems would still easily cope. As Figure 2c shows, the U-value gain from increasing insulation to extreme levels is marginal and probably not economical, to the house buyer and society at large, all factors considered. Considering that the global climate is getting warmer and warmer and extreme weather more likely, rather than designing a house as a heavily insulated thermos, as if the outdoor temperature was minus 10 C all year round, it could be designed for a mild climate and with a small emergency room for the few extreme winter days. Such a room would have to be located near kitchen and bathroom and facilitate both living room and bedroom possibilities in one room. However, this is only to be used for a short time and under extreme weather conditions. Rethinking such possibilities would lead to a more radical approach among house designers and more rational use of materials than the laidback approach we mostly see today.

There is also bound to be a rethink of the value of the extreme insulation achieved for windows and glass doors. The drive towards making better insulated glass have been very successful, but the downside is rarely informed about. In cold climates when the nights are below zero, the sky is cloud free and the humidity in the air is high, combined with no wind, a layer of frost appears on the outside of the glass making it impossible to see through it. It is the dew or condensation that forms a layer on the external side of the glass. It occurs because there is hardly any leakage of heat from the inside of the building through the glass, hence no melting occurs. Many new flat owners wake up on their first frost morning during autumn or winter in their newly acquired, often expensive top flat, only to realize they will not be able to enjoy the view on days like that. This phenomenon is well known among manufacturers and contractors, but not always communicated clearly to the buyers of apartments.

This should also send a signal to planners, including designers of ZEMCH. We must pose the following question: Is this high-quality insulation really worth it when the cost is high cost windows and occasional lost view to the outside? Are there other solutions? Other solutions could be to go for a less insulating window quality and instead allow some more use of energy. If the energy supplied is renewable, the sustainable considerations are covered. If the energy in addition is provided on site and is cheap the total cost for housing and its running costs will decrease or at least stay stable and with less challenges for the views.

Could we instead develop windows with multiple functions, not only as an insulator and an area for view and daylight but also to actually produce energy—passive solar and electricity, PV as shown in Figure 4b?

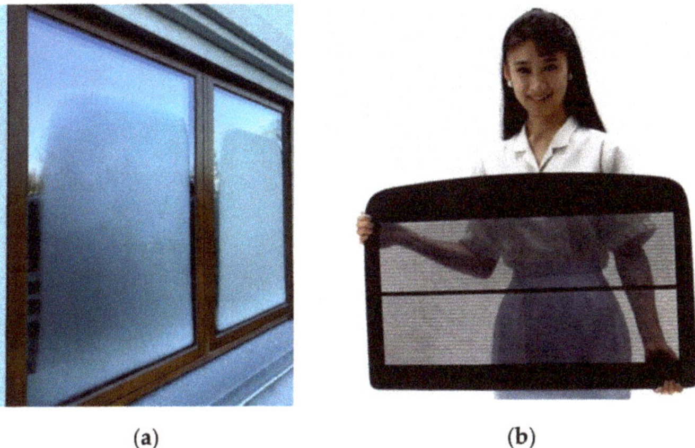

(a)   (b)

Figure 4. After a night of minus 3 degrees C humidity in the outdoor air has frozen to ice on the outside of the windows making them look like frosty glass (**a**). This occurs at regular intervals during the winter when high air humidity is combined with frosty nights and clear sky. The problem is that this frosty glass was not ordered and people that are unaware of this gets shocked when their often expensive views are completely eliminated during most of the following day. The photo is of a brand-new housing development at Ledaal Park, Eiganes, Stavanger, Norway—opened 2020. Windows can also have a double function both being transparent and producing solar electricity PV as shown in this Sanyo product from 1989 (**b**). All photos: the author.

The great concern to insulate buildings better in spite of climate change's consequence, raised temperatures, is the often lack of concern with indoor temperature increase. There are growing g concerns that in the future buildings will no more just be protected from cold environment but avoiding and controlling solar heat gains during the warm season. This challenge will then have to be addressed for example through screens or inner thermal mass.

3.2. Solar Energy

The use of solar energy for all kinds of purposes has expanded fast over the last ten years. Prices of solar PV electricity today are below 1/100th of the level 30 years ago and they keep falling [10]. Solar thermal for buildings is also taking a big share of the market, basically for domestic hot water. The square meter of roof, wall or garden becomes valuable and can be put to use for solar equipment. While in northern Europe, it is possible to get

120 kWh per m^2 of useful electricity out of a PV module per year, the similar figures for evacuated glass tube solar heating system is 500 kWh. Flat plate collectors are slightly below around 300–400 kWh. When using solar in relation to a building, it is crucial that the modules are well integrated and replace other cladding materials in order to save on costs, thus making solar more rational and viable [11] pp. 46–58. The use of passive solar as heat through windows and greenhouses is another way of making use of the sun's energy. However, since this is not a technology, a sellable product, but rather depends on the knowledge of the planner, it is frequently forgotten. However, passive solar is still applicable and the oldest of all solar techniques. Cave dwellers used it and when the forests around Rome were chopped down for log burning purposes, they had to import wood from Caucasus. It became complicated and costly. Emperor Justinian who was ruling from 527 to 565 hence introduced a law ensuring houses' right to sunshine through windows. This was possibly the world's first passive solar law. It was written into the Justinian's Codex [11] (pp. 170–171).

When in the 80s architects tried to design energy autonomous solar housing, they often ended up with very complex combinations of systems combining several energy sources as in the Chanelle case, Figure 1a—as opposed to in the Kolnes case, Figure 3 where only one energy source, solar PV, is used. Nowadays and in the future, it does seem that simpler systems will win. Solar PV is hence a gift because the electricity it can produce can be used as that, as electricity, or it can be used to power a heating system. The challenge is of course energy storage and the solution is a choice between either running two-way on a city- or rural central grid system (grid-connected), a local neighborhood grid system (positive energy district grid—PED) or an in-house electric storage normally battery based, like in the Madla-Revheim case, Figure 5, a new part of town to be housing 10,000 people.

In the latter case and also possibly in the PED case, the huge increase in the number of electric vehicles will allow the sharing of battery capacity between houses and vehicles, as the drawing in Figure 5 shows. Electric vehicles and bikes are often charged at home, but also at work and at public charging points. The vehicles hence often return home with quite full batteries that can be tapped by the household and filled again by the house's solar PV system when there is surplus. In this way, in the future ZEMCH, the battery capacity particularly of electric vehicles has to be considered when designing a total energy balance system. The electric storage buffer will hence be larger than that of the house itself. A more holistic approach-thinking is in demand to respond to the new possibilities offered by a co-operation between buildings and all kinds of vehicles.

In this context, one must follow the development of batteries and push for more ecological or sustainable solutions as regards the exploration of the raw materials used in batteries, the handling of them to protect the workers, the effective use of materials and finally a complete recycling of materials. New materials for battery production will be invented and we already see great developments in this field. In the future, cheaper and more ecological and sustainable batteries are bound to emerge. This will propel the electrification of the vehicle fleet and simultaneously possibly the use of batteries—in houses and vehicles—as electric buffers for individual ZEMCH, groups of housing or whole districts. That option is at least now in sight.

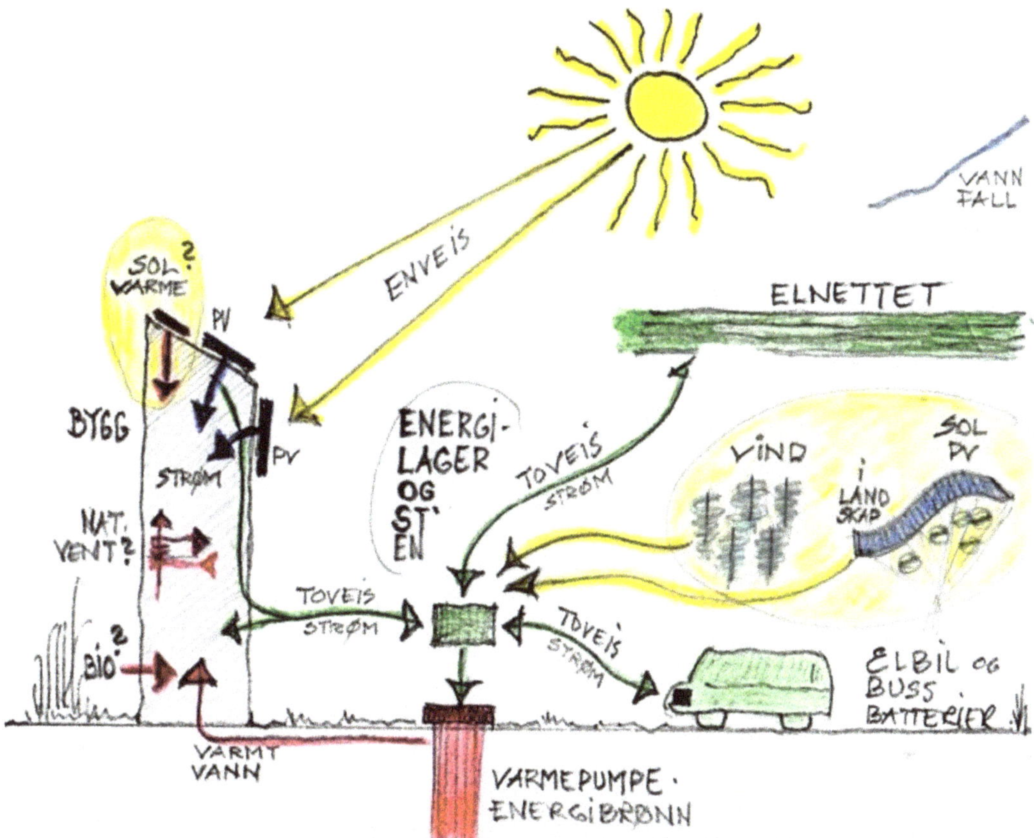

Figure 5. Madla-Revheim, Stavanger, Norway. Energy autonomous new part of town for 10.000 people, where a combination of renewable energy like solar PV, solar thermal on buildings and in the landscape is combined with wind power. This again is connected to running a geothermal heat pump and a stationary central battery bank storage as well as grid-connection when necessary. The transportation sector is linked through this and vehicle batteries provide a united and huge battery bank in addition to the stationary battery bank. The illustration is from the concept design energy system at the Municipality of Stavanger Quality Program for the Madla-Revheim development, designed and drawn by: the author.

It is crucial that solar when applied, is applied with a rational intention behind it. The simple idea that cannot be repeated too often is that the first design step should be to design a building as energy efficient as possible—but without overdoing it, as discussed in the contexts above on insulation values of walls, roofs and windows. It is possible to design a perfectly well-operating building without going to extremes. However, in any climate, we must respond to the local climate. Principles applied in Norway may not be applicable in Sri Lanka or Cyprus. As an example, below is a case of a university building in Cyprus, Figure 6. This building is set in an extremely hot dry climate where earth ducts for cooling the natural ventilation driven air solution was found to be the best option to reduce the enormous cooling need of the building. It was also necessary to first reduce the energy need in order to reduce the size (area demand), capacity and cost of the facade integrated solar PV system intended to run the building. In this kind of setting as much as 80% of the total energy need of the building would typically represents the cooling and ventilation need. Natural ventilation is generally an underrated possibility in our time. Although it can be used both to ventilate and cool buildings in efficient and inexpensive ways if done well. Our knowledge in this age-old method of cooling must be upgraded and literature can help us as the techniques are well-documented [12].

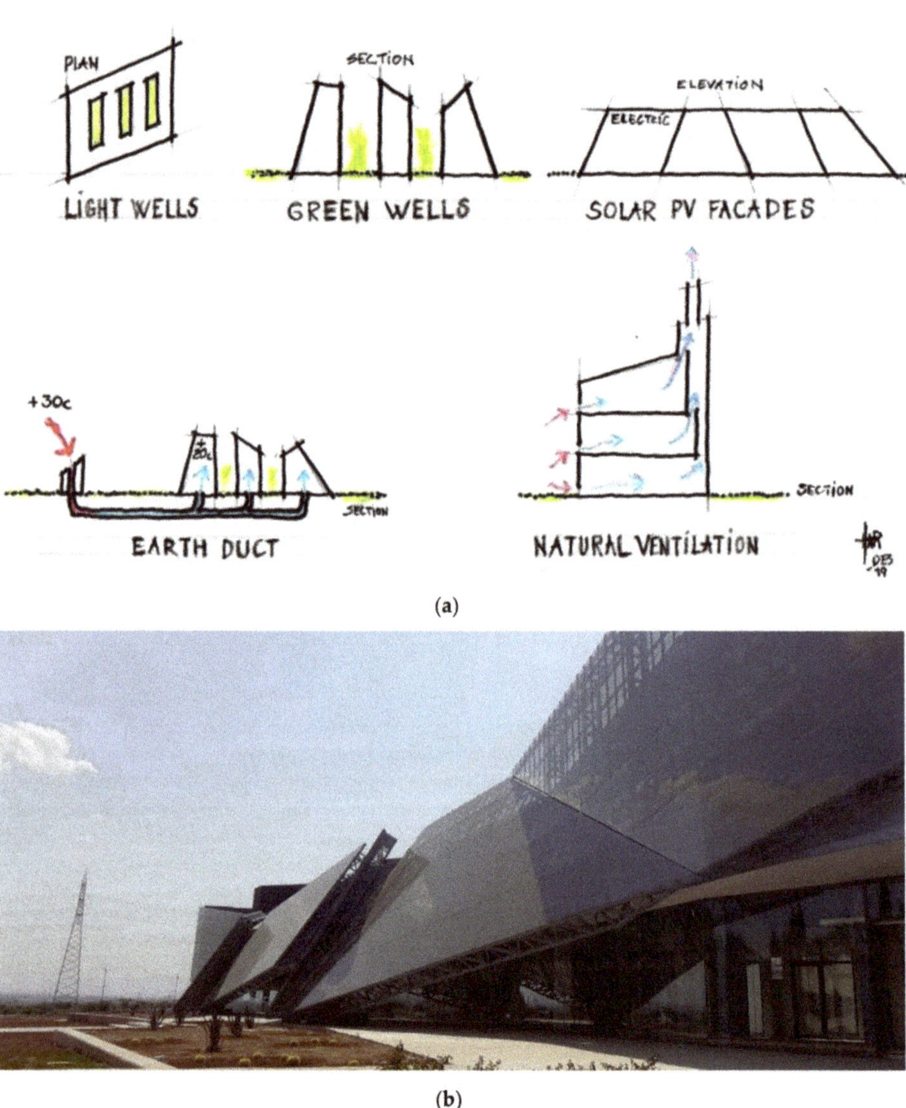

(a)

(b)

Figure 6. Design principles for the International University of Cyprus' Engineering building inaugurated 2019 (**a**). Energy system overall principles (**b**). The author. Natural ventilation design: Max Fordham, London, UK. Architect: Saffet Kaya Ltd. Drawing and photo: the author.

3.3. Burning Logs

The concept of universal design includes considerations to ensure that people with any kind of physical handicap can function as well as possible in society. As a result, planners and architects have become used to dealing with appropriate door widths, low thresholds and ramps to allow people in wheelchairs and blind people to function well in society. However, we have not been so engaged perhaps in ensuring everybody's rights to a clean environment, free from pollution of waters and what connects us all—air. Air pollution is becoming a rising problem in many parts of the world and a lot of it arises from transportation related emissions but also a lot from log fires. The burning of logs in homes has in some areas of the world, at times, created unhealthy air quality. There are two issues here: one is the general international theoretical agreement that wood is climate neutral because trees, while they grow, store CO_2. In an international agreed climate gas accounting system wood, both as building material and for burning to generate heating in homes, is met with positivity. However, the other issue is the consideration of particle emission. Particle matter, PM 2.5 are tiny particles emerging from the burning of wood. Their size allows them to go straight from the lung to the blood vessels and create bad damage. In Norway alone, it is calculated that 1700 people die an early death every year as a result of PM 2.5 from log burning—not from traffic [13]. To give a scale to this, less than 100 people die in traffic accidents in Norway every year. Contrary to earlier beliefs, the health damage to people from log burning is not only caused through outdoor inhaling from chimneys, but also in the indoor rooms. Opening and closing log burner doors to light fires and put logs into a burning fire releases damaging gases into the rooms [14]—so does an open fireplace. The new research shows that both the current log burning technology but also the way we use it is faulty and a health hazard. Research shows that in winter log burning is far more polluting as regards particle (PM 2.5) than traffic fumes. In London half of the PM 2.5 stems from log burning—not traffic [15].

From this follows a need for us all to re-question the whole issue of the applicability of wood in a sustainability context. Two questions arise: One is related to the international agreed assumption that wood is climate neutral. It is of course correct that wood has stored a lot of CO_2 over its lifetime before it is chopped. This is the reason why it is seen as climate neutral. It has done a job in the past and it is valued. However, when trees are chopped for building or firewood purposes an open land area emerges. The catching of CO_2 by trees on that land hence stops. When new trees are planted, it may, depending on tree type, take 30–60 years before trees get old enough to really catch CO_2 in a big way. As the CO_2 catching ability by trees are very small in the younger years and in elder age than in mid-life, chopping should not happen in mid-life as the main catching of CO_2 happens when threes have developed a considerable leaf area. That happens in a tree's mid-life. The lost CO_2 catching time, the negative part of the calculation, while waiting decades for the new, planted trees to develop is not included in the usual CO_2 calculation. Only the positive part, the past catching is calculated [16,17]. This biased calculation leads to a general misconception that the use of timber for construction and burning to generate heat purposes is sustainable, while it is probably not. The debate about these different views have been ongoing among highly qualified people in for example agriculture and many strong opponents to the current CO_2 calculation method are people in the academic and forestry fields. Strange as it seems this discussion has not siphoned into the field of architecture.

Why is this relevant in relation to ZEMCH? It is relevant because in a proper sustainability-based CO_2 counting system timber will probably not keep its remarkably protected position as a do-gooder forever. Its position and the logic behind it are being questioned and it is always wise to be prepared for changes in the CO_2 accounting system before they actually appear.

Having mentioned this, it is of course important to underline the actual qualities of timber. It is beautiful, it can regulate humidity levels in indoor rooms through "breathing", it created a comfortable acoustic environment, it smells good and it is a renewable material.

Sometimes it is even local, although a lot of the timber industry is not at all local and leads to transport over huge distances. All these will be enough good arguments for the use of timber as a construction material, but we cannot ignore its downsides as mentioned above and in addition add the bulkiness of timber constructions and complicated logistics compared to several other materials. Keeping our eyes open for new materials and new applications of old ones is part of the complicated decision puzzle planners and architects will have to engage in in order to ensure sustainable design and construction.

Timber, like any other material will have to undergo LCA analysis or other evaluations in order to uncover too long travel distances for the timber leading to emissions from lorries, calculating refinement processes and their emissions as well as and waste and recyclability possibilities. This applies to individual types of materials as well as ZEMCHs at large.

Finally, it is crucial to understand that in modern, heavily insulated homes, log burning becomes a challenge in that it will occasionally overheat spaces.

4. Ethics

Many architects throughout history have had a tendency to want to serve the rich and mighty, since they were the ones that had the resources to build buildings, and through this they became rich and famous when designing palaces for real kings and queens as well as monetary "kings". Simultaneously, the ancient history of architecture is full of examples of architecture without architects. Modernism, in an attempt at both searching for new expressions and create mass production to provide shelter for the many at a reasonable cost, often stumbled into terribly deep traps. The product they came up with often did not last due to poor workmanship, poor detailing and the use of low-cost materials that hampered durability. Demolishing was sometimes the only way to fix the problem and hence buildings' lives became short. Such an approach, scrapping young buildings, is of course not sustainable and we must look for better solutions with higher quality and—if necessary—more compact and area efficient solutions to create more lasting products.

The ethical responsibilities of architects are supposedly to not different from those of doctors or dentists, tailors and chefs. The questions each one could pose is—who do I want to serve? It is always a dilemma as an architect to make up one's mind as to whether the designing of houses with five bathrooms per family is meaningful when the need for mass produced social housing is so great in most societies. Shifting the priorities away from the mansion sector onto the low-cost sector is a challenge both in terms of fees and prestige, but it is for any profession and ethical issue too. Interesting places like Timbuktu in Mali, Western Africa where the characteristic mud architecture has inspired so many through the beauty of the buildings and the use of local materials. This author had the pleasure to design such buildings in Mali and adding some minor technological gadgets making it possible to light them and run an evaporative cooling device helped by a PV run fan, Figure 7. The function of these were to deliver some electricity miles from any grid by using small solar modules and a battery to provide light. The concept of architecture without architects emerged in in 1974 through the book of Bernard Rudofsky and the great exhibition based on his book "Architecture without Architects", showing at the Design Centre Gallery in The Mall in London the same year [18].

(a)   (b)

Figure 7. (**a**) The book by Rudofsky and other books about the hand-made architecture of Mali has been inspiring many generations of architects. (**b**) The author's solar run and naturally ventilated office and housing building for the Strømme Foundation in Bafoulabe, Mali in 1985 made of local sun-dried mud and covered ion hand-clad mud uses some simple solar PV and thermal technologies to evaporate cool the building. All photos: the author.

The exploration of aged architecture makes a great impression and it inspires most people. Many ask; what can I learn from this? Richard Sennet's great book "Building and Dwelling. Ethics for the City", resounds many architects' own experience where he writes; "As a young urbanist, I was persuaded to the ethics of modern making by reading a book by Bernard Rudofsky" … "who documented how the materials, shapes and siting of the built environment have arisen from the practices of everyday life" … "Rudofsky argued that the making of places had no need of self-conscious artiness" [19] (p. 13).

Coupling this with the work of Hassan Fathy as explained in his book "Architecture for the Poor", there is a good mix of choices eliminated as to who one wants to serve [9]. Fathy's local materials and local production, mass customized as it was at the time, constructed with local builders, self-builders and helpers coupled with his ideas of creating cool indoor spaces through using mass and natural ventilation. This cheap way of temperature- and comfort-control was unbeatable when he added to his mud brick designs using water-filled clay jars in the windowsills and patio fountains removing heat from the air through evaporative cooling.

When this author later got the opportunity to work for Save the Children in Colombo, Sri Lanka to develop all kinds of solar energy-based solutions, they were mostly built from local materials. They were solar stills to produce clean drinkable water from salt or saline water as shown in Figure 8a, cooking by solar heat, drying vegetables and fish to reduce rotting and waste. Later, when designing Sri Lankas's first, and largest, solar electric grid-connected and naturally ventilated building for Worldview Global Media/Young Asia Television (WGM/YATV) in Battaramulla near Colombo and testing out the electric three-wheeler solar taxi in the same city as shown in Figure 3b,c, a combination of solutions were applied. All this happened in the 1990s. It included a steep learning curve on issues like deep poverty, need, compact living and scarcity of materials. It can teach us a lot about the human condition and how different life is in a developed and a developing country. It can teach us a lot about energy, people and the sky above—about the enormous potential for introducing the use of decentralized solar energy in a big way. It is low cost, simple and mostly repairable on the spot. It also teaches us a lesson about lack of repeatability. In many countries like Sri Lanka mass production has not been natural. Instead, hand-made products were dominating. It was only when international construction industries entered the Sri Lankan market through international collaboration that some kind of mass production was introduced. The point is this: if we are going to encourage mass production for houses, other buildings or vehicles we first have to build something in those countries to really understand what the level of workmanship for production and repair is. When we understand that we can start introducing more relevant mass production by preferably using the local work force as a base. It is useless if new industries get entirely based on the flying in of too much expensive foreign staff without interest or knowledge of the local challenge ahead. Development is a slow process and it can only take place and have a chance of succeeding if the conditions on the ground are understood.

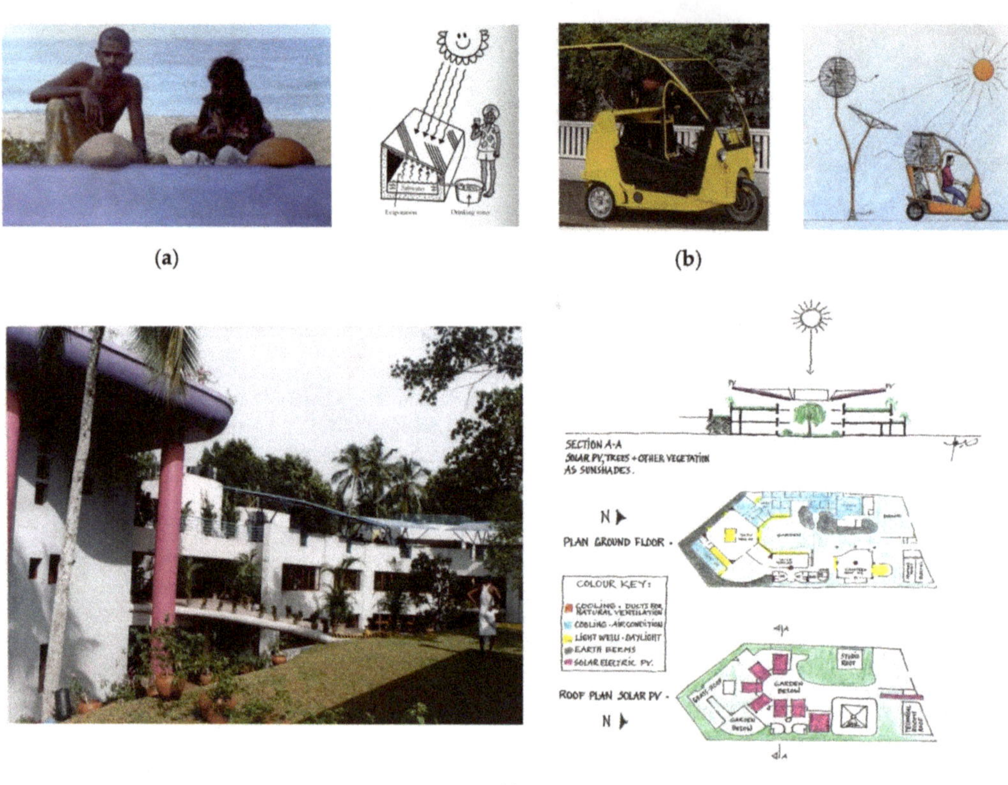

(a)

(b)

(c)

Figure 8. Through working with crucial challenges in developing countries one can learn a lot that can be applied in the developed world, like simplicity, lowering costs and repairability on the spot: simple still to produce drinking water from salt water through evaporation and condensation (**a**), solar run and naturally ventilated World Global Media (WGM)/Young Asia Television (YATV) headquarters (**c**), solar electric three wheeler taxi for cities prototype test. (**b**). All tests performed in and around Colombo, Sri Lanka. The solar electric vehicle design and drawing: the author and Peter Opsvik AS. All other designs and photos: the author.

However, from hot tropical and humid Colombo to cold mountainous Katmandu, Nepal the story is different. People are freezing in the mountain landscapes with its sunny days and extremely cold nights. However, every day, the sun rises and shines bright and warm. The solar potential is gigantic, yet people are chopping down the woods to get firewood for their meals and for heating their crowded, densely populated locally built stone and clay houses. Many knowledgeable scientists and practitioners have observed this and wanted to shout from frustration when seeing our fellow nationals doing development aid but ignoring the huge potential of developing the use of solar energy into a solution and instead, and with huge aid budgets pushing fossil fuels like diesel and gas as the, normally imported, solution or huge centralized hydropower dam projects. Which energy sources are pushed in which country and by whom, is an ethical issue often based on strong economic interests and pure greed?

In the book, Philosophy and Architecture, Christian Illies and Nicholas Ray show many examples of ethical borderlines [10]. Examples of people, both architects and planners that have played a supportive role within regimes that has a long record of human rights violation, are referred to and the overarching theme is that there will always be people that don't see the ethical issues at hand or refuse to see them. The authors refer to philosophical thinkers that have suggested that architects and other professionals should refuse to associate themselves with commissions that obviously end up on the wrong side of the ethical borderlines. Those are similar ideas of double roles and dichotomies that this author developed in the book Corruption the Nobel way [20].

4.1. Being Brave

All planners, architects, suppliers and developers have access to research results. They should be accepted and result in action accordingly. Braveness and long-term goals are crucial to create faith in mankind's own ability to see what is coming. By conscientiously following our internal ethical guidelines and listening to our internal ethical alarm bells, we can better develop our ability to forecast, to see what will be in demand in the future and what will not. When confronted with scary reports on the future, our attitudes and priorities sometimes change. While writing this, an alarming message occurred through The Guardian of dramatic change in our climate; the world is at its hottest in 12,000 years [21]. That kind of research makes a big impact and should accelerate the shift for a more ecological future. This applies to our sustainable future as well as practical responses to it—like being in involved in developing ZEMCHs. There are many kinds of possible futures. It is our hard task to choose which future we believe most in and act to embrace that one according to our holistic conviction after having studied all the parameters involved and trying to look at them without being biased. We must believe in the future—a future we choose to believe in—and be brave.

4.2. Being Humble

It seems everybody is fighting a hard battle, so we should be kind to each other. However, let us not be fooled. There are tough robust people out there and people do not always mean well. There are selfish bullies too and some people steel from need or greed. We should stop them by sticking our necks out and blow the whistle. However, most of the time and above, we should be humble. Many great thinkers were brought up in humble, tiny houses of mediocre quality. It is not the size of the house that decided which thoughts are being thought in there. It is the capacity of the mind, the brain, to fly, that makes the difference.

The philosopher Ludwig Wittgenstein inherited one of the largest fortunes in Europe from his father who was a rich steel magnate in Vienna, Austria. LW gave it all away in order to become a philosopher and he built his modest little 60 m^2 timber house on a mountain shelf in Skjolden, Norway in 1915. It was completed while he spent time as a volunteer in the Second World War on the European continent. In his house, he collected water from the lake 30 m below and he heated his house with logs chopped from the nearby forest as shown in Figure 9a. His 1.5 m^2 unheated toilet shed was 15 m from the house and he had to empty the bucket and bury the content every spring. Candles lit the house after dark. There was, of course—no electricity [22]. However, Wittgenstein was knowledgeable, and he used local craftsmen with local knowledge to help him. In order to become one of the world's leading philosophers ever, he studied many subjects. First, he spent a few years doing engineering at a university in Berlin, then he moved to University of Manchester and spent four year studying aeronautics. He was flying kites there and even patented a fuel injection system for helicopter wings. From there, he moved to University of Cambridge to study philosophy and he spent the rest of his life as a philosophy teacher at the university there, when he was not in Norway and on travels elsewhere [23].

What can we learn from this? That housing size does not matter, really, and that multidisciplinary knowledge is useful? Probably, and there are other examples, like the example of Le Corbusier, the architect. When he built his summer house in the South of France at Cap Martin, it was a 16 m^2 house for him and his wife; 8 m^2 per person. They called it Le Cabanon. He had been inspired by his trips with ocean liners and planes across the Atlantic and learnt how compact the accommodation had become. In his own Cabanon, he applied the same principles, almost like designing a compact cabin for a brain and a pencil only [24]. As shown in Figure 9b, although it did not have a full-size overdone kitchen because there was a restaurant next door where they had all their meals, the compactness is striking and impressive. Perhaps we can learn from it that bare essentials are enough for many, most of the time. At least it was enough for these two giants, Wittgenstein and Corbusier. They were certainly showing their humble side in the way they selected their accommodation.

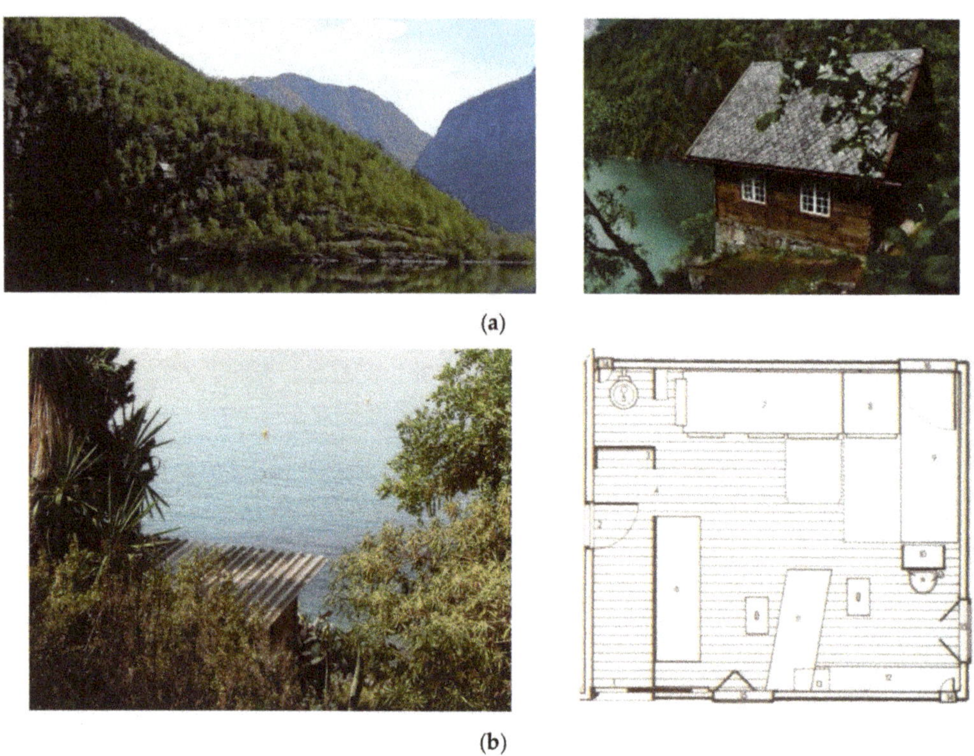

(a)

(b)

Figure 9. Ludwig Wittgenstein's 60 m^2 remotely located cottage outside the village Skjolden, near the end of Norway's longest fjord Sognefjorden and by a lake higher up. Note the string on the drawing by which a bucket was lowered to get drinking water from the lake (**a**). Le Corbusier's 16 m^2 Le Cabanon, Cap Martin, France (**b**). Photos and drawings: the author.

This leads to the big question: are there still things to learn for us from other industries than the housing industry. Can we draw experience and learning from the compact caravan or even vehicle industry, in the way they mass produce and make the solutions compact? The caravan sector is certainly a field to be studied, both from the perspective of insulation, compact living—kitchen, bath, living- and sleeping quarters. Although it is not applicable to all, that is not the point, it could be applicable to many and it could provide many more with reasonable, although humble, low-cost and mass-produced solutions on a global scale as shown in Figure 10.

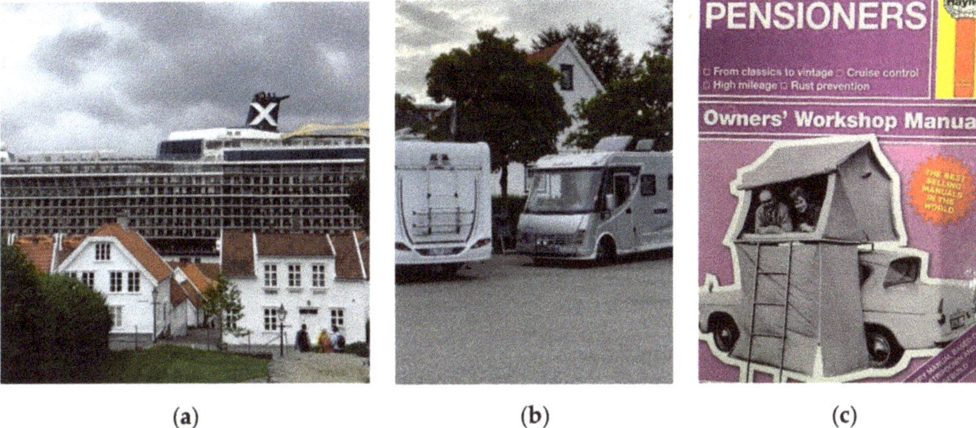

(a) (b) (c)

Figure 10. There are useful zero energy mass custom homes (ZEMCH) lessons to be learnt from other industries, like cruises where thousands of passengers are transported in relatively compact living quarters, although the rest of the floating hotels are full of huge shared spaces (**a**). Compact living as a camper can end in many physical expressions (**b**,**c**). All photos from Stavanger city: the author.

If the concept of ZEMCH is going to develop in a big way, we need to look at it from all angles and try out many more solutions than the on-site, part on site, part factory mass production now going on. We must look for even more ingenious solutions, and perhaps we have to go back 70 years, when Europe was rebuilding after the Second World War and mass production became necessary in order to house the homeless. At that time, housing was produces in huge numbers—and fast. The result was often mediocre quality. In an earlier ZEMCH conference paper from 2012 this author covered that side of the issue [9] (pp. 83–95).

5. Forecasting

Most of us are poor forecasters and we fail to accept this fact. Most of us are too entangled in our own ideas and preconceived positions in the art of debate, so that we cement our attitudes rather than stay open to learn about new positions and possibilities. This hampers our personal development and developments at large—privately, in the family, socially and at work. That is a great loss. Super-forecasting or the art of prediction has been extensively dealt with by Wharton professor Philip Tetlock and journalist Dan Gardener through the release of their book Superforecasting: The Art and Science of Predictions [25]. They argued that everything we do involves some kind of forecasting about how the future will unfold. In a twenty-year study they registered actual predictions and tested them against what actually happened twenty years later. They found that the actual experts were only slightly better than the average person doing guesswork. They also found that some of the most visible experts used as commentators by the media were lousy at predicting the future. Contrary to our general belief in experts, the study uncovered that the absolutely best predictors were quite ordinary people, from former ballroom dancers to retired computer programmers. They had an extraordinary ability to predict the future twenty years ahead with an accuracy of 60 percent greater than the average. Tetlock's study found that independence seemed to be the most valuable quality at being a great predictor and that the ability can be trained and improved. The main point was that most people are not independent. Our predictions tend to lean towards what we have sympathy for, general interest in or vested interest in. When we find ourselves in double roles, we tend to be lousy forecasters. We hence need to distance ourselves from our interests and sympathies and look cooler at the questions ahead in order to improve our prediction abilities.

What Tetlock's study showed us was really that when we listen to other people's predictions, we need to question why they lean the way they do. When an employee in an oil company forecasts the oil age to be lasting hundreds of years, we should treat that prediction with caution and ask about vested interests as well as concrete, factual arguments, why oil will be dominating that long. Similarly, when a solar expert forecasts the solar age, ask for data and concrete arguments, not wishful thinking, and when a designer of zero energy housing argues that this is the future, ask for data as to why. Is it the future because it is a wish or are there logical explanations? If so, which?

These are the test questions that help us understand better, both our fellow men and women, but also our future. We can only predict the future if we can argue for why a certain direction in the future will "win" or dominate. We must have some basic logical data at hand, only then can we begin from there to use our imagination and use our gut feeling, our experience and only if we manage to stay independent can we pose reliable predictions. If not, we are just chatters, like so many of the commentators: big talkers, lousy predictors, useless forecasters, and time wasters.

We don't want to be like that, so from here on we must ask some hard questions and if necessary, turn everything upside down in order to possibly arriving at some new "truths". As a vehicle, as theme, we can use ZEMCH and try to forecast the future of ZEMCH and how it should develop in order to endure.

Covid

The Covid-19 pandemic has changed a lot. It is a disastrous experience for most people although some are hit harder than others. If only we could learn enough from it to justify the damage and loss of life, it would be easier to accept that nature from time to time hits at us in this way. Throughout history, pandemics have occurred from time to time and they will occur again in the future. This author has (during the pandemic) slightly changed some of the courses taught at the university to study the actual learning outcome of the pandemic so far. By doing that, students have been able to concentrate on events actually ongoing and discuss them in light of sustainability related issues and draw learning from the pandemic so far. One must of course be cautious of such an approach, since we have not yet seen the end of the pandemic, as this is written. Surprises may yet occur, but the following have been experienced and noted:

Traffic naturally decreased during the pandemic because people stayed at home. As a result, CO_2 emissions fell. The world appeared more sustainable and less polluting, however:

In many parts of the world, people fired more with logs while working from home during daytime and the particle pollution (PM 2.5), due to increased use of log fires compensated for the air pollution gains from less traffic. The air was in many places almost as polluted as before the pandemic [16].

Due to the increased cleanliness caused by hand-washing and social distancing, the occurrence of other respiratory diseases, like flu and colds, fell during the pandemic. In most countries large numbers of people die every winter from flu and the number varies from year to year depending on how tough the virus is. In Norway, a country of 5.4 million people an average of 900 people dies, every winter, because of flu. In a particularly hard winter four years ago, 1.600 people dies. The variation over the last decades has been between 200 and 1.600. As a comparison, 550 people have dies from the Covid-19 pandemic so far in Norway from mid-March to end December 2020. During all of 2020, 94 people lost their life in traffic accidents. That is the scale of the log fire problem on human health. It is important to add that the Norwegian society have been practically closed down for months to keep the pandemic death toll down so the cost to society has been extremely high as it caused unemployment, loneliness and mental challenges for many.

From these examples we can at least draw two conclusions and learning experiences:

We can make life better for many, if we increase the level of hygiene in the future.

We can reduce congestion, climate gas emissions and air pollution if those that are able to work from home, does that say one day a week or more. Or we can reduce morning and afternoon congestion by working from home an hour or two in the morning before travelling. In this way, the number of job travels are reduced or shifted timewise and hence pollution is reduced. The world becomes more sustainable through such examples of behavior change. Such measures are also so much easier to implement than establishing huge monitors for toll road payments and road prizing. Behavior change does not need any new infrastructure that needs excavation by air polluting heavy machinery, production and transport of toll road equipment. All it takes to change behavior is knowledge, motivation and the right attitude.

The pandemic has also reminded of us the health side of our lives and could lead to an upgrade of issues like healthy buildings and environments. Planners will possibly find new challenges to be addresses like congestion in lifts and shared ventilation systems circulating air in buildings. The issue of more decentralized air supply and removal as well as issues related to recirculation of air will have to be addressed. Air filters will have to have a better quality and a better servicing routine to make sure they are changed regularly so that quality and efficiency is not compromised. High efficiency particulate air (HEPA) filters will have to be explored and applied. All this will lead to a new way of thinking about how human beings share spaces and share air. The pandemic might already have changed our way of thinking in so many ways. Many of the changes are wise, ecological and long-term winners.

6. The Next Step and Conclusions

There will always be a next step, but the biggest mistake we can make is to ignore facts. When something is not working, we must step back, have a look at it again and either fix it or scrap it and start again. Not accepting being on the wrong track can lead to big damage. Progress is only progress if we accept that we sometimes do mistakes. Learning from mistakes is crucial. The process of innovation is sometimes cruel when having to kill a good idea—a "darling". However, this is a necessary process in order to proceed with some kind of chance of success.

The above also applies to sustainability and ZEMCH. It is crucial that we constantly keep a critical eye on the development of new knowledge, preferably research-based knowledge, and make sure it is addressed and somehow included in the way we deal with innovation in everyday life. This is as crucial as it is to define where one builds and why. The world has many climatic zones—some are cold, and some are hot, there are many different cultures as well as many different economic levels. The range is so wide it makes it almost impossible to write or talk about housing people without defining exactly the context and the human condition on the ground. If the concept of ZEMCH is going to succeed even more, it needs to go through this motion too; clearly define the context as challenges differ in London, Calcutta and Stockholm [26].

What's next then? The simple answer is we have no idea that can be clearly cut in stone apart from this: The best forecasters and the ones we normally listen to, see on screens or read about are lousy forecasters. As shown in the chapter on forecasting they fail to deliver exact forecasters that are useful. Many "ordinary" people though do, basically because many of them are more unbiased. They do not have so many positions, earlier own forecasts or vested interests to defend. The late Beatle John Lennon, in his song and for his son "Beautiful Boy" wrote: "Life is what happens to you when you're busy making other plans" [27]. Lennon, while making plans, was shot dead. Humankind, while making plans were suddenly faced with a pandemic that has changed a lot of routines and of our thinking about the future. Seemingly impossible ideas become realistic and former realistic ideas impossible. Our main goal and the main message and conclusion from this contribution is hence that the best we can do is to stay open and receptive to new ideas, always, and hence expose ourselves to the difficulty of an insecurity as regards future vision. Although we have ideas, we must realize that we have no exact idea of what the future holds. However, what we at least can do is to try to stay on top of the flow of new information, stay updated, build on new research and always question cemented truths.

In conclusion, the ZEMCH issue could be turned on its head and we could look at it from the perspective of a homeless person. How, if it is a wish that is, can a homeless person find better shelter and how can we that work on ZEMCHs help? That is a challenge possibly more complicated and meaningful than many tasks we are posed with. As Victor Papanek argued in his book "Design for the Real World", we must make sure we use our capabilities and competencies to the best of mankind and design useful stuff [28]. As the illustrations in Figure 11 below indicate, life on the street is rough. Some have chosen it, most have not. Can we, through very small measures be of assistance and apply our imagination as the examples below show? Can huge advertising structures in cities house homeless in the space between a building and the front board? The space is available, the construction up. It takes minor readjustment to ensure proper ladder access and separate spaces. There could even be a legal requirement in cities, the "Thick-Ad Law", using thick advertising boards as shelter. It is a wild idea from the 1987 UN Shelter for the Homeless international competition. That year UN set the goal of abolishing homelessness by 2000. A useless, beautiful dream that is not anywhere near coming through. However, we must keep trying, like UN does [29,30].

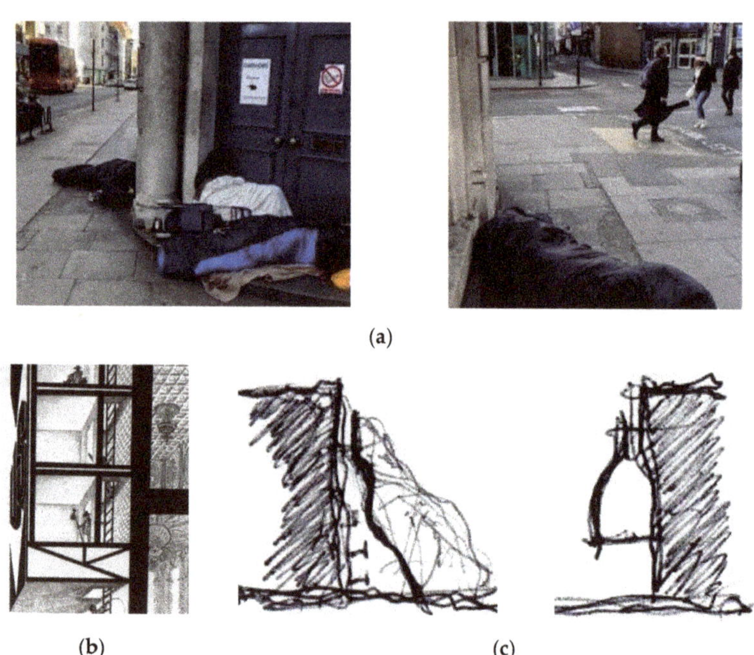

Figure 11. Targeting the client. Who are ZEMCHs for? Homeless asleep on a frosty London February morning in 2018 (**a**). Can we use the spaces "in between" for safe shelters (**b**)? Can huge advertising structures in cities house homeless in the space between a building and the front board. The space is available, the construction up. It takes minor readjustment to ensure proper ladder access and separate spaces. It is a wild idea from the 1987 UN Shelter for the Homeless international competition (**c**). All photos and illustrations: the author.

Funding: This research received no external funding.

Institutional Review Board Statement: Not applicable.

Informed Consent Statement: Not applicable.

Conflicts of Interest: The author declare no conflict of interest.

Entry Link on the Encyclopedia: https://encyclopedia.pub/9078.

References

1. IPCC. Global Warming of 1.5 °C. 2020. Available online: https://ipcc.ch/sr15/ (accessed on 1 March 2021).
2. Mernild, S. IPCC Author. Interview VG, Oslo 30.12.2020. Available online: https://www.vg.no/nyheter/i/qAJyp1/varmeste-aar-noensinne-og-varmere-skal-det-bli (accessed on 1 March 2021).
3. Røstvik, H.N. Does research results have an impact? *Adv. Build. Energy Res.* **2013**, *7*, 1–19. [CrossRef]
4. Hastings, R.; Wall, M. *Sustainable Solar Housing. Strategies and Solutions*; Earthscan/James & James: London, UK, 2007; ISBN 9781844073252.
5. Carson, R. *Silent Spring*; Houghton & Mifflin: Boston, MA, USA, 1962; ISBN 9780618253050.
6. Toffler, A. *Future Shock*; Pan Books Ltd.: London, UK, 1971; ISBN 0-330-02861-8.
7. The UN Commission on Environment and Development Our Common Future. 1987. Available online: https://sustainabledevelopment.un.org/content/documents/5987our-common-future.pdf (accessed on 10 January 2021).
8. ISES. 50 Years. 1970–2020 on UN Rio Conference. 1992. Available online: https://www.uis.no/nb/solenergipioneren (accessed on 21 December 2020).
9. Noguchi, M. (Ed.) *Proceedings of the ZEMCH 2012 International Conference, Glasgow, UK, 20–22 August 2012*; ZEMCH Network: Glasgow, UK, 2012; Available online: http://www.zemch.org/proceedings/2012/files/assets/basic-html/page83.html (accessed on 17 March 2021).
10. Gholami, H.; Røstvik, H.N. Economic analysis of BIPV systems as a building envelope material for building skins in Europe. *Energy* **2020**, *7*, 117931. [CrossRef]
11. Røstvik, H.N. *The Sunshine Revolution*; SunLab Publishers: Stavanger, Norway, 1992; ISBN 82-91052-03-4.
12. European Commission. DG Energy Altener program. In *Natural Ventilation in Buildings: A Design Handbook*; Allard., F., Santamouris, M., Eds.; James&James: London, UK, 1998; ISBN 1873936729.
13. The Guardian. Available online: https://www.theguardian.com/cities/2016/jan/19/log-fires-traffic-fumes-cause-bergen-air-pollution (accessed on 27 February 2021).
14. The Guardian. Available online: https://www.theguardian.com/environment/2020/dec/18/wood-burners-triple-harmful-indoor-air-pollution-study-finds (accessed on 27 February 2021).
15. Available online: https://www.newscientist.com/article/2119595-wood-burners-london-air-pollution-is-just-tip-of-the-iceberg (accessed on 25 February 2021).
16. Available online: https://www.nrk.no/norge/vedfyring-pa-hjemmekontor-forurenser-lufta-ute-1.15257570 (accessed on 25 February 2021).
17. Horne, T. *Den Store Klimaguiden*; Forlaget Press: Oslo, Norway, 2020; pp. 238–239. ISBN 978-82-328-0315-6.
18. Rudofsky, B. *Architecture Without Architects*; Academy Editions: London, UK, 1974; ISBN 0902620789.
19. Sennet, R. *Building and Dwelling. Ethics for the City*; Penguin Random House: London, UK, 2018; ISBN 978-0-713-99875-7.
20. Røstvik, H.N. *Corruption the Nobel Way: Dirty Fuels & The Sunshine Revolution*; Kolofon: Oslo, Norway, 2015; ISBN 978-82-300-1240-6.
21. The Guardian. Available online: https://www.theguardian.com/environment/2021/jan/27/climate-crisis-world-now-at-its-hottest-for-12000-years (accessed on 1 March 2021).
22. Fortidsminneforeningen. Available online: https://fortidsminneforeningen.no/aktuelt/har-gjenreist-skrivehuset-til-filosofen-ludwig-wittgenstein-i-sogn (accessed on 1 March 2021).
23. Røstvik, H.N. *Building Histories: Proceedings of the Fourth Conference of the Construction History Society, Cambridge, UK, 7–9 April 2017*; Campbell, J.W.P., Ed.; Queens College, University of Cambridge: Cambridge, UK, 2017; pp. 505–514. ISBN 9780992875138.
24. Chiambretto, B. *Le Cabanon*; Parentheses: Marseilles, France, 1987; ISBN 2863640461.
25. Tetlock, P.; Gardner, D. *Superforecasting*; Cornerstone: London, UK, 2015; ISBN 9781847947147.
26. Taylor, B.B. *Geoffrey Bawa*; Revised edition; Thames and Hudson: London, UK, 1995; ISBN 050027858-X.
27. Lennon, J. Beautiful Boy. LP/CD. In *Double Fantasy*; Geffen Records: Santa Monica, CA, USA, 1980.
28. Papanek, V. *Design for the Real World. Human Ecology and Social Change*; Thames and Hudson: London, UK, 1985; ISBN 0500273588.
29. Fathy, H. *Architecture for the Poor*; The University of Chicago Press: Chicago, IL, USA, 1973; ISBN 0-226-23915-2.
30. Illies, C.; Ray, N. *Philosophy of Architecture*; Cambridge Architectural Press: Cambridge, UK, 2014.

encyclopedia

MDPI

Entry

Solar Chimney Applications in Buildings

Haihua Zhang, Yao Tao * and Long Shi *

School of Engineering, RMIT University, Melbourne, VIC 3004, Australia; s3637231@student.rmit.edu.au
* Correspondence: yao.tao@rmit.edu.cu (Y.T.); shilong@mail.ustc.edu.cn (L.S.); Tel.: +61-03-9925-3215

Definition: A solar chimney is a renewable energy system used to enhance the natural ventilation in a building based on solar and wind energy. It is one of the most representative solar-assisted passive ventilation systems attached to the building envelope. It performs exceptionally in enhancing natural ventilation and improving thermal comfort under certain climate conditions. The ventilation enhancement of solar chimneys has been widely studied numerically and experimentally. The assessment of solar chimney systems based on buoyancy ventilation relies heavily on the natural environment, experimental environment, and performance prediction methods, bringing great difficulties to quantitative analysis and parameterization research. With the increase in volume and complexity of modern building structures, current studies of solar chimneys have not yet obtained a unified design strategy and corresponding guidance. Meanwhile, combining a solar chimney with other passive ventilation systems has attracted much attention. The solar chimney-based integrated passive-assisted ventilation systems prolong the service life of an independent system and strengthen the ventilation ability for indoor cooling and heating. However, the progress is still slow regarding expanded applications and related research of solar chimneys in large volume and multi-layer buildings, and contradictory conclusions appear due to the inherent complexity of the system.

Keywords: natural ventilation; solar chimney; Trombe wall; renewable energy; passive ventilation; building application

Citation: Zhang, H.; Tao, Y.; Shi, L. Solar Chimney Applications in Buildings. *Encyclopedia* **2021**, *1*, 409–422. https://doi.org/10.3390/encyclopedia1020034

Academic Editor: Antonio Frattari

Received: 29 March 2021
Accepted: 24 May 2021
Published: 27 May 2021

Publisher's Note: MDPI stays neutral with regard to jurisdictional claims in published maps and institutional affiliations.

1. Introduction to Solar Chimneys

Due to the potential benefits of passive ventilation systems in economic and energy conservation and resistance against noise and carbon dioxide emission [1–3], more research has focused on exploring and improving passive ventilation in the past 20 years. Passive ventilation strategies have been extensively studied over the years. According to local climate conditions and building characteristics, passive ventilation systems show different airflow characteristics and temperature distributions. Simultaneously, some passive ventilation systems also have heat dissipation and heat acquisition functions for space cooling and heating apart from providing adequate ventilation [3–9]. Most modern buildings rely entirely on mechanical ventilation, i.e., active ventilation systems, to satisfy indoor comfort. The majority of the energy supply is used for those active ventilation systems, occupying usable space due to its relatively large volume and structural complexity. Additionally, buildings with mechanical ventilation are often highly airtight to minimize energy consumption and heat loss, resulting in an inadequate fresh air supply [2].

Passive ventilation systems are increasingly being advocated as low-energy alternatives and low-cost solutions for energy conservation buildings. According to the pressure difference sources, typical modes of passive ventilation are referred to as wind-induced, buoyancy-driven, and hybrid ventilation [10,11]. Corresponding air movement is caused by wind pressure, temperature difference, or both of the above, and humidity difference [12]. It has been found that natural ventilation has the potential to provide adequate capacity for thermal regulation and satisfying indoor air quality in available climatic conditions without reliance on mechanical systems [10,13,14].

Passive ventilation systems rely on natural physical mechanisms, which make many uncertainties occur during operation. Wind-induced ventilation systems are solely dependent on prevailing wind speed and incident angle. The stochastics of wind direction and wind intensity bring significant challenges to system performance evaluation and design [15]. Buoyancy-driven ventilation builds upon the air intensity difference caused by the internal and external temperature difference, ventilating the space even in windless conditions. However, under extremely hot and humid climatic conditions (the temperature difference is insignificant), the system is probably not working properly. Not every passive ventilation system has the dual function of heating and cooling space driven by natural forces. The natural ventilation system can remove the stale warm airflow indoors by accelerating the air movement to provide a space cooling effect. Achieving heating usually requires collecting and storing heat gain and releasing heat when needed to increase the indoor temperature. As the most representative buoyancy ventilation system, the solar chimney has attracted researchers' attention because of its simultaneous ventilation, heating, and cooling functions.

A typical solar chimney is presented in Figure 1. It consists of an absorption wall, a glazing wall, tuyeres/vents, and heat-insulating materials. Airflow is affected by the air density difference between the internal and external environment and the external wind [16–19]. Stale air escapes from the purpose-built openings by the thermo-siphoning effect. The solar chimney components can employ direct or indirect solar energy to drive the airflow in the space. Quesada et al. [20,21] comprehensively reviewed the research on transparent and translucent solar facades in the past ten years based on theory and experiment and explored its development and applicability. The solar façade absorbs and reflects incident solar radiation and converts direct or indirect solar energy into usable heat. Jiménez-Xamán et al. [22] verified that a roof-top solar chimney applied to a single room for cooling purposes could increase the ventilation rate by 1.16–45.0%. The numerical code was generated to solve the conjugate turbulent heat transfer in a single room equipped with a solar chimney based on the coupling of CFD and global energy balances.

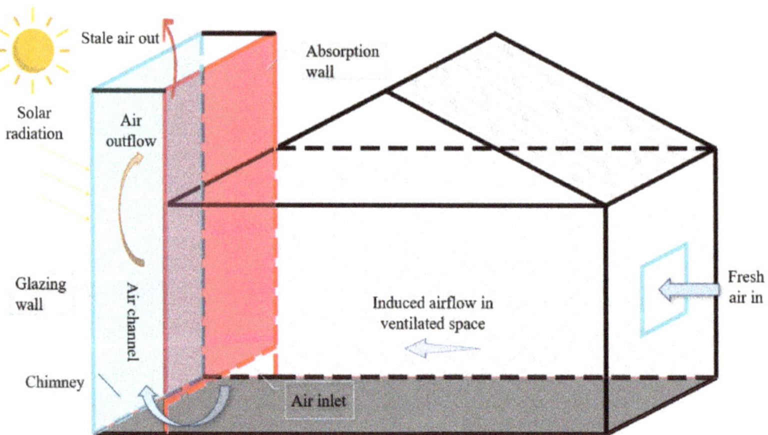

Figure 1. A schematic diagram of a solar chimney.

Solar chimneys stand out among many natural ventilation systems not only because of the convenience of their structural features when they are integrated into buildings or in conjunction with other ventilation systems but also because the solar chimney has heating and cooling modes through the cooperation of damping and openings, which makes the structure more sustainable. Figure 2 presents two modes that a solar chimney can achieve in the cooling season and heating season. In order to improve thermal comfort and enhance the applicability of natural ventilation, Monghasemi et al. [6] summarized the existing combined passive ventilation system based on solar chimneys and investigated the thermal regulation of the selected systems and their ability to improve ventilation efficiency.

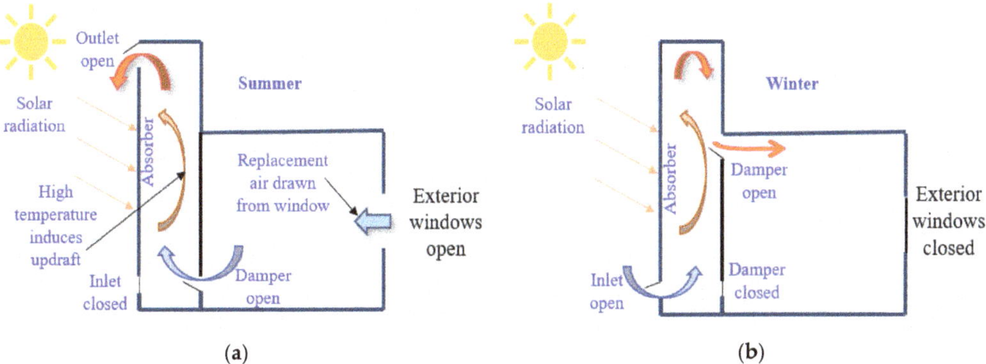

Figure 2. Schematic of solar chimney under heating and cooling modes: (**a**) cooling mode of a solar chimney; (**b**) heating mode of a solar chimney. Revised from [23].

2. Current Research Methods of the Solar Chimney

According to the review of Chen et al. [24,25], the existing Computer-Aided Prediction Models involve numerical models (usually assisted with CFD), analytical models, empirical models, full-scale or small-scale experimental models, multizone network models, and zonal models. A considerable part of the research on natural ventilation methods is based on experimental methods coupled with numerical simulations. Because of their complexity and the limitations of modeling methods, most previous studies are without detailed parametric analysis or merely focus on validating analytical models in corresponding experiments. According to the ref. [1], only 31% of the research used analytical methods, 10% of which used pure analytical methods, and the combination of analytical methods and experiments accounted for 13%. The remaining part used analytical methods combined with a numerical value. Based on simplified assumptions, different analytical models are capable of solving a series of energy-balance equations under specific scenarios, which in turn would limit the establishment of universal analytical models and solutions.

A prevailing method in passive ventilation uses computational fluid dynamics (CFD) to predict the air distribution and temperature distribution. Using CFD modeling can save time in simulating natural convection and predicting ventilation performance, allowing researchers to quickly develop optimal strategies for improving ventilation performance [26]. CFD can also analyze the multiple flow regimes caused by different driving forces (such as Laminar and turbulent) from an intuitive multi-dimensional perspective [27]. With the aids of CFD simulation, Kong et al. [28] determined the optimal inclination angles for the single-chamber roof solar chimney to achieve optimal ventilation performance. Sundar et al. [29] examined eighteen cases under different heat flux intensities and geometrical parameters with the aid of CFD and an experimental method for an inclined solar chimney. Nguyen and Wells [30] used CFD models to predict the ventilation rate and thermal efficiency of wall solar chimneys with four types of adjacent walls and different chimney configurations, and the heat source location. Salari et al. [31] developed a three-dimensional quasi-steady CFD model of a compound solar chimney with the photovoltaic module and phase change material and verified the three types of combined system performance.

With the aid of CFD-based computational methods, previous studies tried to find the optimal solution to enhance natural ventilation through geometric modification of the chimney configuration and the prediction of the flow pattern in the space connected to the solar chimney. The accuracy of CFD's prediction of the performance of solar chimneys applied to natural ventilation in buildings has been widely recognized. Section 3 further discusses the influencing parameters of solar chimney performance in previous studies.

3. Influencing Factors of Solar Chimney Performance

Researchers have conducted many parametric studies to find the influencing parameters directly related to the ventilation rate. Shi et al. [32] summarised four groups of influencing factors (a total of thirteen influencing parameters) on solar chimney performance, including the configurations of solar chimney and buildings, installation methods, material properties, and the external environment. Changing the configuration parameters of the system is the most convenient and effective way, which has been reflected in many documents in detail, so this part is not presented in detail here. The influence of external environmental factors is usually related to solar radiation intensity and wind effect. With the continuous emergence of new materials, the properties of different materials also directly affect the performance of the solar chimney. The aggregated influencing factors are listed in Figure 3.

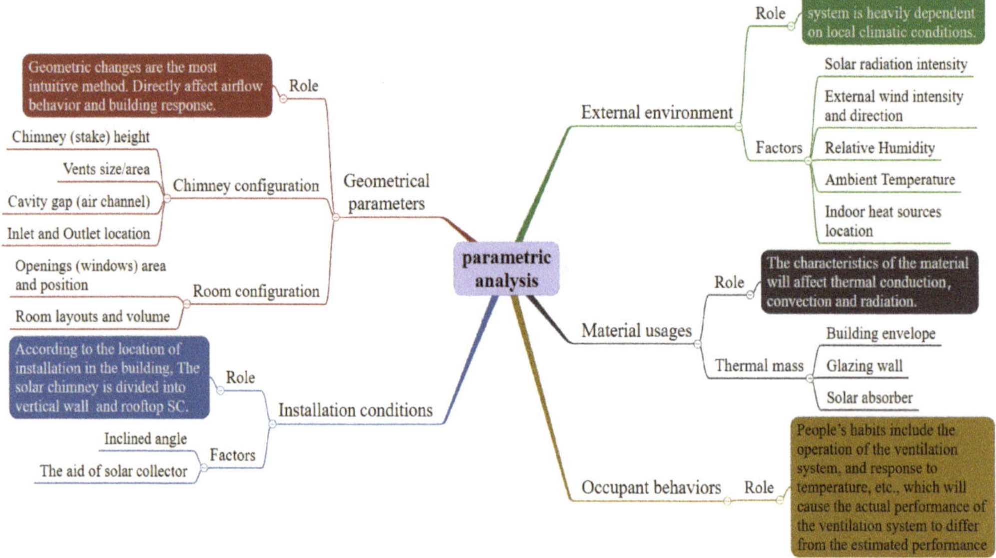

Figure 3. The influencing factors on the performance of the solar chimney.

The stack effect intensity is associated with aperture positions and the air density difference between inlet and outlet. Solar radiation intensity as one of the external environmental factors can result in a straight improvement in buoyancy-driven ventilation efficiency without considering the external wind effect. Menchaca-Brandan et al. [33] emphasized the importance of considering radiative effects in theoretical models or numerical simulations and experimental setup. The correct establishment of the radiation model will affect the accuracy of the prediction of ventilation efficiency and the accuracy of assessing the thermal comfort of indoor residents. Due to the buoyancy effect brought by the temperature difference, it is known that buoyancy-driven ventilation is preferable in temperate climates [34].

Excepting solar radiation, wind plays an equally critical role in influencing airflow patterns and thermal behavior. Some studies also compared the concurrent influences of solar radiance and external wind on the solar chimney system, and sometimes one driving force could be dominant under a particular condition. Additionally, the effect of wind factors on stack ventilation may be beneficial or unfavorable, depending on the pressure difference caused by the combination of wind pressure and air density difference at the air inlet, whether positive or negative. Shi et al. [35] investigated the interaction between solar chimney performance and external wind numerically and theoretically. When considering the influence of wind, the performance of solar chimneys depends not only on the wind speed but also on the incident angle. Based on the theoretical model developed, the critical wind speed was proposed. Wang et al. [36] verified that the dependence of solar chimney performance and airflow characteristics on solar radiation was reduced when the external wind was taken into account. It was found that keeping the vent size constant under the influence of wind, the optimal design value of the chimney cavity is 0.4–0.5 m, which is different from the previous optimal value of 0.2–0.3 m, which only considers the effect of buoyancy. Even at low wind speeds, solar chimney performance has been enhanced.

The coupling effect of the thermal mass and the buoyancy effect of naturally ventilated buildings has also attracted researchers' attention. Yang and Guo [37] analyzed the non-linearity of the coupling effect between the thermal mass and the stack effect under an external heat source theoretically. The thermal behavior and ventilation rate fluctuations caused by coupling effects were discussed. Thermal-mass-integrated PCM provides an effective method of thermal storage [38]. Through the phase transition, combined with the thermal mass of PCM, naturally ventilated space maintains a relatively uniform temperature. It can significantly reduce the cooling load of the building in mild weather. Especially in the case of night ventilation, the role of PCM becomes more prominent [39,40]. Vargas-Lopez et al. [41] presented an extensive review of transient mathematical models based on global energy balance models of solar chimneys with/without PCMs. It is further developed the mathematical models for a double-channel solar chimney integrated with PCMs.

Figure 4 shows the controlling parameters used to evaluate the performance of passive ventilation systems. There are two main parameters used to evaluate and characterize the efficiency of passive ventilation systems from ventilation rate perspectives: air exchange efficiency and ventilation effectiveness [18]. Additionally, thermal efficiency is usually related to indoor temperature regulation and humidity control. For the specific values of each indicator, refer to ref. [42].

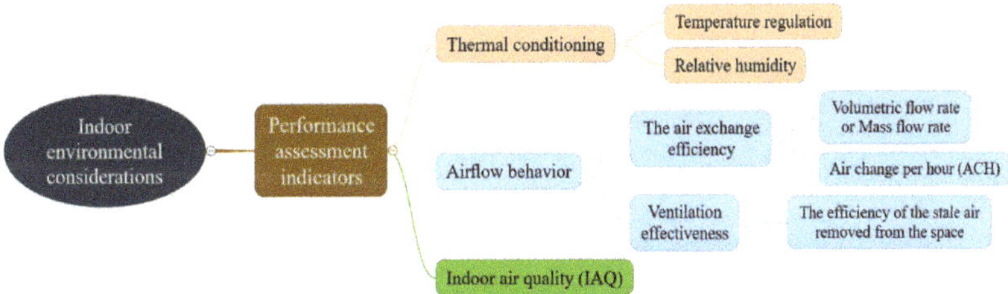

Figure 4. Indicators of performance.

Based on the analysis of the influence of control parameters, researchers have carried out extensive research on enhancing stand-alone solar chimney performance. The performance of the solar chimney can be significantly improved by selecting the optimal design parameters. Khanal and Lei pointed out [1] that previous studies related to solar chimneys aim to find optimum design solutions for enhancing natural ventilation, considering different design parameters. Most of the research is to optimize the system performance by changing the configuration of the solar chimney components and room openings from the geometric aspect. Enhancing natural ventilation will first consider changing the system's configuration, mainly by changing the geometric parameters to determine the design parameters corresponding to the optimal ventilation efficiency [32]. Second, the sensitivity of system performance to external environmental parameters is considered to verify the applicable conditions of the system [17,32]. Thirdly, based on the analysis of the response, ventilation rate, and thermal behavior of the naturally ventilated building, the desired effect can be achieved by changing the system or building envelope materials, such as solar absorptance of the heat storage materials, heat transmittance, and solar reflectance of glazing cover [17,43]. Shi et al. [32] comprehensively summarized the correlation between different controlling parameters and solar chimney performance. The diversity of influencing parameters also brings variables to the design of solar chimneys according to local conditions. Zhang and Shi [44] clarified the optimal design value to improve the performance of the solar chimney by reviewing three factors: chimney configurations, installation conditions (chimney installation angle), and material properties.

In addition to geometrically changing the configuration parameters of the solar chimney and the ventilated room, recent studies have shown that some attempts to add components into the chimney cavity can also improve ventilation efficiency. Sheikhnejad et al. [45] employed a passive vortex generator (VG) in the air channel to enhance heat transfer. The effectiveness of this configuration improvement was successfully verified by using simulation technology. Among many studies on improving the heat storage capacity of absorption walls, it is worth mentioning that PCM integration into the solar chimney has been verified to affect ventilation rates and thermal comfort positively. By integrating PCM, a solar chimney can effectively store solar energy during the day and release heat at night, thereby improving the solar chimney performances and making the indoor temperature uniform [46]. Dordelly et al. [47] investigate the influence of integrating a PCM on the performance of two laboratory prototypes of solar chimneys and verified the effectiveness of integrating solar chimney and PCM to improve ventilation efficiency.

Most previous parameterization studies have been based on predicting ventilation rates and building response through geometric modifications. Critical considerations for improving natural ventilation performance include introducing different configurations, such as the inclination position of the roof-top solar chimney, window-to-wall ratio, cavity gaps, stack height, and orifices areas. In addition, there are some studies on the influence of thermal mass and the transmittance and reflectivity of glass materials on ventilation performance. Some parameter analysis also focuses on developing various mathematical models to examine the correlation between airflow or thermal performance and the design variables.

4. Works Have Been Done at RMIT University, Australia

Figure 5 presents the contributions of RMIT University in the passive ventilation strategies over the years. Researchers at RMIT University have explained and presented a relationship between controlling parameters and ventilation capacity in buoyancy-driven ventilation. Referring to the parameterization studies on the various aspects mentioned in Section 2, researchers verified the influence of control parameters on the performance of independent solar chimneys and extended the solar chimney practice to the field of fire prevention and smoke control. Shi et al. [48,49] developed empirical models to predict the airflow rate through the air inlets under natural ventilation and smoke exhaustion. The effect of interaction mechanisms between the air inlet and the room window on the performance of the solar chimney was investigated.

Additionally, Shi et al. [50] also promoted the practical application of solar chimneys, which laid the conditions for future field measurement. Cheng et al. [51] identified the factors that affect the natural ventilation efficiency and smoke suppression of a typical solar chimney. Regarding the dual functions of the solar chimney on energy conservation and fire safety, Shi et al. [50] developed a consistency coefficient to evaluate the correlation between controlling parameters and the solar chimney performance under the two modes in actual practice in Melbourne, Australia. The effectiveness of the solar chimney as a means of ventilation and smoke extraction was confirmed. Based on this actual project, Shi et al. [52] further conducted an optimization design of a solar chimney. Simultaneously, the realization of solar chimney ventilation and smoke exhaust in the tunnel is related to traffic safety and environmental impact. Cheng et al. [14] investigated four influential factors that govern solar chimney effectiveness in tunnel applications by developing a numerical model. Chimney height and the cavity gap become the dominant factors affecting the performance of solar chimneys in the two modes.

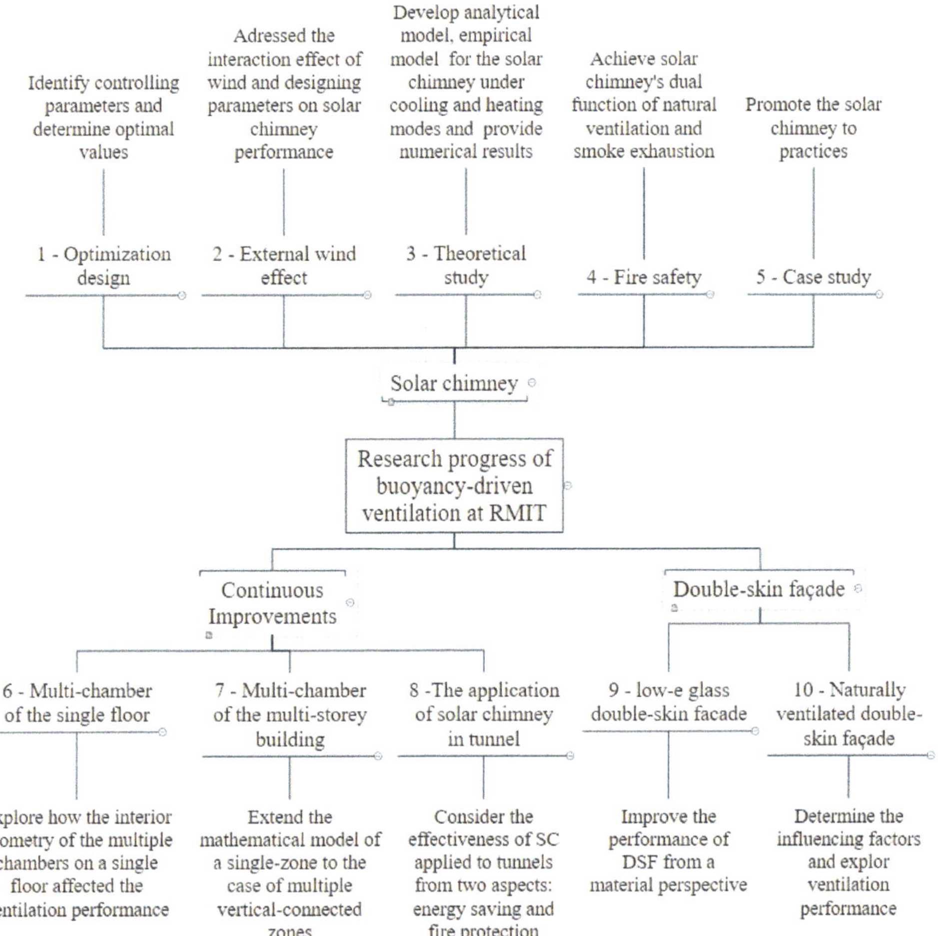

Figure 5. Research contributions on the solar chimney at RMIT University.

Through different research methods, Dr. Shi has made significant contributions to solar chimney research and promotion. Shi et al. [48] conducted the parametric analysis considering installation methods, room/chimney configuration, and cavity materials, thus developing the empirical models with the assistance of a fire dynamics simulator (FDS) to predict typical solar chimney performance. In another study, aiming at roof solar chimney, Shi et al. [19] not only summarized the previous mathematical models, the input parameters of testing ranges, and their experimental tests but also established the empirical models based on the experimental data of various general test rigs to predict the flow rate of roof-top solar chimneys and pointed out the configuration parameter values required to enhance ventilation rate. Shi et al. [53] developed theoretical models on heating and cooling modes of the wall solar chimney to predict the volumetric flow rate and temperature. Four solar chimney types were analyzed theoretically, considering room and chimney configurations and different fresh air supply methods. Consequently, the theoretical and experimental research on applying solar chimneys in single-chamber rooms tends to be consummated.

The emergence of the double-skin façade (DSF) with an air channel as one of the strategies for enhancing natural ventilation makes buoyant ventilation cater more to modern buildings' functional and aesthetic needs due to its excellent acoustic insulation, thermal insulation, and transparent appearance. Solar-assisted passive ventilation systems attached to the building envelope have similar operational mechanisms and structural composition. Using solar-assisted thermal convection and stack effects, the buoyant airflows moving through the openable channels interconnected flow paths can ventilate the occupant spaces. Tao et al. [54] explored the correlation between the airflow behavior of the double-layer façade and the room configuration, coupled with the environmental factor on buoyancy-driven natural ventilation, by adopting experimentally validated numerical models. It revealed the best cavity gap and exhaust port height based on the simulation results and found the influence of the room window's size and position on the ventilation rate. The numerical analysis of the naturally ventilated double-skin façade (NVDSF) confirmed the ability of NVDSF to induce indoor airflow and its energy-saving potential. Tao et al. [55] proposed a low-emissivity glass double-skin façade to reduce the indoor heating demand, considering solar radiation and natural convection. Additionally, the study also obtained the optimal air channel value of 0.15–0.3 m.

Research to improve natural ventilation mechanisms in sustainable buildings is ongoing. Researchers at RMIT used experiments, analysis, and simulation methods to conduct extensive research on thermosyphon air channels represented by solar chimneys with thermal buoyancy as the main driving force for air movement. The studies involve many considerations in designing the essential elements of solar chimneys and similar passive ventilation systems.

5. Potential Trends and Challenges of the Solar Chimney

The theoretical studies of solar chimneys in multiple chambers are scarce to support the view that it is feasible to use buoyancy ventilation in buildings with more rooms. Research on the application of independent solar chimneys in a single space has been extensively studied. However, with the increase in volume and structural complexity of modern buildings, the performance of solar chimneys in the multi-chamber is insufficient. The lack of theoretical research has led to the limitations of the solar chimney in the practical application of multi-chamber spaces. The studies on buoyancy-driven ventilation in a multi-chamber building can be divided into two types: the multi-chamber of the single floor and the multi-chamber of the multi-storey building. In recent years, only a few studies have involved the application of solar chimneys in multi-storey buildings. Punyasompun et al. [56] established a three-storey prototype building with a wall-integrated solar chimney and validated the analytical model by experimental data. The optimal structure of the multi-layer solar chimney is verified. That is, only one outlet is set at the top. It was found that the indoor temperature of a multi-story building using solar chimneys was 4–5 °C lower than that of a building without solar chimneys. By establishing an analytical model and conducting the numerical simulation, Yang and Li [57,58] compared the ventilation efficiency of a 10-story building prototype with/without a wall-integrated solar chimney with buoyancy ventilation and mixed ventilation mode and deduced a dimensionless expression and further determined the value of control parameters.

Moreover, the necessary conditions to ensure adequate natural ventilation include natural forces available, building characteristics, exterior envelope air infiltration and airtightness, and occupant behaviors [59–61]. Among them, the envelope's properties with the most direct contact with the outside environment have become the leading research subjects. The coupling effect between the thermal mass of building envelopes and the stack effect has been projected from the improvement methods of the buoyancy-driven ventilation efficiency. However, such coupling effects studies are insufficient for the practice of solar chimneys in previous studies, whether through experimental verification or numerical simulation. Significantly, the practice of solar chimneys in conjunction with different building forms and climatic conditions requires sufficient case studies and theoretical guidance.

Although some studies draw attention to the attempts of combined systems over the years, parameterization research and theoretical analysis are far from enough to meet practical application needs. In a given environment, a coupling of multiple ventilation systems can compensate for each other's performance and increase the service life of a single system. Figure 6 shows the independent passive systems combined with solar chimneys based on the selected literature review. Zhang et al. [42] categorized and reviewed the cross-application and combination potentials between independent natural ventilation systems. Compared to the conventional stand-alone passive ventilation system, considerably less research has been conducted on the dynamics and performance of multiple passive-assisted ventilation technologies in sustainable buildings. The study exhaustively recorded considerable attempts in natural ventilation in the past 20 years. Researchers mostly combined solar chimneys with other stand-alone systems and applied them to actual structures in the previous literature.

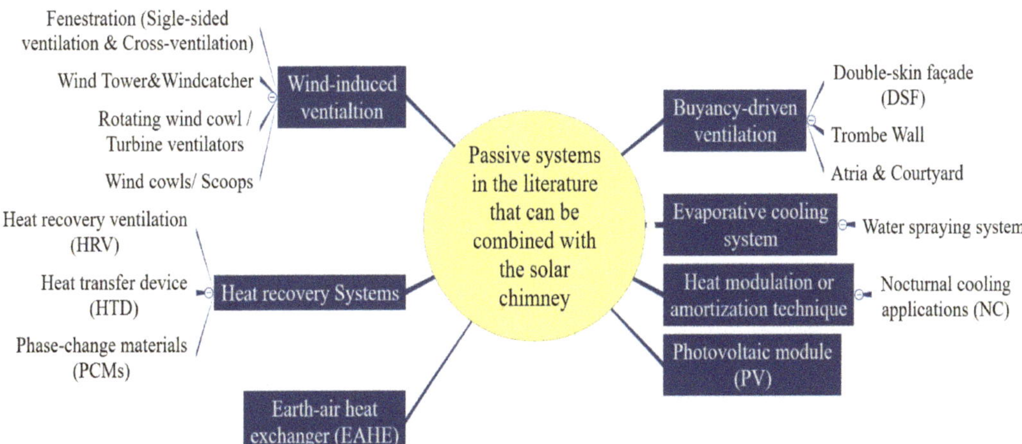

Figure 6. The coupling of passive systems and a solar chimney.

Based on structural superiority and economic and environmental benefits, the solar chimney has attracted growing interest as an excellent member of passive ventilation systems and advocated for as an auxiliary or alternative system for mechanical ventilation. The current research trend is more focused on combine PV modules with solar chimneys by coordinating components. Sivaram et al. [62] pointed out the commercial applicability of building-integrated passive solar energy technology in which photovoltaic modules and solar still are integrated into a single building configured with a solar chimney. The research of combined systems also shows diversification: the combination of multiple systems to maximize energy saving. Chandavar [63] explored the effectiveness of coupling the solar chimney and PV modules by field tests. As two self-sustained systems, the two components complement each other in their work, thus improving thermal efficiency. With the increasing demand for energy-saving, the combination of different ventilation systems is not only limited to double combinations, or triple combinations will get more attempts. Sakhri et al. [64] integrated a windcatcher, solar chimney, and earth-to-air heat exchanger into a test room and conducted a full-scale experiment. Ahmed et al. [13] suggested that a potential solution to combat the heatwaves in warm climates is a combination of a solar chimney, a windcatcher, and evaporative cooling. The related renewable energy systems that could be combined with a solar chimney are listed in Figure 6.

However, the passive ventilation system's performance prediction depends on the local climatic conditions, the experimental settings, and the selection of the prediction method. Therefore, it is challenging and unpredictable to make quantitative comparisons between different passive ventilation combinations. In order to formulate the accessible strategy according to local conditions, not only indispensable theoretical support is needed, but also sufficient case analysis and experimental data are ensured for well-designed combinations, so as to determine the correlation and dependence between different controlling factors and weaken the unexpected correlation of different parameters.

6. Discussion and Conclusions

It is evident from extensive literature that a solar chimney is an excellent passive ventilation strategy used to enhance natural ventilation and provide thermal comfort. The effectiveness of the solar chimney on the improvement of ventilation rate and thermal regulation has been studied by various researchers numerically and experimentally. Table 1 summarizes the ventilation efficiency of independent solar chimneys and passive systems combined with a solar chimney. The indicators for characterizing ventilation efficiency have an extensive range of changes due to different study conditions, such as local climate, experimental settings, prediction tools, and design parameter selection (i.e., geometry modification of solar chimneys and adjacent spaces). It can also be seen from Table 1 that the combined system significantly improves temperature regulation and ventilation rate.

Table 1. The performance of a solar chimney and combined systems.

Performance	Air Temperature Adjustment (°C)	Volumetric/Mass Flow Rate (m³/s, m³/h, or kg/s)	Air Change per Hour (ACH)	Energy-Saving Percentage (%)	Indoor Relative Humidity (%)
Stand-alone Solar chimney	Drop 1.0–5 °C [16,29,56,65–67]; 16.7 °C [68]	50–374 m³/h [16,65,69]; 0.019–0.033 m³/s [70]; 0.55–44.44 kg/s [22,29,71]; 70.6 m³/h~1887.6 m³/h [72];	0.16–15 [22,65,66,69,73,74]; 27.11 [68]; 30 [75];	12–50% [72,76]	
Integrated systems based on solar chimneys	**Drop**: 2.0–14 °C with water spaying [66,77]; 6.7–11.5 °C with evaporative cooling [78–81]; 3.2–9 °C with EAHE [82,83]; 5.2 °C with windcatcher [84]; 10–13 °C with a windcatcher and EAHE [64]; 8 °C with a Trombe wall and water spraying system [85]; with ; **Raises**: 14 °C with Trombe wall and PV [86]; 6.4 °C with EAHE [87].	1.4 kg/s with wind tower [88]; 0.0184 m³/s with EAHE [87]; 0.038–0.144 m³/s or 130.5 m³/h or 414 m³/h with evaporative cooling [78,80,81]	35–73 with a wind tower [88]; 12 with PV [89]; 2.42–4.33 with evaporative cooling [79]; 9 with windcatcher [84]	50% with a windcatcher and EAHE [64]; 75–90% with windcatcher [84]	Increases 17% with Trombe wall and water spraying system [85]; increases 28–45% [77]

An increasing number of studies tend to combine multiple energy-saving systems with SCs with vertical or inclined thermosiphon air channels as an effective passive ventilation enhancement strategy. The strengths of solar chimneys come from providing the desired airflow rate economically, with a simple structure, less space requirement, and ease of integration into existing building facades. Although the research on the application of solar chimneys in single chambers has been going on for a long time, there is no unified theoretical basis and sufficient experimental results for its application in multi-story buildings. Additionally, the significant differences in the research results in the literature indicate that solar chimneys have not yet been thoroughly studied and developed. More theoretical support and experimental verification are needed to enhance the universality of solar chimneys to be popularized in practical applications. Simultaneously, previous research results have revealed the effectiveness of the multi-purpose application of solar chimneys (i.e., heating and cooling spaces, fire prevention, and smoke control), which also makes the research of solar chimneys very promising.

This article investigates the current development of solar chimneys used in buildings to enhance natural ventilation. The working mechanism and typical structure of solar chimneys are briefly presented. The influencing parameters related to ventilation efficiency, temperature distribution, air pattern, and IAQ have been confirmed. This article explores current research gaps of solar chimneys and the various possibilities for enhanced performance. It was observed that a stand-alone solar chimney used in a single room and ideal climatic conditions could effectively provide indoor thermal regulation and enhanced ventilation rate, while a combined system based on solar chimneys can cope with more diverse climatic conditions and provide buildings with functional energy-saving solutions. Previous research has focused on achieving the optimal design strategy based on parameter analysis, including geometrically improving the configuration of the ventilation system and the subsequent response analysis of adjacent spaces. Some of the research hotspots of buoyancy-driven ventilation systems represented by solar chimneys also include the impact of building opening positions on natural ventilation and the coupling effect of the building's thermal mass on buoyancy-driven ventilation. However, since natural ventilation is heavily dependent on local climatic conditions and experimental settings under different working conditions, this brings challenges to quantitative analysis and parallel comparison of system performance. How to improve the stability of the system to cope with different climatic conditions and prolong the durability of the system has become a problem that needs to be solved. In addition, although there have been some attempts at hybrid natural ventilation, the practical application of passive-assisted ventilation systems in a building is limited due to the lack of a large amount of experimental and theoretical support for the coupling system. As the population continues to increase, it is not surprising that multiple rooms have natural ventilation scenarios. At present, only a small amount of research involves the application of solar chimneys in multi-chamber spaces, including single-story buildings with multi-chambers and multi-storey buildings. However, due to the increase in the amount of spaces, the dimensionality of the variables is undoubtedly increased, thereby increasing the difficulty of theoretical derivation. So far, there is no unified design guide for solar chimneys to enhance natural ventilation in multi-chamber scenarios.

Author Contributions: Conceptualization, L.S.; methodology, H.Z. and L.S.; formal analysis, H.Z. and Y.T.; investigation, H.Z.; writing—original draft preparation, H.Z.; writing—review and editing, L.S. and Y.T.; supervision, L.S. and Y.T.; funding acquisition, L.S. All authors have read and agreed to the published version of the manuscript.

Funding: This research was funded by the Australian Government through the Australian Research Council (DE200100892).

Data Availability Statement: Data sharing not applicable.

Conflicts of Interest: The authors declare no conflict of interest.

Entry Link on the Encyclopedia Platform: https://encyclopedia.pub/11130.

References

1. Khanal, R.; Lei, C. Solar chimney—A passive strategy for natural ventilation. *Energy Build.* **2011**, *43*, 1811–1819. [CrossRef]
2. Geetha, N.; Velraj, R. Passive cooling methods for energy efficient buildings with and without thermal energy storage—A review. *Energy Educ. Sci. Technol. Part A* **2012**, *29*, 913–946.
3. Santamouris, M.; Kolokotsa, D. Passive cooling dissipation techniques for buildings and other structures: The state of the art. *Energy Build.* **2013**, *57*, 74–94. [CrossRef]
4. Ascione, F. Energy conservation and renewable technologies for buildings to face the impact of the climate change and minimize the use of cooling. *Sol. Energy* **2017**, *154*, 34–100. [CrossRef]
5. Bhamare, D.K.; Rathod, M.K.; Banerjee, J. Passive cooling techniques for building and their applicability in different climatic zones-The state of art. *Energy Build.* **2019**, *198*, 467–490. [CrossRef]

6.   Monghasemi, N.; Vadiee, A. A review of solar chimney integrated systems for space heating and cooling application. *Renew. Sustain. Energy Rev.* **2018**, *81*, 2714–2730. [CrossRef]
7.   Sameti, M.; Kasaeian, A. Numerical simulation of combined solar passive heating and radiative cooling for a building. *Build. Simul.* **2015**, *8*, 239–253. [CrossRef]
8.   Zhai, X.Q.; Wang, R.Z.; Dai, Y.J. Solar chimney combined with underground vent for natural ventilation in energy saving of buildings. In Proceedings of the 3rd International Symposium on Heat Transfer Enhancement and Energy Conservation, Guangzhou, China, 16 January 2004; pp. 1117–1123.
9.   Hughes, B.R.; Chaudhry, H.N.; Ghani, S.A. A review of sustainable cooling technologies in buildings. *Renew. Sustain. Energy Rev.* **2011**, *15*, 3112–3120. [CrossRef]
10.   Chartier, Y.; Pessoa-Silva, C. *Natural Ventilation for Infection Control in Health-Care Settings*; World Health Organization: Geneva, Switzerland, 2009.
11.   Awbi, H. Basic concepts for natural ventilation of buildings. In Proceedings of the CIBSE BSG Seminar: Natural and Mixed-Mode Ventilation Modelling, London, UK, 24 May 2010.
12.   Etheridge, D.W.; Sandberg, M. *Building Ventilation: Theory and Measurement*; John Wiley & Sons: Chichester, UK, 1996; Volume 50.
13.   Ahmed, T.; Kumar, P.; Mottet, L. Natural ventilation in warm climates: The challenges of thermal comfort, heatwave resilience and indoor air quality. *Renew. Sustain. Energy Rev.* **2021**, *138*, 110669. [CrossRef]
14.   Cheng, X.; Shi, Z.; Nguyen, K.; Zhang, L.; Zhou, Y.; Zhang, G.; Wang, J.; Shi, L. Solar chimney in tunnel considering energy-saving and fire safety. *Energy* **2020**, *210*, 118601. [CrossRef]
15.   Horan, J.M.; Finn, D.P. Sensitivity of air change rates in a naturally ventilated atrium space subject to variations in external wind speed and direction. *Energy Build.* **2008**, *40*, 1577–1585. [CrossRef]
16.   Arce, J.; Jimenez, M.J.; Guzman, J.D.; Heras, M.R.; Alvarez, G.; Xaman, J. Experimental study for natural ventilation on a solar chimney. *Renew. Energy* **2009**, *34*, 2928–2934. [CrossRef]
17.   Lee, K.H.; Strand, R.K. Enhancement of natural ventilation in buildings using a thermal chimney. *Energy Build.* **2009**, *41*, 615–621. [CrossRef]
18.   Zhai, X.Q.; Song, Z.P.; Wang, R.Z. A review for the applications of solar chimneys in buildings. *Renew. Sustain. Energy Rev.* **2011**, *15*, 3757–3767. [CrossRef]
19.   Shi, L.; Zhang, G.M.; Cheng, X.D.; Guo, Y.; Wang, J.H.; Chew, M.Y.L. Developing an empirical model for roof solar chimney based on experimental data from various test rig. *Build. Environ.* **2016**, *110*, 115–128. [CrossRef]
20.   Quesada, G.; Rousse, D.; Dutil, Y.; Badache, M.; Halle, S. A comprehensive review of solar facades. Opaque solar facades. *Renew. Sustain. Energy Rev.* **2012**, *16*, 2820–2832. [CrossRef]
21.   Quesada, G.; Rousse, D.; Dutil, Y.; Badache, M.; Hallé, S. A comprehensive review of solar facades. Transparent and translucent solar facades. *Renew. Sustain. Energy Rev.* **2012**, *16*, 2643–2651. [CrossRef]
22.   Jiménez-Xamán, C.; Xamán, J.; Gijón-Rivera, M.; Zavala-Guillén, I.; Noh-Pat, F.; Simá, E. Assessing the thermal performance of a rooftop solar chimney attached to a single room. *J. Build. Eng.* **2020**, *31*, 101380. [CrossRef]
23.   Asfour, O. Natural ventilation in buildings: An overview. In *Natural Ventilation: Strategies, Health Implications and Impacts on the Environment*; Nova Science Pub Inc.: New York, NY, USA, 2015; pp. 1–26.
24.   Chen, Q.; Lee, K.; Mazumdar, S.; Poussou, S.; Wang, L.; Wang, M.; Zhang, Z. Ventilation performance prediction for buildings: Model assessment. *Build. Environ.* **2010**, *45*, 295–303. [CrossRef]
25.   Chen, Q.Y. Ventilation performance prediction for buildings: A method overview and recent applications. *Build. Environ.* **2009**, *44*, 848–858. [CrossRef]
26.   Jiru, T.E.; Bitsuamlak, G.T. Application of CFD in Modelling Wind-Induced Natural Ventilation of Buildings—A Review. *Int. J. Vent.* **2010**, *9*, 131–147. [CrossRef]
27.   Khanal, R.; Lei, C.W. Flow reversal effects on buoyancy induced air flow in a solar chimney. *Sol. Energy* **2012**, *86*, 2783–2794. [CrossRef]
28.   Kong, J.; Niu, J.; Lei, C. A CFD based approach for determining the optimum inclination angle of a roof-top solar chimney for building ventilation. *Sol. Energy* **2020**, *198*, 555–569. [CrossRef]
29.   Sundar, S.; Prakash, D.; Surya, V. Analysis and Optimization of Passive Wall Solar Chimney through Taguchi's Technique. *Appl. Sol. Energy* **2020**, *56*, 397–403. [CrossRef]
30.   Nguyen, Y.; Wells, J. A numerical study on induced flowrate and thermal efficiency of a solar chimney with horizontal absorber surface for ventilation of buildings. *J. Build. Eng.* **2020**, *28*, 101050. [CrossRef]
31.   Salari, A.; Ashouri, M.; Hakkaki-Fard, A. On the performance of inclined rooftop solar chimney integrated with photovoltaic module and phase change material: A numerical study. *Sol. Energy* **2020**, *211*, 1159–1169. [CrossRef]
32.   Shi, L.; Zhang, G.; Yang, W.; Huang, D.; Cheng, X.; Setunge, S. Determining the influencing factors on the performance of solar chimney in buildings. *Renew. Sustain. Energy Rev.* **2018**, *88*, 223–238. [CrossRef]
33.   Menchaca-Brandan, M.A.; Espinosa, F.A.D.; Glicksman, L.R. The influence of radiation heat transfer on the prediction of air flows in rooms under natural ventilation. *Energy Build.* **2017**, *138*, 530–538. [CrossRef]
34.   Chan, H.-Y.; Riffat, S.B.; Zhu, J. Review of passive solar heating and cooling technologies. *Renew. Sustain. Energy Rev.* **2010**, *14*, 781–789. [CrossRef]
35.   Shi, L. Impacts of wind on solar chimney performance in a building. *Energy* **2019**, *185*, 55–67. [CrossRef]

36. Wang, Q.; Zhang, G.; Li, W.; Shi, L. External wind on the optimum designing parameters of a wall solar chimney in building. *Sustain. Energy Technol. Assess.* **2020**, *42*, 100842. [CrossRef]

37. Yang, D.; Guo, Y.H. Fluctuation of natural ventilation induced by nonlinear coupling between buoyancy and thermal mass. *Int. J. Heat Mass Transf.* **2016**, *96*, 218–230. [CrossRef]

38. Kośny, J. *PCM-eNhanced Building Components: An Application of Phase Change Materials in Building Envelopes and Internal Structures*; Springer: Amsterdam, The Netherlands, 2015.

39. Piselli, C.; Prabhakar, M.; de Gracia, A.; Saffari, M.; Pisello, A.L.; Cabeza, L.F. Optimal control of natural ventilation as passive cooling strategy for improving the energy performance of building envelope with PCM integration. *Renew. Energy* **2020**, *162*, 171–181. [CrossRef]

40. Prabhakar, M.; Saffari, M.; de Gracia, A.; Cabeza, L.F. Improving the energy efficiency of passive PCM system using controlled natural ventilation. *Energy Build.* **2020**, *228*, 110483. [CrossRef]

41. Vargas-López, R.; Xamán, J.; Hernández-Pérez, I.; Arce, J.; Zavala-Guillén, I.; Jiménez, M.J.; Heras, M.R. Mathematical models of solar chimneys with a phase change material for ventilation of buildings: A review using global energy balance. *Energy* **2019**, *170*, 683–708. [CrossRef]

42. Zhang, H.; Yang, D.; Tam, V.W.; Tao, Y.; Zhang, G.; Setunge, S.; Shi, L. A critical review of combined natural ventilation techniques in sustainable buildings. *Renew. Sustain. Energy Rev.* **2021**, *141*, 110795. [CrossRef]

43. Barbosa, S.; Ip, K. Perspectives of double skin facades for naturally ventilated buildings: A review. *Renew. Sustain. Energy Rev.* **2014**, *40*, 1019–1029. [CrossRef]

44. Zhang, G.; Shi, L. Improving the performance of solar chimney by addressing the designing factors. In Proceedings of the IOP Conference Series: Earth and Environmental Science, Banda Aceh, Indonesia, 26–27 September 2018; p. 012010.

45. Sheikhnejad, Y.; Nassab, S.A.G. Enhancement of solar chimney performance by passive vortex generator. *Renew. Energy* **2021**, *169*, 437–450. [CrossRef]

46. Tiji, M.E.; Eisapour, M.; Yousefzadeh, R.; Azadian, M.; Talebizadehsardari, P. A numerical study of a PCM-based passive solar chimney with a finned absorber. *J. Build. Eng.* **2020**, *32*, 101516. [CrossRef]

47. Dordelly, J.C.F.; El Mankibi, M.; Roccamena, L.; Remiona, G.; Landa, J.A. Experimental analysis of a PCM integrated solar chimney under laboratory conditions. *Sol. Energy* **2019**, *188*, 1332–1348. [CrossRef]

48. Shi, L.; Zhang, G. An empirical model to predict the performance of typical solar chimneys considering both room and cavity configurations. *Build. Environ.* **2016**, *103*, 250–261. [CrossRef]

49. Shi, L.; Cheng, X.; Zhang, L.; Li, Z.; Zhang, G.; Huang, D.; Tu, J. Interaction effect of room opening and air inlet on solar chimney performance. *Appl. Therm. Eng.* **2019**, *159*, 113877. [CrossRef]

50. Shi, L.; Ziem, A.; Zhang, G.; Li, J.; Setunge, S. Solar chimney for a real building considering both energy-saving and fire safety—A case study. *Energ Build.* **2020**, *221*, 110016. [CrossRef]

51. Cheng, X.D.; Shi, L.; Dai, P.; Zhang, G.M.; Yang, H.; Li, J. Study on optimizing design of solar chimney for natural ventilation and smoke exhaustion. *Energy Build.* **2018**, *170*, 145–156. [CrossRef]

52. Shi, L. Optimization Design of Solar Chimney in GH Soppett Pavillion Mentone Reserve for Kingston City Council in Melbourne. 2018. Available online: https://www.researchgate.net/publication/326370433_Optimization_design_of_solar_chimney_in_GH_Soppett_Pavillion_Mentone_Reserve_for_Kingston_City_Council_in_Melbourne (accessed on 25 May 2021).

53. Shi, L. Theoretical models for wall solar chimney under cooling and heating modes considering room configuration. *Energy* **2018**, *165*, 925–938. [CrossRef]

54. Tao, Y.; Zhang, H.; Zhang, L.; Zhang, G.; Tu, J.; Shi, L. Ventilation performance of a naturally ventilated double-skin façade in buildings. *Renew. Energy* **2021**, *167*, 184–198. [CrossRef]

55. Tao, Y.; Zhang, H.; Huang, D.; Fan, C.; Tu, J.; Shi, L. Ventilation performance of a naturally ventilated double skin façade with low-e glazing. *Energy* **2021**, *229*, 120706. [CrossRef]

56. Punyasompun, S.; Hirunlabh, J.; Khedari, J.; Zeghmati, B. Investigation on the application of solar chimney for multi-storey buildings. *Renew. Energy* **2009**, *34*, 2545–2561. [CrossRef]

57. Yang, D.; Li, P. Natural ventilation of lower-level floors assisted by the mechanical ventilation of upper-level floors via a stack. *Energy Build.* **2015**, *92*, 296–305. [CrossRef]

58. Yang, D.; Li, P. Dimensionless design approach, applicability and energy performance of stack-based hybrid ventilation for multi-story buildings. *Energy* **2015**, *93*, 128–140. [CrossRef]

59. Pomponi, F.; Piroozfar, P.A.E.; Southall, R.; Ashton, P.; Farr, E.R.P. Energy performance of Double-Skin Façades in temperate climates: A systematic review and meta-analysis. *Renew. Sustain. Energy Rev.* **2016**, *54*, 1525–1536. [CrossRef]

60. Al Assaad, D.; Ghali, K.; Ghaddar, N. Effect of flow disturbance induced by walking on the performance of personalized ventilation coupled with mixing ventilation. *Build. Environ.* **2019**, *160*, 106217. [CrossRef]

61. Fernandez-Aguera, J.; Dominguez-Amarillo, S.; Alonso, C.; Martin-Consuegra, F. Thermal comfort and indoor air quality in low-income housing in Spain: The influence of airtightness and occupant behaviour. *Energy Build.* **2019**, *199*, 102–114. [CrossRef]

62. Sivaram, P.; Premalatha, M.; Arunagiri, A. Computational studies on the airflow developed by the building-integrated passive solar energy system. *J. Build. Eng.* **2021**, *39*, 102250. [CrossRef]

63. Chandavar, A.U. Quantifying the performance advantage of using passive solar air heater with chimney for photovoltaic module cooling. *Int. J. Energy Res.* **2021**, *45*, 1576–1586. [CrossRef]

64. Sakhri, N.; Moussaoui, A.; Menni, Y.; Sadeghzadeh, M.; Ahmadi, M.H. New passive thermal comfort system using three renewable energies: Wind catcher, solar chimney and earth to air heat exchanger integrated to real-scale test room in arid region (Experimental study). *Int. J. Energy Res.* **2021**, *45*, 2177–2194. [CrossRef]
65. Bansal, N.; Mathur, R.; Bhandari, M.J.B. Solar chimney for enhanced stack ventilation. *Build. Environ.* **1993**, *28*, 373–377. [CrossRef]
66. Chungloo, S.; Limmeechokchai, B. Application of passive cooling systems in the hot and humid climate: The case study of solar chimney and wetted roof in Thailand. *Build. Environ.* **2007**, *42*, 3341–3351. [CrossRef]
67. Gan, G.H. Simulation of buoyancy-induced flow in open cavities for natural ventilation. *Energy Build.* **2006**, *38*, 410–420. [CrossRef]
68. Villar-Ramos, M.; Macias-Melo, E.; Aguilar-Castro, K.; Hernández-Pérez, I.; Arce, J.; Serrano-Arellano, J.; Díaz-Hernández, H.; López-Manrique, L. Parametric analysis of the thermal behavior of a single-channel solar chimney. *Sol. Energy* **2020**, *209*, 602–617. [CrossRef]
69. Khedari, J.; Mansirisub, W.; Chaima, S.; Pratinthong, N.; Hirunlabh, J. Field measurements of performance of roof solar collector. *Energy Build.* **2000**, *31*, 171–178. [CrossRef]
70. Saleem, A.A.; Bady, M.; Ookawara, S.; Abdel-Rahman, A.K. Achieving standard natural ventilation rate of dwellings in a hot-arid climate using solar chimney. *Energy Build.* **2016**, *133*, 360–370. [CrossRef]
71. Layeni, A.T.; Waheed, M.A.; Adewumi, B.A.; Bolaji, B.O.; Nwaokocha, C.N.; Giwa, S.O. Computational modelling and simulation of the feasibility of a novel dual purpose solar chimney for power generation and passive ventilation in buildings. *Sci. Afr.* **2020**, *8*, e00298. [CrossRef]
72. Zha, X.Y.; Zhang, J.; Qin, M.H. Experimental and Numerical Studies of Solar Chimney for Ventilation in Low Energy Buildings. In Proceedings of the 10th International Symposium on Heating, Ventilation and Air Conditioning, Ishvac2017, Jinan, China, 19–22 October 2017.
73. Mathur, J.; Bansal, N.; Mathur, S.; Jain, M. Experimental investigations on solar chimney for room ventilation. *Sol. Energy* **2006**, *80*, 927–935. [CrossRef]
74. Khedari, J.; Boonsri, B.; Hirunlabh, J. Ventilation impact of a solar chimney on indoor temperature fluctuation and air change in a school building. *Energy Build.* **2000**, *32*, 89–93. [CrossRef]
75. Imran, A.A.; Jalil, J.M.; Ahmed, S.T. Induced flow for ventilation and cooling by a solar chimney. *Renew. Energy* **2015**, *78*, 236–244. [CrossRef]
76. Miyazaki, T.; Akisawa, A.; Kashiwagi, T.J. The effects of solar chimneys on thermal load mitigation of office buildings under the Japanese climate. *Renew. Energy* **2006**, *31*, 987–1010. [CrossRef]
77. Rabani, R.; Faghih, A.K.; Rabani, M.; Rabani, M. Numerical simulation of an innovated building cooling system with combination of solar chimney and water spraying system. *Heat Mass Transf.* **2014**, *50*, 1609–1625. [CrossRef]
78. Swiegers, J.J. Inlet and Outlet Shape Design of Natural Circulation Building Ventilation Systems. Master's Thesis, Stellenbosch University, Stellenbosch, South Africa, 2015.
79. Maerefat, M.; Haghighi, A.P. Natural cooling of stand-alone houses using solar chimney and evaporative cooling cavity. *Renew. Energy* **2010**, *35*, 2040–2052. [CrossRef]
80. Abdallah, A.S.H.; Yoshino, H.; Goto, T.; Enteria, N.; Radwan, M.M.; Eid, M.A. Integration of evaporative cooling technique with solar chimney to improve indoor thermal environment in the New Assiut City, Egypt. *Int. J. Energy Environ. Eng.* **2013**, *4*, 45. [CrossRef]
81. Abdallah, A.S.H.; Hiroshi, Y.; Goto, T.; Enteria, N.; Radwan, M.M.; Eid, M.A. Parametric investigation of solar chimney with new cooling tower integrated in a single room for New Assiut city, Egypt climate. *Int. J. Energy Environ. Eng.* **2014**, *5*, 92. [CrossRef]
82. Ramírez-Dávila, L.; Xamán, J.; Arce, J.; Álvarez, G.; Hernández-Pérez, I. Numerical study of earth-to-air heat exchanger for three different climates. *Energy Build.* **2014**, *76*, 238–248. [CrossRef]
83. Serageldin, A.A.; Abdeen, A.; Ahmed, M.M.; Radwan, A.; Shmroukh, A.N.; Ookawara, S. Solar chimney combined with earth to-air heat exchanger for passive cooling of residential buildings in hot areas. *Sol. Energy* **2020**, *206*, 145–162. [CrossRef]
84. Moosavi, L.; Zandi, M.; Bidi, M.; Behroozizade, E.; Kazemi, I. New design for solar chimney with integrated windcatcher for space cooling and ventilation. *Build. Environ.* **2020**, *181*, 106785. [CrossRef]
85. Rabani, M.; Kalantar, V.; Dehghan, A.A.; Faghih, A.K. Empirical investigation of the cooling performance of a new designed Trombe wall in combination with solar chimney and water spraying system. *Energy Build.* **2015**, *102*, 45–57. [CrossRef]
86. Elghamry, R.; Hassan, H. Experimental investigation of building heating and ventilation by using Trombe wall coupled with renewable energy system under semi-arid climate conditions. *Sol. Energy* **2020**, *201*, 63–74. [CrossRef]
87. Elghamry, R.; Hassan, H. Impact a combination of geothermal and solar energy systems on building ventilation, heating and output power: Experimental study. *Renew. Energy* **2020**, *152*, 1403–1413. [CrossRef]
88. Bansal, N.K.; Mathur, R.; Bhandari, M.S. A Study of Solar Chimney Assisted Wind Tower System for Natural Ventilation in Buildings. *Build. Environ.* **1994**, *29*, 495–500. [CrossRef]
89. Sivaram, P.; Mande, A.B.; Premalatha, M.; Arunagiri, A. Investigation on a building-integrated passive solar energy technology for air ventilation, clean water and power. *Energy Convers. Manag.* **2020**, *211*, 112739. [CrossRef]

Entry

SEM-PLS Approach to Green Building

Nasim Aghili [1,*] **and Mehdi Amirkhani** [2,*]

[1] Faculty of Health, Engineering, and Sciences, School of Civil Engineering and Surveying, University of Southern Queensland, Brisbane 4350, Australia

[2] ZEMCH EXD Lab, Faculty of Architecture, Building and Planning, The University of Melbourne, Melbourne 3010, Australia

* Correspondence: nasim71.aghili@gmail.com (N.A.); mehdi.amirkhani@unimelb.edu.au or mehdi@abiandesign.com.au (M.A.)

Definition: Green buildings refer to buildings that decrease adverse environmental effects and maintain natural resources. They can diminish energy consumption, greenhouse gas emissions, the usage of non-renewable materials, water consumption, and waste generation while improving occupants' health and well-being. As such, several rating tools and benchmarks have been developed worldwide to assess green building performance (GBP), including the Building Research Establishment Environmental Assessment Method (BREEAM) in the United Kingdom, German Sustainable Building Council (DGNB), Leadership in Energy and Environmental Design (LEED) in the United States and Canada, Comprehensive Assessment System for Built Environment Efficiency (CASBEE) in Japan, Green Star in Australia, Green Mark in Singapore, and Green Building Index in Malaysia. Energy management (EM) during building operation could also improve GBP. One of the best approaches to evaluating the impact of EM on GBP is by using structural equation modelling (SEM). SEM is a commanding statistical method to model testing. One of the most used SEM variance-based approaches is partial least squares (PLS), which can be implemented in the SmartPLS application. PLS-SEM uses path coefficients to determine the strength and significance of the hypothesised relationships between the latent constructs.

Keywords: energy management; green building performance; office building; SEM-PLS

Citation: Aghili, N.; Amirkhani, M. SEM-PLS Approach to Green Building. *Encyclopedia* **2021**, *1*, 472–481. https://doi.org/10.3390/encyclopedia1020039

Academic Editor: Nikos Salingaros

Received: 19 April 2021
Accepted: 17 June 2021
Published: 18 June 2021

Publisher's Note: MDPI stays neutral with regard to jurisdictional claims in published maps and institutional affiliations.

1. Introduction

The building sector is one of the most energy-intensive sectors, irrespective of its geographical location [1]. It is estimated that buildings' energy consumption accounts for approximately 32% of global energy use, and buildings are responsible for about 40% of the total energy-related carbon dioxide emissions [2]. This is due to the rapid depletion and inefficient use of natural resources and energy and increasing waste production in the building sector. However, buildings can conserve energy by appropriate resource management (RM) practices at the design, construction, and operation stages (see Figure 1).

Research suggests that implementing energy-efficiency strategies at the design stage can significantly influence building performance and reduce energy usage [3–5]. As such, there has been an increasing amount of literature on zero energy building construction and renewable energy sources [6]. These studies have investigated, proposed, and improved different parameters, indices, and approaches to improving building performance [7–9]. They primarily focus on enhancing building systems and integrating renewable energy at the design phase using various building performance simulation (BPS) tools. The most commonly used approach to applying BPS tools during the design phase is based on evaluating the simulation outcomes, and if not satisfactory, the design is altered until the desired results are achieved. One of the limitations of this approach is that it uses direct modelling and simulation workflow in which the impacts of modifying the values of parameters on energy performance are examined one at a time, without considering the combined impacts of parameters [10]. An accurate and simplified prediction method of users' comfort in buildings is another major challenge of using BPS tools. Several parameters, such as human behaviour in buildings, their interventions in design conditions, resource usage, and management, can significantly negatively impact the predicted energy saving. In practice, it is a challenge for building designers to assess the possibility of achieving predicted energy usage at the design stage. There is also little published data on user interventions and RM during the operation phase of the buildings and their impact on energy usage. Figure 1 gives a brief description of RM.

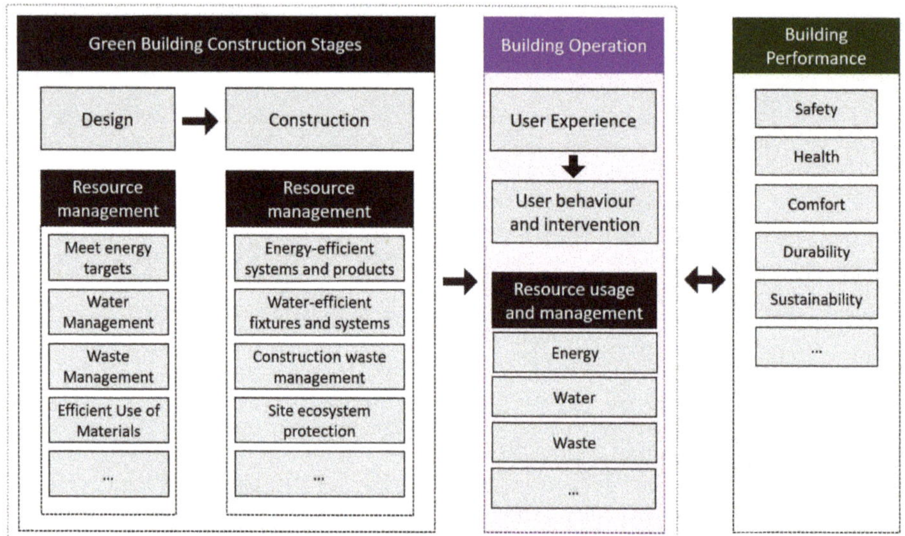

Figure 1. Building recourse management at different stages.

2. Resource Management

RM means attaining more with less. It is the management of natural usage resources by humans to provide the greatest advantage to current generations while retaining the capacity to meet future generations' requirements [11]. The current study declares challenges encountered in RM to improve building performance. According to Graham [12], significant amounts of natural resources are used by the building industry, and therefore many of the initiatives pursued to create ecology-sustaining buildings are concentrated on boosting the efficiency of resource consumption.

While there are several existing tools to optimise resource usage and management during the design and construction stages, the building occupants' experience and behaviour have received minimal attention. Studies have shown that harmful user interventions in green building design conditions resulting from experiencing discomfort could destroy predicted energy savings up to 75% [13,14]. Appropriate RM during building operation could improve building performance and user experience, reducing operating costs and enhancing end-users' health and well-being. The following sections explain RM's different factors, including water management, waste management, and energy management during building operation.

2.1. Water Management

Water management is one of the main elements of sustainable development. The world is facing water scarcity, and water shortages have become an issue worldwide. Additionally, water availability and quality are crucial throughout the building life cycle, and it is one of the most common natural resources used in buildings [15]. Building industries contribute to about 16% of freshwater consumption; hence greater emphasis was given to water efficiency and conservation in almost all the sustainable building rating schemes worldwide. Water management is one of the world's most significant challenges due to competition for limited resources, increasing global water need, regional disparities in water supply and affluence, aquifer depletion, pollution, and climate change-induced water stress. Accordingly, integrated sustainable water resource management requires innovation, progress, and international cooperation in the coming decades [16]. While water management in buildings, including water treatment, is so expensive, it is the best solution to conserve and protect future generations' water resources [17].

2.2. Waste Management

Waste is a pressing environmental, social, and economic issue, and it is one of the biggest challenges faced by every urban area in the world [18]. The waste is categorised into different types based on its:

- physical state (e.g., solid waste, liquid waste, gaseous waste),
- source (e.g., household/domestic waste, industrial waste, agricultural waste, commercial waste, demolition and construction waste, mining waste), and
- environmental impact (hazardous waste and non-hazardous waste) [19].

A large amount of construction waste is generated every year. Construction waste refers to damaged and surplus materials as the result of building activities, such as demolition, new construction, and renovations. While the construction industry has a significant impact on economic development, rising construction waste has become a serious global issue [20]. Therefore, due to the increase in construction waste production, there is a need to implement waste management in order to ensure the protection of the environment and natural resources for future generations.

Waste management has been widely recognised as a technical problem strongly influenced by various political, legal, socio-cultural, environmental, and economic factors and by the resources available to tackle it. The primary purpose of waste management is to practice optimisation with a broader resource conservation goal [18]. Construction waste management helps redirect reusable materials to appropriate sites and redirect recyclable resources back to the manufacturing process. Project waste should be recognised as an integral part of overall materials management [21].

During the operation phase of a building, a considerable amount of waste is produced by the occupants. According to Illankoon and Lu [22], waste generated during the operational phase of buildings consists of:

1. Food waste.
2. Cardboard and paper.
3. Plastics (including bottles and other containers).
4. Glass (including green, brown, and clear).
5. Metals (including aluminium cans and tin cans).

Waste management plans in the operation phase of buildings is generally divided into the following stages [23]:

1. Occupier source segregation.
2. Occupier deposit and storage.
3. Bulk storage and on-site management.
4. On-site treatment and off-site removal.
5. End destination of wastes.

Briefly, organised and accurate waste management implementation decreases negative environmental impacts (e.g., litter and, to a lesser extent, contamination of soil and water, etc.) [23].

2.3. Energy Management

Energy Management's (EM) principal objectives are the preservation of resources, preventing climate change, and cutting costs, as well as guaranteeing simple and ingrained access for all to the energy spectrum [24]. According to Danish [25], building EM is a complex and multifaceted function that depends on various factors based on the type of building. It is the constant process of managing devices that consume energy to improve building energy performance while minimising energy usage. Building EM is feasible through desirable building design and management. It is a branch of building services engineering [26]. Besides the economic costs of EM, energy requires further expenses, mostly environmental and societal, associated with its resource and waste depletion and contribution to climate change.

EM programs consist of three processes: energy auditing, energy targeting, and energy planning. Energy auditing is the process of profiling energy usage and identifying the energy-saving opportunities critical in a systematic approach for decision-making in EM. A process that can be used to determine the percentage of energy efficiency and save energy, energy targeting, is possible from energy audit results. Energy planning is the process of making decisions regarding energy saving in the preliminary stage, including support for energy policies, organisational structure, and implementation [27].

Buildings consume much energy during the operation phase [28]. Green building tools evaluate environmental designs and buildings' performance during the design phase to predict energy savings during building operation and achieve sustainable development [29]. However, poor EM and/or harmful human interventions in design conditions during the operation phase could significantly reduce predicted energy savings in green buildings [13]. Appropriate EM could help diminish building users' harmful interventions and achieve green building objectives. Therefore, it has persisted as a significant and contemplative issue for scholars to reduce increasing energy usage in recent decades [30].

This study focuses on the influence of EM on Green Building Performance (GBP) during the operation phase. It also uses and analyses data collected in previous studies in Malaysia by authors. The SmartPLS approach is used to illustrate the impact of EM on the performance of green office buildings. The current study develops a building performance framework based on these previous investigations. Additionally, various measurement items include benchmarking tools for specific building sectors. Table 1 shows EM items in the current study, which are adopted from the Canada Green Building rating [31]. The following section explains different factors of GBP.

Table 1. Energy management measurement.

No	Measurement Item	Description
1	EM vision	the organisation develops a vision statement that has the unambiguous target of reducing energy usage through specific activities and conveys them to all staff.
2	EM Current	the organisation evaluates the existing state of EM, set objectives, and defines specific and measurable actions for continuous enhancement.
3	EM Appointed	the organisation has appointed an energy manager.
4	EM Audit	the organisation conducts an energy audit exercise.
5	EM Advantage	the organisation takes advantage of utility and government incentive programs concerning energy.
6	EM Track	the organisation tracks project-level resource savings according to the International Performance Measurement and Verification Protocol (IPMVP) framework.
7	EM Performance	the organisation reports energy performance data to the management committee.
8	EM Provide	the organisation offers training and/or develops customised training programs to address capacity gaps in energy management.

3. Green Building Performance (GBP)

In British Standard (BS) 5240, building performance is described as a behaviour that can be used to illustrate a building's physical performance characteristics and its parts. Therefore, it is related to a building's capability of contributing to fulfilling the functions of its envisioned utilisation. Building performance can play a crucial role in understanding how the building behaves [32]. Regarding building performance, numerous scenarios exist, with higher complexity than ever. Modern society is increasingly demanding efficient, functional, environmentally friendly, robust, adaptable, durable, beautiful, comfortable, and healthy buildings. The environment is degrading at a worrying rate that puts continuous pressure on facility managers, engineers, builders, and architects to design and construct buildings capable of performing their best in every given condition [33]. As a result, obtaining and sharing knowledge concerning building performance is of high significance.

Green building is one of the measures that can relieve buildings' effects on the environment, society, and the economy [34]. Green building refers to the design and construction of buildings that positively impact the environment [18]. Green buildings primarily aim to enhance the residents' comfort and satisfaction while reducing environmental impacts and costs. GBP evaluates the green buildings' performance at different baselines such as transportation, energy, water, operations and maintenance, occupant satisfaction, waste generation, and recycling [35]. According to Foliente and Becker [36] and Hitchcock [37], there are several assessments of GBPs to achieve sustainable development. A review of the most frequently used items available now is presented and compares the items related to GBP in the Malaysian context. The current study develops a framework that includes safety, health and hygiene, comfort, durability, and sustainability (see Table 2).

Table 2. Building performance dimensions.

Measurement Items	Description	References
Safety	Buildings must enjoy safety. This issue is of high importance for architects, engineers, facilities, and building managers. To appraise the building performance regarding the building's safety, we must determine those directly affected by safety conditions.	[36–38]
Health and Hygiene	The term "healthy building" implies the influence that the building can impose on the occupants. The sick building syndrome (SBS) issue resulted in an increasing demand for healthier buildings for occupants. Different researchers have recognised the significance of health in destroying the performance of a building.	[36,37,39]
Comfort	Indoor comfort is a crucial parameter for enhancing building performance. A building must arrange for an internal environment with an acceptable comfortability level, particularly in thermal, visual, and acoustics comfort areas.	[36,37,40]
Durability	This covers the building materials' durability, which is a significant aspect of the building performance.	[36,37,40]
Sustainability	Sustainability leads to diminishing energy usage and CO_2 emissions.	[36,37]

4. Impact of EM on GBP

This study uses data collected by Aghili [41] and interpreted them differently. The data was gathered from Malaysia green buildings to understand EM and building performance. Since data were gathered from a particular source, it is essential to check the standard method variance. As such, we conducted Harman's single factor test by entering all the primary constructs in a component factor assessment.

The present study used sampling information due to easy access, geographical proximity, and availability at a specific time, and willingness to participate in the study. The respondents were Malaysian green building experts, managers, and facilitators certified by the green building index. A total of 89 certified green building managers, experts, and facilitators participated in this study. A structured survey questionnaire was developed to collect data.

Structural equation modeling (SEM) is a commanding statistical approach to model testing [42]. SEM is valuable since it helps the researcher test a number of hypothesised relationships simultaneously, makes available an indication of fit between the hypothesised model and actual data, and evaluates the alternative models [43]. SEM combines two approaches to model testing, i.e., factor analysis and multiple regression analysis. The regression analysis deals with the relationship between a criterion variable and predictor variables, while the factor analysis attempts to find a set of latent variables (i.e., factors). It explains the common variance that exists amongst a set of observed variables. The factor analysis is most commonly applied to determining the factor structure that underlies the scores in a set of questionnaire items [42,44]. Researchers consider two methods when SEM is applied: variance-based partial least squares and covariance-based techniques [45,46]. One of the most commonly used SEM variance-based approaches is Partial Least Squares (PLS) [46]. AMOS and LISREL applications could be used to exemplify covariance-based SEM analysis. PLS Graph and Smart-PLS could be used to implement SEM-PLS analysis. Accordingly, Smart-PLS software has been employed to achieve the research goal based on the research objective to investigate the relationship between variables and predict key target constructs. Furthermore, in this research, the sample size ($N = 89$) is small, and the research data is secondary as well. Therefore, SEM-PLS is well suited to this study. In this way, the current research examined the measurement model (validity and reliability) and the structural model (testing the relationship among variables).

4.1. Measurement Model

We analysed convergent validity and discriminant validity to evaluate the measurement model. The measurement model includes the unidirectional predictive relationships between each latent construct and its associated observed indicators.

4.1.1. Convergent Validity

Convergent validity refers to the degree that multiple items use in the research to measure the same concepts that are in agreement [44,47]. For reflective scale measurements, convergent validity is assessed through factor loadings of the items, average variance extracted (AVE), and composite reliability (CR) [48]. The results show that the factor loading of all the items is 0.5 or above, the AVEs of all the variables are greater than 0.5, and CR is above 0.7. Therefore, the convergent validity for scale analysis is achieved (Table 3). That way, it is feasible to determine the estimated model fit.

Table 3. The results of convert validity.

Variable	Items	Factor Loadings	CR	AVE	Cronbach's Alpha
EM	EM Audit	0.500	0.890	0.618	0.845
	EM Advantage	0.617			
	EM Appointed	0.806			
	EM Current	0.723			
	EM Performance	0.794			
	EM Provide	0.715			
	EM Track	0.750			
	EM Vision	0.566			
GBP	Comfort	0.819	0.857	0.552	0.791
	Durability	0.870			
	Health and Hygiene	0.762			
	Safety	0.530			
	Sustainability	0.690			

4.1.2. Discriminant Validity

Discriminant validity is the degree to which items distinguish between constructs. Two criteria are used to test discriminant validity in the current study, including Fornell–Larcker and cross-loadings [36,41,42]. Using the Fornell–Larcker criterion, results indicate that the square root of the average variance extraction is greater than the inter-construct correlations. Regarding the cross-loadings criterion, the factor loadings of each item or indicator must be greater than the rest of its cross-loadings to ensure the discriminant validity of the construct [44,49]. Table 4 illustrates the results of the discriminant validity measurements. The value on the diagonals was greater than the inter-construct correlation in its respective row. Thus, there is no issue of discriminant validity in the measurement model. Additionally, the results indicate that the item is producing the highest loading on its respective constructs. Thus, discriminant validity has been established.

Table 4. The results of discriminant validity analysis.

	EM	GBP
EM	0.786	
GPB	0.549	0.743

Note: Square root of average variance extraction.

4.2. Structural Model

To assess the structural model (path relationship), the R^2 value, standard beta, t-values through a bootstrapping process with a resample of 5000, the predictive applicability (Q^2), and the effect sizes (f^2) were considered [50]. As SEM-PLS's goal is to explain the endogenous latent variance, the key target is to have a higher R^2. The R^2 value is the exogenous variable that can explain the variance in the percentage representatives of the research model's predictive power and its values ranging between 0 and 1. The greater the value, the better the explanatory power of the model [51]. Cohen [52] argued that the values of R^2 ranging between 0.02 and 0.12 could be considered weak, values ranging between 0.13 and 0.25 could be considered moderate, and values equal to or greater than 0.26 could be considered substantial as a rule of thumb. According to Hair and Ringle [46], R^2 appropriateness depends on the research context. The results obtained in the present restudy show that the R^2 value was 0.358. This indicates that 35% of the variation in green building performance is caused by EM.

SEM-PLS uses path coefficients to define the strength and importance of the hypothesised relationships between the latent construct. These path coefficients could also be considered standardised beta coefficients [49]. Typically, in PLS-SEM, the bootstrapping technique is used to analyse the t-value for the path coefficients to assess the importance of hypothesised connections [46,53,54]. The standardised range of the path coefficient values is between -1 and $+1$. The standard estimate path coefficients near to $+1$ signify a strong positive linear association and vice versa for negative values [44]. In theory, the p-value is a constant measure of evidence, but it is usually dichotomised approximately into highly important, marginally important, and not statistically important at conventional levels, with cut-offs at $p \leq 0.01$, $p \leq 0.05$, and $p > 0.10$ [55]. Table 5 shows a significant relationship between EM and GBP (path coefficient between energy management and GBP was 0.623; t-value was 7.851; $p \leq 0.01$). Cohen's f^2 is used to recognise an appraisal of local effect size. Effect size evaluates the strength, size, or magnitude of the relationship between the latent variables. Based on the f^2 value, the omitted construct's effect size for an endogenous construct could be defined such that 0.02, 0.15, and 0.35 illustrate small, medium, and significant effects, respectively [56]. The results illustrate that the effective size is 0.557, demonstrating a significant correlation between EM and GBP. This means that the hypothesis was supported.

Table 5. The results of the structural model.

Hypothesis	Relationship	Standard Beta	Standard Deviation	t-Value	f^2	R^2
H1	EM $\leftrightarrow$ GBP	0.623	0.076	7.851	0.557	0.358

5. Conclusions and Prospects

The current study developed a building performance framework that included safety, health and hygiene, comfort, durability, and sustainability based on previous investigations. It has examined the direct influence of EM on the office building's overall performance. The findings have revealed that EM plays a significant role in improving GBP. The study highlights the importance of EM in achieving sustainable development in developing countries. It encourages green building managers, experts, and facilitators to apply EM to improve GBP. That way, EM improves the performance of the green office building in conserving and protecting energy resources and cuts down on bills to achieve sustainable development.

Further research can discern other management practices and evaluate each parameter of GBP's effects using comprehensive questionnaires that include closed and open questions. Furthermore, future researchers can use interviewing as a tool to gather the required data. The research recommendation is to employ empirical studies to further investigate the relevance between a variety of dimensions of green building management key practice and green building performance.

Author Contributions: Conceptualisation, N.A. and M.A.; methodology, N.A.; software, N.A.; validation, N.A. and M.A.; formal analysis, N.A.; investigation, N.A. and M.A.; resources, N.A. and M.A.; data curation, N.A.; writing—original draft preparation, N.A. and M.A.; writing—review and editing, M.A.; visualisation, M.A. and N.A.; project administration, N.A. and M.A.; All authors have read and agreed to the published version of the manuscript.

Funding: This research received no external funding.

Conflicts of Interest: The authors declare no conflict of interest.

Entry Link on the Encyclopedia Platform: https://encyclopedia.pub/12025.

References

1. Hamdaoui, S.; Mahdaoui, M.; Allouhi, A.; El Alaiji, R.; Kousksou, T.; El Bouardi, A. Energy demand and environmental impact of various construction scenarios of an office building in Morocco. *J. Clean. Prod.* **2018**, *188*, 113–124. [CrossRef]
2. Jin, X.; Wu, J.; Mu, Y.; Wang, M.; Xu, X.; Jia, H. Hierarchical microgrid energy management in an office building. *Appl. Energy* **2017**, *208*, 480–494. [CrossRef]
3. Abediniangerabi, B.; Makhmalbaf, A.; Shahandashti, M. Deep learning for estimating energy savings of early-stage facade design decisions. *Energy AI* **2021**, *5*, 100077. [CrossRef]
4. Fan, Y.; Ito, K. Integrated building energy computational fluid dynamics simulation for estimating the energy-saving effect of energy recovery ventilator with CO_2 demand-controlled ventilation system in office space. *Indoor Built Environ.* **2014**, *23*, 785–803. [CrossRef]
5. Shen, J.; Zhang, X.; Yang, T.; Tang, L.; Wu, Y.; Pan, S.; Wu, J.; Xu, P. The early design stage of a novel Solar Thermal Façade (STF) for building integration: Energy performance simulation and socio-economic analysis. *Energy Procedia* **2016**, *96*, 55–66. [CrossRef]
6. Rezaee, R.; Vakilinezhad, R.; Haymaker, J. Parametric framework for a feasibility study of zero-energy residential buildings for the design stage. *J. Build. Eng.* **2021**, *35*. [CrossRef]
7. Hu, M. *Net Zero Energy Building: Predicted and Unintended Consequences*; Routledge: Abingdon, UK, 2019.
8. Lechner, N.; Wallace, C. *Heating, Cooling, Lighting: Sustainable Design Methods for Architects*, 4th ed.; Wiley: Hoboken, NJ, USA, 2015.
9. Garg, V.; Mathur, J.; Bhatia, A. *Building Energy Simulation: A Workbook Using DesignBuilder*; CRC Press: Boca Raton, FL, USA, 2017.
10. Kerdan, I.G.; Raslan, R.; Ruyssevelt, P.; Gálvez, D.M. An exergoeconomic-based parametric study to examine the effects of active and passive energy retrofit strategies for buildings. *Energy Build.* **2016**, *133*, 155–171. [CrossRef]
11. Kusimo, H.; Oyedele, L.; Akinade, O.; Oyedele, A.; Abioye, S.; Agboola, A.; Mohammed-Yakub, N. Optimisation of resource management in construction projects: A big data approach. *World J. Sci. Technol. Sustain. Dev.* **2019**. [CrossRef]
12. Graham, P. *Building Ecology: First Principles for a Sustainable Built Environment*; John Wiley & Sons: Hoboken, NJ, USA, 2009.
13. Amirkhani, M. *Innovative Integrated Window Design with Electric Lighting Design System to Reduce Lighting Intervention in Office Buildings*; Queensland University of Technology: Brisbane, Australia, 2018.

14. Amirkhani, M.; Garcia-Hansen, V.; Isoardi, G.; Allan, A. Innovative window design strategy to reduce negative lighting interventions in office buildings. *Energy Build.* **2018**, *179*, 253–263. [CrossRef]

15. Al-Qawasmi, J.; Asif, M.; El Fattah, A.A.; Babsail, M.O. Water efficiency and management in sustainable building rating systems: Examining variation in criteria usage. *Sustainability* **2019**, *11*, 2416. [CrossRef]

16. Gray, S.; Semiat, R.; Duke, M.; Rahardianto, A.; Cohen, Y. Reference module in earth systems and environmental sciences. In *Seawater Use Desalination Technology*; Elsevier: Amsterdam, The Netherlands, 2011; pp. 73–109.

17. Kumar, D.P.; Satish, A.; Asadi, S. An analytical approach for evaluation of land resources management in construction industry—A model study. *Int. J. Civ. Eng. Technol.* **2018**, *9*, 105–114.

18. Ribić, B.; Voća, N.; Ilakovac, B. Concept of sustainable waste management in the city of Zagreb: Towards the implementation of circular economy approach. *J. Air Waste Manag. Assoc.* **2017**, *67*, 241–259. [CrossRef]

19. Amasuomo, E.; Baird, J. The concept of waste and waste management. *J. Mgmt. Sustain.* **2016**, *6*, 88. [CrossRef]

20. Lu, W.; Chi, B.; Bao, Z.; Zetkulic, A. Evaluating the effects of green building on construction waste management: A comparative study of three green building rating systems. *Build. Environ.* **2019**, *155*, 247–256. [CrossRef]

21. Kubba, S. *Green Construction Project Management and Cost Oversight*; Butterworth-Heinemann: Oxford, UK, 2010.

22. Illankoon, I.C.S.; Lu, W. Cost implications of obtaining construction waste management-related credits in green building. *Waste Manag.* **2020**, *102*, 722–731. [CrossRef] [PubMed]

23. Byrne, I. Construction Phase & Operational Phase Waste Management Plan—GlencairnSHD Residential Development. 2018. Available online: http://glencairnshd.ie/wp-content/uploads/2018/09/Glencairn-Waste-Managment-Plan-R5-21.08.18.pdf (accessed on 16 June 2021).

24. Islam, M.; Hasanuzzaman, M. Introduction to energy and sustainable development. In *Energy for Sustainable Development*; Elsevier: Amsterdam, The Netherlands, 2020; pp. 1–18.

25. Danish, M.S.S.; Senjyu, T.; Ibrahimi, A.M.; Ahmadi, M.; Howlader, A.M. A managed framework for energy-efficient building. *J. Build. Eng.* **2019**, *21*, 120–128. [CrossRef]

26. Moss, K. *Energy Management in Buildings*; Taylor & Francis: Abingdon, UK, 2006.

27. Ahmad, A.S.; Hassan, M.Y.; Abdullah, H.; Rahman, H.A.; Majid, M.S.; Bandi, M. Energy efficiency measurements in a Malaysian public university. In Proceedings of the 2012 IEEE International Conference on Power and Energy (PECon), Kota Kinabalu, Malaysia, 2–5 December 2012; pp. 582–587. [CrossRef]

28. Amaral, R.E.; Brito, J.; Buckman, M.; Drake, E.; Ilatova, E.; Rice, P.; Sabbagh, C.; Voronkin, S.; Abraham, Y.S. Waste Management and Operational Energy for Sustainable Buildings: A Review. *Sustainability* **2020**, *12*, 5337. [CrossRef]

29. Mohammad, I.S.; Zainol, N.N.; Abdullah, S.; Woon, N.B.; Ramli, N.A. Critical factors that lead to green building operations and maintenance problems in Malaysia. *Theor. Empir. Res. Urban Manag.* **2014**, *9*, 68–86.

30. Shaikh, P.H.; Nor, N.B.M.; Sahito, A.A.; Nallagownden, P.; Elamvazuthi, I.; Shaikh, M. Building energy for sustainable development in Malaysia: A review. *Renew. Sustain. Energy Rev.* **2017**, *75*, 1392–1403. [CrossRef]

31. Natural Resources Canada. *Energy Management Best Practices Guide for Commercial and Institutional Buildings*; Natural Resources Canada: Ottawa, ON, Canada, 2015.

32. Douglas, J. Building performance and its relevance to facilities management. *Facilities* **1996**. [CrossRef]

33. Fleming, D. Facilities management: A behavioural approach. *Facilities* **2004**. [CrossRef]

34. Zuo, J.; Zhao, Z.-Y. Green building research–current status and future agenda: A review. *Renew. Sustain. Energy Rev.* **2014**, *30*, 271–281. [CrossRef]

35. Henderson, J.; Fowler, K. Assessing green building performance a post occupancy evaluation of 14 air force buildings. *Rep. Prep. US Dep. Energy* **2014**. [CrossRef]

36. Foliente, G.C.; Becker, R. CIB PBBCS proactive programme—Task 1. In *Compendium of Building Performance Models*; CSIRO Building; Construction and Engineering: Victoria, Australia, 2001.

37. Hitchcock, R.J. High performance commercial building systems program, Element 2 Project 2.1-Task 2.1.2. In *Standardized Building Performance Metrics*; Final Report; Building Technology Department, Lawrence Berkeley National Laboratory: Berkeley, CA, USA, 2002.

38. Preiser, W.F.; Schramm, U. Intelligent office building performance evaluation. *Facilities* **2002**. [CrossRef]

39. Preiser, W.F.E.; Vischer, J.C. *Assessing Building Performance*; Butterworth-Heinemann: Oxford, UK, 2005.

40. Douglas, J. Developments in appraising the total performance of buildings. *Struct. Surv.* **1994**. [CrossRef]

41. Aghili, N. *Green Building Management Practices Model for Malaysia Green Building*; Universiti Teknologi Malaysia: Skudai, Malaysia, 2018.

42. Biddle, S.J.H.; Markland, D.; Gilbourne, D.; Chatzisarantis, N.L.D.; Sparkes, A.C. Research methods in sport and exercise psychology: Quantitative and qualitative issues. *J. Sports Sci.* **2001**, *19*, 777–809. [CrossRef]

43. Bolino, M.C.; Varela, J.A.; Bande, B.; Turnley, W.H. The impact of impressio—Management tactics on supervisor ratings of organizational citizenship behavior. *J. Organ. Behav. Int. J. Ind. Occup. Organ. Psychol. Behav.* **2006**, *27*, 281–297.

44. Hair, J.F., Jr.; Hult, G.T.M.; Ringle, C.; Sarstedt, M. *A Primer on Partial Least Squares Structural Equation Modeling (PLS-SEM)*; Sage Publications: London, UK, 2016.

45. Chin, W.W. The partial least squares approach to structural equation modeling. *Mod. Methods Bus. Res.* **1998**, *295*, 295–336.

46. Hair, J.F.; Ringle, C.M.; Sarstedt, M. PLS-SEM: Indeed a silver bullet. *J. Mark. Theory Pract.* **2011**, *19*, 139–152. [CrossRef]

47. Ramayah, T.; Lee, J.W.C.; In, J.B.C. Network collaboration and performance in the tourism sector. *Serv. Bus.* **2011**, *5*, 411–428. [CrossRef]
48. Hair, J.; Black, W.; Babin, B.; Anderson, R.; Tatham, R. *Multivariate Data Analysis*, 6th ed.; Pearson Prentice Hall: Upper Saddle River, NJ, USA, 2006.
49. Götz, O.; Liehr-Gobbers, K.; Krafft, M. Evaluation of structural equation models using the partial least squares (PLS) approach. In *Handbook of Partial Least Squares*; Springer: Berlin/Heidelberg, Germany, 2010; pp. 691–711.
50. Hair, J.F., Jr.; Sarstedt, M.; Hopkins, L.; Kuppelwieser, V.G. Partial least squares structural equation modeling (PLS-SEM): An emerging tool in business research. *Eur. Bus. Rev.* **2014**, *26*, 106–121. [CrossRef]
51. Huang, C.-C.; Wang, Y.-M.; Wu, T.-W.; Wang, P.-A. An empirical analysis of the antecedents and performance consequences of using the moodle platform. *Int. J. Inf. Educ. Technol.* **2013**, *3*, 217. [CrossRef]
52. Cohen, J. *Statistical Power Analysis for the Behavioral Sciences*, 2nd ed.; Erlbaum Associates: Hillsdale, MI, USA, 1988.
53. Efron, B. Bootstrap methods: Another look at the jackknife. In *Breakthroughs in Statistics*; Springer: Berlin/Heidelberg, Germany, 1992; pp. 569–593.
54. Yung, Y.F.; Bentler, P.M. Bootstrap—Corrected ADF test statistics in covariance structure analysis. *Br. J. Math. Stat. Psychol.* **1994**, *47*, 63–84. [CrossRef] [PubMed]
55. Rice, W.R. Analyzing tables of statistical tests. *Evolution* **1989**, *43*, 223–225. [CrossRef] [PubMed]
56. Wong, K.K.-K. Partial least squares structural equation modeling (PLS-SEM) techniques using SmartPLS. *Mark. Bull.* **2013**, *24*, 1–32.

Entry

Domestic Environmental Experience Design

Sajal Chowdhury [1,*], **Masa Noguchi** [1] **and Hemanta Doloi** [2]

[1]   ZEMCH EXD Lab, Faculty of Architecture, Building and Planning, The University of Melbourne, Melbourne, VIC 3010, Australia; masa.noguchi@unimelb.edu.au
[2]   Smart Villages Lab, Faculty of Architecture, Building and Planning, The University of Melbourne, Melbourne, VIC 3010, Australia; hdoloi@unimelb.edu.au
*   Correspondence: sajal.chowdhury@unimelb.edu.au; Tel.: +61-401-543-154

Definition: The term 'domestic environmental experience' was defined as users' experiences of cognitive perceptions and physical responses to their domestic built environments. Domestic environments can be enriched through the implementation of environmental experience design (EXD) by combining users' environmental, spatial and contextual factors that may accommodate occupants' needs and demands as well as their health and wellbeing. Here, an EXD theoretical concept has been developed based on the 'User-Centred Design' thematical framework.

Keywords: domestic environment; occupant experience; environmental design; health and wellbeing

1. Introduction

Generally, people spend most of their time in indoor environments [1]. In the domestic setting, occupants' living experiences are diverse, and they have numerous preferences related to their spatial needs and demands [2,3]. These preferences are connected to the environment of a domestic setting and are perceived through occupants' household experiences [4]. Every design component of a domestic environment has a negative or positive impact on occupants' psychological responses [2,4,5]. Several studies identified that surrounding environments stimulate occupants' mediative capacities in their living environments [6]. Therefore, it is necessary to explore occupants' perceptions and experiences in their domestic settings that may enhance their psychological health and wellbeing [4]. Occupants' way-of-living conditions may be modified or changed through the adjustment of their subjective perceptions and attachment to their domestic environment given—i.e., "Human experience" [5,7]. Particularly, in developing countries such as Bangladesh, lower- and middle-income families have the minor capacity to change or modify their existing domestic living conditions due to their socioeconomic limitations [8]. Most of these families live in small or congested urban domestic spaces where only physical architectural design configurations are considered today's architectural design practice [8,9]. Such limited design considerations alone may not be sufficient in addressing occupants' wellbeing to improve the quality of their living [8,10,11]. Domestic indoor environmental attributes have been studied extensively, the research on the occupants' domestic environmental experiences is still marginal to date [4]. The architectural environmental design concept may be enriched by integrating human perceptions and experiences in different spaces of the built environment [12]. The key aim of this study is to explore an architectural design concept for the domestic living environment through the occupants' household experiences that may influence their mental health and wellbeing.

2. Context of the Study

In order to explore the stipulation of this particular study, as well as to enlighten the concept to the reader with the limitation or gap of which the motivation for this study depends, it is important to provide a study contextualisation in which it explains. In this study, middle-income families are considered as the focus group locally and globally and

Citation: Chowdhury, S.; Noguchi, M.; Doloi, H. Domestic Environmental Experience Design. *Encyclopedia* **2021**, *1*, 505–518. https://doi.org/10.3390/encyclopedia1020042

Academic Editor: Nikos Salingaros

Received: 8 April 2021
Accepted: 15 June 2021
Published: 21 June 2021

Publisher's Note: MDPI stays neutral with regard to jurisdictional claims in published maps and institutional affiliations.

their domestic living environment particularly in Dhaka, Bangladesh, has been described as an example of a developing country's urban housing scenario to explore the concept of domestic environmental experiences that may affect occupants' health and wellbeing.

2.1. World Income Group Projections

Since 2000, the number of middle-income people in the world's urban areas has been steadily increasing [13,14]. Sustainable development goals (SDG) estimate that by 2030, the number of middle-income groups will be about 35–45 percent of the total world population [15]. Middle-income groups contribute significantly to the worlds' socioeconomic growth and are increasing faster in Asia than anywhere else compared to other world regions [16]. The middle-income groups are the driving forces of the world economy [16]. Bangladesh is the second-fastest rising economy in South Asia according to the report of World Bank entitled 'South Asia Economic Focus, Fall 2019: Making (De) centralization Work' jointly with the Office of the Chief Economist for the South Asia Region (SARCE) and the Macroeconomics, Trade and Investment (MTI) Global Practice, in 2019 [17].

2.2. Income Group Scenarios of Bangladesh as an Example

Bangladesh is the fastest urbanising emerging country in the world. In Bangladesh, young generations are the key contributor to the country's economy and significantly influence the country's socioeconomic growth [18,19]. Around 47.6 million (approximately 29 percent) of the total population has been identified as young generations of Bangladesh, with an age limit between 18 and 35 years old [18]. Millions of people from this young community in Bangladesh belong to middle-income families and they are the core economic contributor to the country [8]. At present, these income groups have become the most trained and educated part of the country's population [20].

2.3. Health and Wellbeing Problems in Bangladesh

According to the World Health Organization (WHO), a larger portion of the young generation in Bangladesh is suffering from numerous health and wellbeing problems. It is noteworthy that mental health and wellbeing issues are gaining global attention [21]. The National Mental Health Survey of Bangladesh (2019) identified that lack of entertainment, natural environment and socialisation are the main contributing factors to young people's mental illness. It is also identified that urban youth are affected (approx. 40.4 percent) more than rural ones [22]. This young community mainly belongs to middle-income families in Bangladesh who are usually living in highly dense urban housing. According to the WHO, UN and BRAC reports, these middle-income families suffer from inadequate, unaffordable housing facilities in their daily lives and face various types of health and wellbeing difficulties [8,20,23].

2.4. Dhaka City and Middle-Income Families

Dhaka is the capital and main urban business centre of Bangladesh [24]. Dhaka's population has increased rapidly from 1.37 million to about 21 million from 1970 to the present 2020. By 2030, Dhaka will have a population of about 28 million [25,26]. Dhaka is the densest urban area globally, with a density of about 49,182 people per square kilometre in the core region [24,27]. To find new livelihood opportunities every day, approximately 2000–2500 people migrate to Dhaka City from different areas across the country [8,24,25]. Because of rapid population growth, Dhaka faces extreme challenges in the housing sector to grapple with increasing housing demands [8,26]. In Dhaka, the proportion of middle-income people is above 50 percent compared to other income groups (Figure 1) [8]. They are the most skilled part of the country's population, leading to driving the country's economic growth [20]. Nonetheless, according to the Household Income and Expenditure Survey (HIES) in Bangladesh by 2005, 2010 and 2016, the urban household size (e.g., family members) gradually decreased because of their socioeconomic limitations (e.g., monthly income) (Figure 1) [8,28].

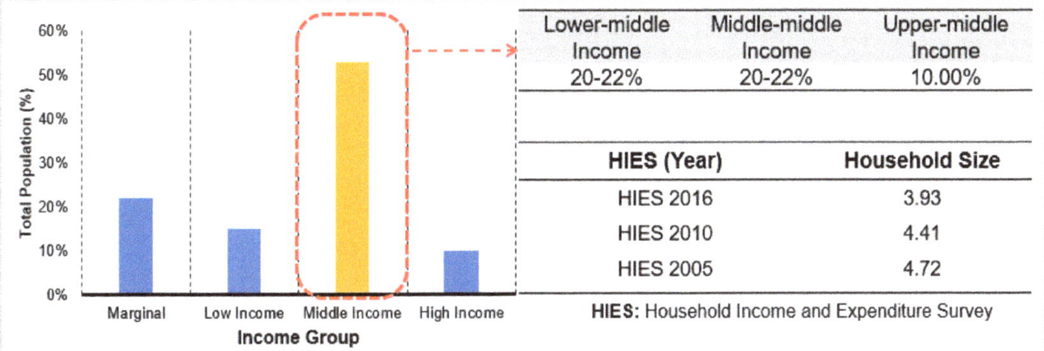

Figure 1. Different income groups in Dhaka City, Bangladesh. Source: [8,9,28].

According to the housing survey and BRAC reports in 2017, about 70–80 percent of middle-income families in Dhaka metropolitan areas are renters [8]. In Dhaka, house prices are increasing disproportionately compared to the housing needs of middle-income families [8,11,20]. Therefore, almost 70–78 percent of these middle-income families, particularly lower-middle and middle-income groups, cannot afford their own houses or apartments in the current market price of housing in Dhaka City, Bangladesh (Figure 2) [8,11,20,23].

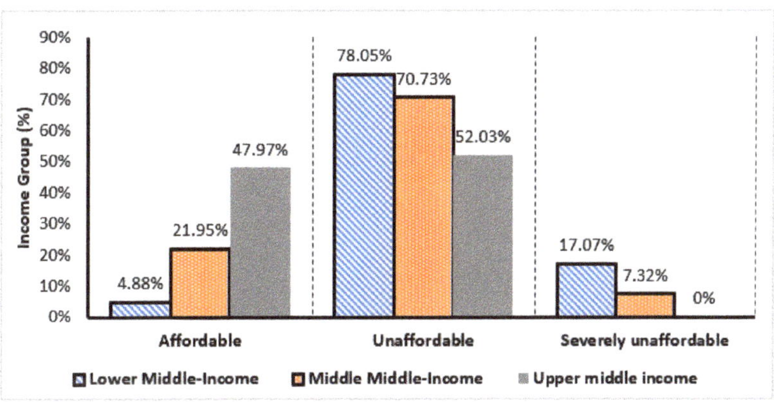

Figure 2. Affordability level of middle-income groups, Dhaka, Bangladesh. Source: [8,20,23].

The report entitled 'State of Cities 2017: Housing in Dhaka' published by BRAC Institute of Governance and Development (BIGD) stated that approximately 61 percent of homeowners from middle-income families depend on their personal or family savings to buy housing units in urban areas. Due to excessive house rents and apartment prices, lower-middle and middle-income families need to adjust their other daily expenditures to afford a dwelling unit. Seemingly, middle-income households reduce their clothing, entertainment, food, education and everyday expenses to cope with their excessive housing prices (Figure 3) [8].

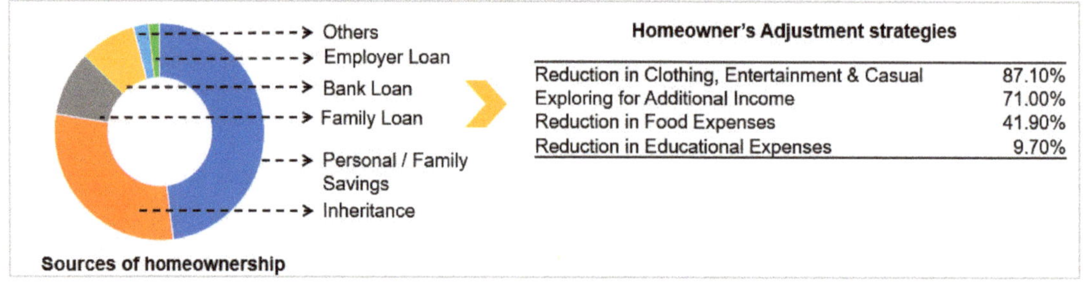

Figure 3. Source of homeownership and adjustment strategies in Dhaka City. Source: [8].

2.5. Overview of Housing Issues in Dhaka

The housing sector of Dhaka City is mainly reliant on public and private providers [8,9,27]. The public sectors supply approximately 7–10 percent of the total housing that is predominantly to the government's public servants. The rest of the housing sector relies on the private sectors where about 70 percent of private real estate developers are targeting the higher- and upper-middle-income groups that include houses and apartments with a size greater than 1000 sqft (92 sqm) [8,9,11]. It is also estimated that there is a shortage of about 0.5 million housing units by 2020, whereas only 25,000 housing units are supplied yearly by private sectors. However, the supply of housing is insufficient so far, and it is expected that by 2035, the backlog of private sector housing units will be approximately 0.7 million [8,9,23].

Consequently, it is increasingly becoming difficult for middle- (i.e., lower-middle and middle-middle) income families to buy or afford decent residential space (e.g., apartment or dwelling unit) within the apartment size greater than 1000 sqft or more in comparison of their income level in an urban area such as Dhaka, especially when only single-earning members support families (Figure 4) [8,20].

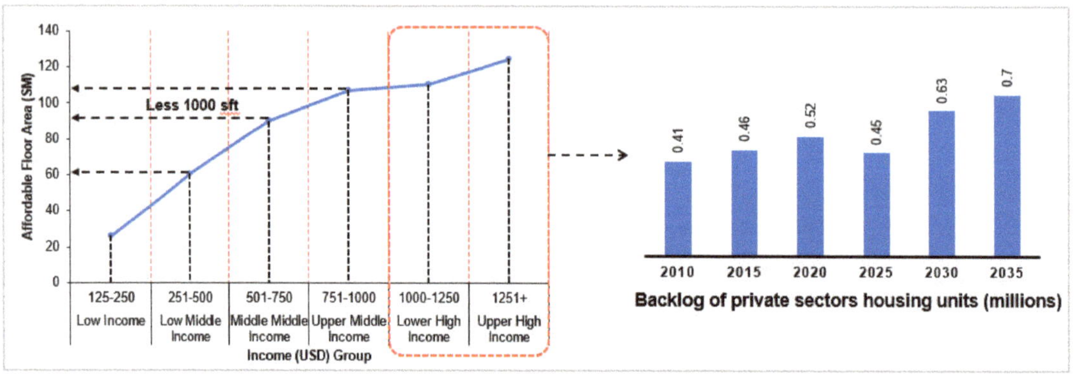

Figure 4. Affordable housing floor area and market supply trend in Dhaka, Bangladesh. Source: [8,9,20,23].

To address the situation, housing sectors (both public and private) are now changing their target group from higher and upper-middle income to lower-middle and middle-middle income groups by developing high-density small-sized flats or apartments (Figure 5) [8,9,29].

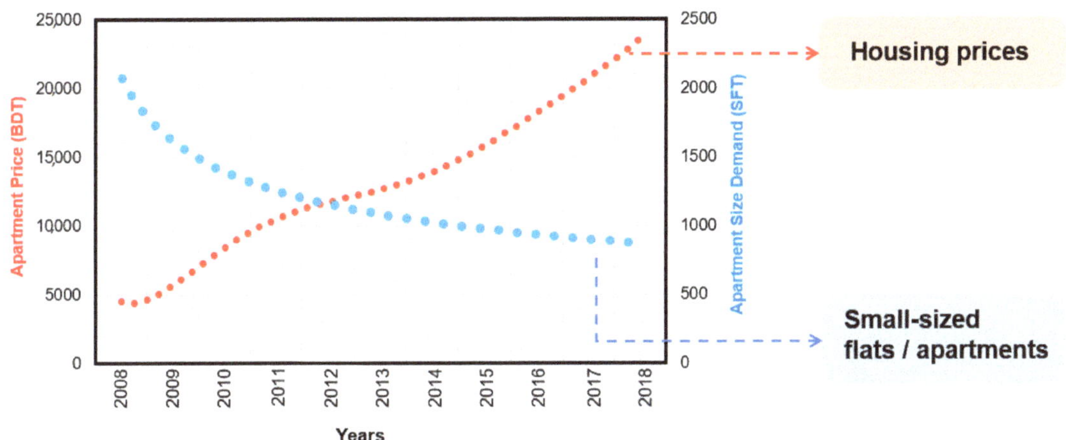

Figure 5. Apartment price vs. apartment size scenario in Dhaka, Bangladesh. Source: [8,9,20,23,29].

Consequently, in recent years, the demand for vertical expansion of high-density housing development in Dhaka has increased exponentially as the horizontal growth is difficult because of buildable land shortage in the metropolitan areas [24,27]. Due to limited budget restrictions and comparatively high rents, the middle-income groups in Dhaka usually live in tiny or compact domestic environments [4]. Notably, most lower-middle and middle-middle income families are living in small or congested domestic spaces in high-density apartments due to housing unaffordability where physical design elements (e.g., room numbers, sizes, configurations and layout) are the main consideration in the architectural design process (Figure 6) [8,9,20,26].

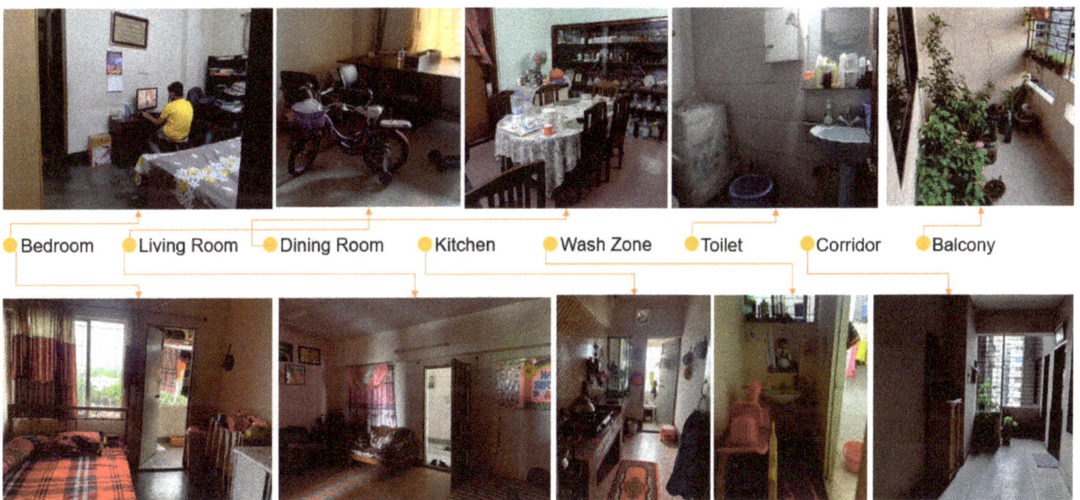

Figure 6. An example of domestic living environments in Dhaka, Bangladesh. Source: (Residents took photos from high-density urban housings, Dhaka, Bangladesh).

These compact accommodations are typically crowded with a lack of privacy, ventilation, daylighting, air qualities and other environmental attributes which may create adverse living conditions and consequently, may directly or indirectly impact occupants' wellbeing [8,30]. More than 50 percent of Dhaka City residents suffer from numerous physical and mental health problems [31]. These problems are prominent among middle-income families, mainly those who are living in high-density urban housing. Particularly, these families have little capability to change or modify their existing domestic living circumstances because of their socioeconomic boundaries [8,23,26]. Such limited architectural design considerations for the high-density urban housing may not be sufficient to address occupants' wellbeing problems in the local context [4].

3. Study Rationale for Local and Global Context

Due to socioeconomic restrictions, most lower-middle and middle-middle income families are living in small or congested dwelling spaces or apartments not only in Dhaka, Bangladesh, but also globally, where physical design components in the architectural design practice are the primary consideration of today's local housing sectors [4]. These small living conditions may create clumsiness leading to the deterioration of indoor environmental quality (i.e., less or no privacy) and indoor air quality (i.e., heavy CO_2 concentration and $PM_{2.5}$) [32]. Thus, the situations may affect occupants' mental wellbeing in their domestic environments [32]. Occupants' mental wellbeing can be enhanced by changing or modifying their domestic living environments that are related to household experiences in addition to the existing physical environments of buildings (Figure 7) [5]. However, modifying the existing design elements or components could be difficult because of middle-income families' socioeconomic limitations. Alternatively, their way of living environments can be changed by adjusting occupants' subjective perception, behaviour and place attachment in their domestic settings given—i.e., 'Human Experience' [4].

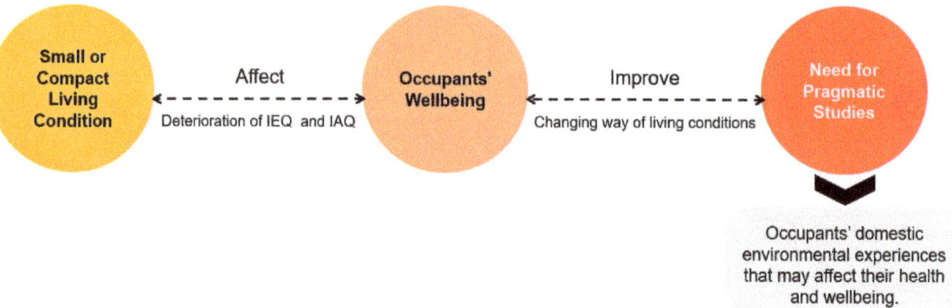

Figure 7. The rationale for the study.

Though residential indoor environmental qualities have been studied broadly, the study on the occupants' domestic environmental experiences for families with socioeconomic limitations is still not well rationalised to date. Without a clear perception of occupants' domestic experiences, environmental design solutions in architectural design decisions may not be enhancing occupants' health and wellbeing.

4. Theoretical Background

4.1. Built Environment and Human Responses

The built environment has diverse terms and ideologies and most spatial design elements formed by humans are the elements of built environments [33]. The studies of environmental design and psychology elaborate that each design element of the built environment directly or indirectly affects users' physical, biological and psychological health and wellbeing factors [7]. Psychologists have examined the correlations between human brain and cognitive interactions in built environments that trigger humans' factors through vision, hearing, smell, touch and movement where 'human perception' is cognitive and 'human emotional responses' are physical (Figure 8) [34,35]. These human factors affect individual perceptions (e.g., feelings, emotions and moods) in the built environment where human emotions are complex phenomena to define precisely [34,35]. Generally, human psychological response concepts or theories describe the core feelings of humans' subjective emotional situations [36]. Mehrabian (1974) elaborated three primary human emotional responses such as pleasure, arousal and activation that can be perceived through experiences [36]. Several studies focusing on indoor environmental qualities such as noise, lighting, material, air, odours and colour examine that environmental psychology bridges the correlation between built environmental design and human responses at indoor and outdoor scales [7]. Kaplan's 'attention restoration theory (ART)' proposed a framework that differentiates between stress and attentional components of human experiences in built environments. Emphasising the critical role of natural environments, this integration contributes towards human–environmental interaction. According to Kaplan, experiencing a natural environment reduces human stress [37]. In addition, Ulrich's 'stress reduction theory (SRT)', focusing on nature in wellbeing, indicates from an evolutionary perspective that natural experiences have an immediate benefit to human mental wellbeing. Ulrich emphasises the affective and aesthetic human responses to the natural environment [38]. Both Kaplan and Ulrich identify natural settings or environments as stimulating components for human wellbeing in numerous ways. Lawlor describes the interconnectedness of human emotions and feelings with the architectural design components based on humans' spiritual perceptions in their living environment [39]. Evans argues that every element of the built environment directly or indirectly affects occupants' mental health and wellbeing [5]. Ergan examined that humans' emotional reactions to colour, light, noise, air quality and crowding are distinctive and momentary in their living environments [40]. Ambient environmental interactions with architectural design components positively impact human physical, biological and psychological factors [41]. Several studies found that spatial ergonomics affect occupants' emotional perceptions such as relaxation and pleasantness within built environments [2]. Pallasmaa revisited human emotional adaptation processes and interactions within the built environment [42]. Humans observe their near environments through their senses [43–45]. Diverse environmental design attributes influence humans' sensory perceptions in numerous ways within their built environments [3,6,36].

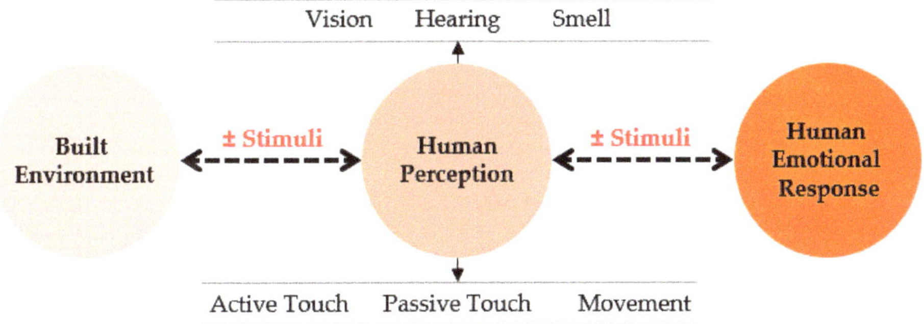

Figure 8. Relationship between the built environment and human responses. Modified from Source: [2–6,39,40,42–44].

4.2. Home as Domestic Environment and Conceptual Meaningfulness

The domestic setting has both indoor and outdoor environmental factors that reflect occupants' needs and demands [43]. The domestic setting protects occupants from unwanted bad weather, while occupants' psychological comfort or satisfaction needs to be well maintained to enhance their wellbeing [2,33,43]. The term "domestic environment" refers to the space where occupants live and correlate with sociocultural and environmental factors that may affect their health and wellbeing [5,6]. Cooper (1974) indicated that home is the mother of the entire environment of security and love [46]. Mallett (2004) addressed the notion of home as a dwelling place of interaction between humans, places and things; considering sociocultural aspects. It can be associated with occupants' numerous feelings, moods and emotions [47]. Stokols (1972) described that a home is a place where people can meet all their psychological, physical and social needs and demands [48]. Hayward (1977) suggested the psychological concept of a home according to the meaning of individual human factors [49]. Dovey (1985) emphasises that home is a series of connections between a person and the world and orients people and connects all with the physical environment and social world [50]. The domestic environment influences occupants' psychological responses and behaviours that have impacts on human health and wellbeing [7]. Designing these settings may need to be developed with due consideration of occupants' sociocultural and psychological aspects.

4.3. Domestic Environment and Occupants' Psychological Responses

Home has multilayered characteristics integrating memories, desires, intimacy, privacy, identity, function and even language [50]. The domestic environments reflect occupants' behaviours because of their numerous physical, psychological and social contextual experiences (Figure 9). The domestic environment has various purposes and meaningfulness related to occupants' spatial and environmental factors. Generally, a domestic environment is the combination of three types of spatial zones (e.g., public, semipublic and private) distribution [43]. The domestic indoor environment can be connected with the external outdoor environment. Each area of a domestic environment has its different spatial and environmental factors that may accelerate occupants' psychological responses [4,51,52]. These factors in the domestic setting are associated with occupants' preferences (e.g., needs and demands). Moore (2000) illustrated that domestic environments reflect diversities of occupants' preferences due to their living experiences [53]. Caan (2011) described that human experiences are essential in indoor living spaces to mediate tangible and intangible design factors [2]. Graham (2015) addresses the different issues of human perceptions related to the domestic environment [4,52]. Sussman (2015) and Hollander (2021) emphasises the need for understanding human cognitive experiences in the creation of living environments [44,45]. There are different types of human experiences (e.g., physical, mental, emotional, spiritual, social, subjective and virtual) in built environments. Occupants' living experiences are different in every step of their life activities [2,3,54]. From the literature review, it has been identified that human factors e.g., privacy, variety, identity, control, order, security, choice, sociability and aesthetics in different living spaces of a domestic environment influence occupants' psychological experiences relating to their spatial and environmental needs and demands [2,7,55]. From the literature, it has been identified that the overall indoor environmental components such as light, colour, temperature, ventilation, materials, layout, shape, size, height, ergonomics, opening and greenery affect occupants' psychological responses negatively or positively [2,41]. Other indoor air qualities such as dust, mould, moisture, odours, pollutants and dampness are directly related to occupants' physical and mental wellbeing [2,5,41]. Occupants may prefer to control these issues according to their preferences. The outside environmental relationship through openings, window, balcony and toilet attachment may create a spiritual connection with occupants' minds [3,6]. Domestic satisfaction or comfort depends on the occupants' needs and preferences closely related to their spatial and environmental factors in living environments [43].

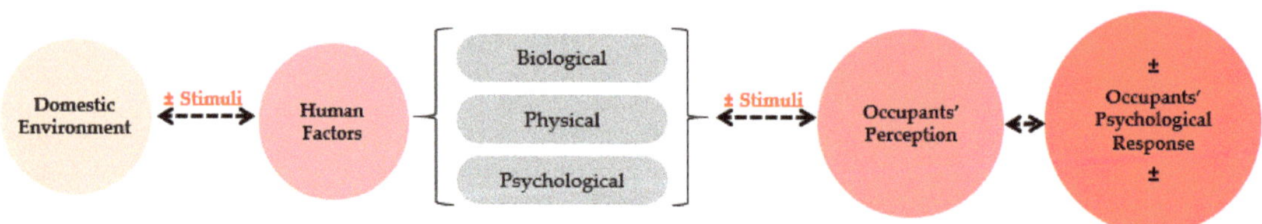

Figure 9. Domestic environment and occupants' psychological responses. Source: [2,4,5,39,40,43,56].

Consequently, occupants may change or modify their domestic environments' physical characteristics to create more comfortable settings. Occupants in their domestic spaces have individual human factors regarding meaningfulness, attachment and perceptions [2,3,5,6]. Occupants' sociocultural contextual factors may impact different spaces in a domestic setting where the spatial character to be related to different sociocultural preferences. Today's architectural design approaches do not adequately address the relationship between users' context, spatial and environmental design factors as well as occupants' psychological satisfaction and comfort [43].

4.4. Defining Domestic Environmental Experience

Several psychological and phenomenological studies have been conducted to define the meaning of the domestic environment [4]. According to Pallasmaa, the phenomenology of the domestic environment (as home) is not just an architectural effort. It has an aesthetic view considering physical, psychological and sociocultural phenomena. Pallasmaa believes that the domestic environment has multilayered characteristics integrating memories, desires, intimacy, privacy, identity, function and even language [42,51]. Continuing the exploration, the domestic environment becomes an essential feature of 'self-identification' for the middle class where privacy, comfort and domesticity are the occupants' core achievements [4,53,56,57]. In general terms, each element of the domestic environment has the meditative power of occupants' emotions and feelings [4]. Considering several studies, the term 'domestic environmental experience' has been defined as " … *user experiences of cognitive perceptions and physical responses to their domestic built environment* … " with a diversity of occupants' daily household activities [4]. In short, domestic environmental experience connects occupants' physical, psychological and social needs and demands that are correlated with different factors of the built environment (Figure 10) [4].

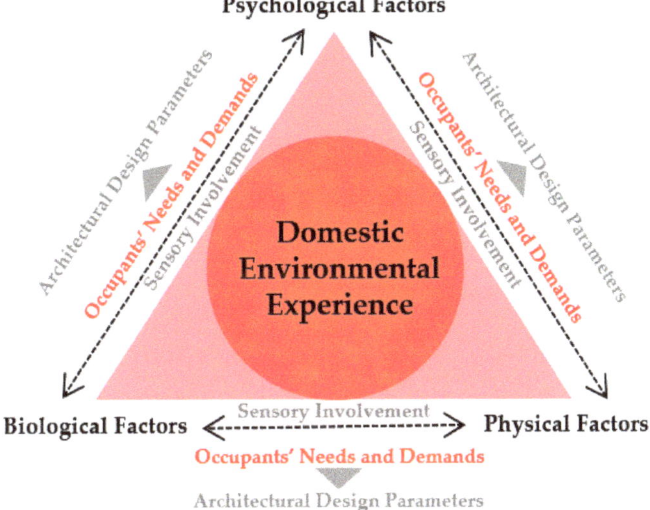

Figure 10. Thematic diagram of 'domestic environmental experience'. Modified from Source: [4,43].

5. Toward Domestic 'Environmental Experience Design'

Today's architectural design approaches or practices create a gap between users' spatial, environmental and psychological needs and demands [43]. According to Lawrence (1990), a study gap exists between two current theories, namely the theory of 'environmental deterministic' and 'social constructivism' [4,58]. The environmental deterministic theory defines the physical environment's impacts on human behavior. The scope of explanation about users' sociocultural contexts is limited in this theory. On the other hand, the social constructivism theory describes users' social and cultural perceptions where consideration of built environmental effects is limited [59]. The position of 'environmental experience design (EXD)' between the two spectra that derive from the user-centred theory and address the user's social, cultural, spatial and environmental aspects (Figure 11) [43].

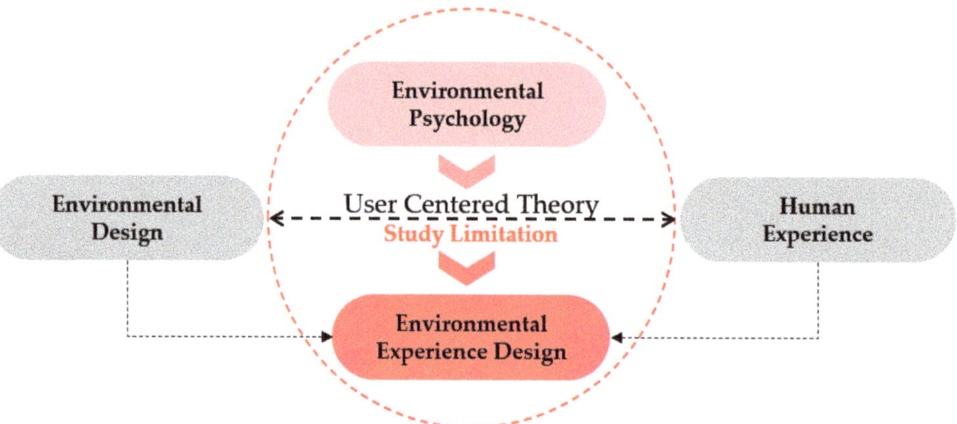

Figure 11. Research gap and future experience design direction. Modified from Source: [4,43].

Kling (1977) coined the term 'user-centred design (UCD)', a person-centric philosophical design approach that focused on the human cognitive interaction with objects, products or things [60]. The concept of UCD became widely popular as 'user experience (UX) design' due to the publication of a book entitled 'User-Centered System Design: New Perspectives on Human-Computer Interaction' by Donald A. Norman in 1986 at the University of California, San Diego [61]. According to Norman (1988),

"Human-centered design is a design philosophy. It means starting with a good understanding of people and the needs that the design is intended to meet. This understanding comes about primarily through observation, for people themselves are often unaware of their true needs, even unaware of the difficulties they are encountering."

Additionally, in the book entitled 'The Design of Everyday Things', Norman expanded the concept of 'experience design' based on the industrial design domain where the author elaborated the concept of human psychology behind design practice and its' importance in everyday lives, considering usability and usefulness [4,62]. The concept of EXD has hardly been applied to architectural design decisions before in the built environmental design solutions [12]. This concept is derived from the 'user experience (UX)' associated with the industrial design domain that mainly focuses on product interfaces for usability. Reflecting the notion of experience design approaches, Ma et al. (2017) and Noguchi et al. (2018) introduced the term and concept of environmental experience design (EXD) [4,12,63]. Following the EXD concept, Chowdhury et al. (2020) defined the term "domestic environmental experience" and a conceptual correlation of the domestic environmental parameters was illustrated based on occupants' perceptions [4]. The environmental experience design approach may combine occupants' preferences (needs and demands) and environmental design factors (EDF) through their domestic living experiences that may improve health and wellbeing [4]. This new design application may change the architectural design concept from a technology-driven idea to a human-centric design decision (Figure 12).

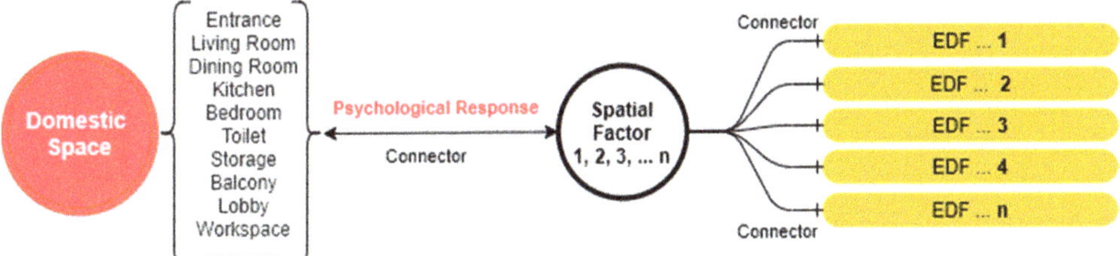

Figure 12. Conceptual parametric relationship of occupants' domestic experiences. Modified from Source: [4,43,64].

6. Domestic Environmental Experience Design Hypothetical Construct

An EXD theoretical concept has been developed based on this 'user-centred design' thematical framework (Figure 13) [12,63]. In this theoretical concept, the user experience's core aspects are summarised in three separate components: spatial design factors, environmental factors and user contextual experiences. Without understanding occupants' experiences, it may be difficult to identify architectural environmental design solutions to enhance occupants' mental health and wellbeing or satisfaction in their living environment not only in locally e.g., Dhaka, Bangladesh but also globally.

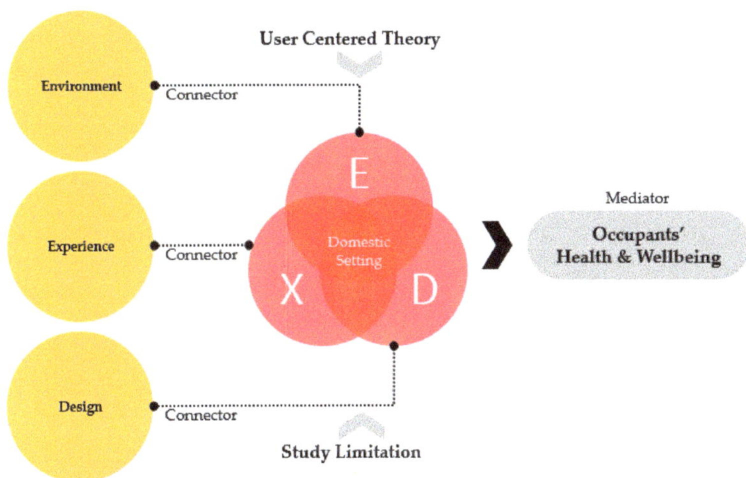

Figure 13. Domestic 'environmental experience design'.

The first component, which concerns environmental design and qualities (e.g., temperature, humidity, air, noise, light, colour, smell, material, texture and nature) deals with the user's needs and demands with existing environmental design aspects. It encompasses indoor environmental elements but also the psychological part of occupants' comfortable feelings (e.g., indoor comfort, air quality, noise and visual comfort) or satisfaction. The second component deals with user contextual experiences related to their preferences and restrictions in their domestic living environment that may shape the interaction between design and environmental factors. This component also deals with user sociocultural contextual factors related to their preferences and restrictions in their living environment. This sociocultural contextual factor varies according to occupants' different domestic living contexts (e.g., age, sex, education, religion, income, restriction, needs and demands). The third component considering the perceived spatial design factors (e.g., privacy, variety, identity, control, order, security, choice, sociability and aesthetics) focuses on users' spatial experiences or requirements. It may be linked to users' physical, psychological and biological needs and demands in their domestic living environments. Thus, this theoretical concept may be applied to extract the 'environmental experience design (EXD)' methodological framework for the domestic environment not only in Dhaka, Bangladesh, as well as globally to enhance occupants' health and wellbeing, as suggested in Figure 13 [4,64].

7. Conclusions

This study explored a concept of domestic environmental experiences design following a research question: what is the notion of architectural design for domestic settings through occupants' experiences that may have impacts on occupants' mental health and wellbeing? Human perception and phenomenology, environmental design and psychology, residential environment and design, health and wellbeing and user experience are the main thematic areas of this study. An extensive literature review was conducted using a combination of different keywords related to domestic environmental experiences through Scopus, Science-Direct, PubMed, Google Scholar, Mendeley, Research Gate and other academic databases. This study found the significant impacts of domestic environments on human perceptions and behaviours that to some extent influence occupants' mental health and wellbeing. Every design element of the built environment contributes to stimulating occupants' psychological responses positively or negatively. Considering the literature, the term domestic environmental experience can be defined as " ... users' experiences of cognitive perceptions and physical responses to their domestic built environment ... " Without a pragmatic perception of occupants' domestic household experiences, the environmental design solutions in architectural practice may be difficult to implement and enhance their mental health and wellbeing who have socioeconomic constraints. Nonetheless, limited architectural design considerations for the high-density urban housing in Bangladesh and other developing countries may not be sufficient to address occupants' wellbeing problems in the local context. Domestic environments can be enriched through the implementation of 'environmental experience design (EXD)' that may accommodate occupants' preferences and restrictions as well as their needs and demands. Therefore, exploring the notion of domestic environmental experience design will be the future research direction in the architectural design domain to enhance occupants' health and wellbeing in high-density urban housing locally (e.g., Bangladesh) and globally.

Author Contributions: S.C. led overall research activities, conceptualisation, methodology, literature reviews, formal analysis, research outcomes and manuscript drafting; M.N. contributed to principal research supervision and refinement of manuscript development; and H.D. contributed to partial research supervision. All authors have read and agreed to the published version of the manuscript.

Funding: This research received no external funding.

Acknowledgments: The authors would like to express their sincere gratitude to Melbourne School of Design (MSD), Faculty of Architecture, Building and Planning, the University of Melbourne for providing access to the facilities required for this research activity as well as a full PhD scholarship given to the first author of this paper.

Conflicts of Interest: The authors declare no conflict of interest.

Entry Link on the Encyclopedia Platform: https://encyclopedia.pub/12068.

References

1. Bluyssen, P.M. Towards an integrated analysis of the indoor environmental factors and its effects on occupants. *Intell. Build. Int.* **2019**, *12*, 1–9. [CrossRef]
2. Caan, S. *Rethinking Design and Interiors: Human Beings in the Built Environment*; Laurence King: London, UK, 2011.
3. Mallgrave, H.F. *From Object to Experience: The New Culture of Architectural Design*; Bloomsbury Publishing: London, UK, 2018.
4. Chowdhury, S.; Noguchi, M.; Doloi, H. Defining Domestic Environmental Experience for Occupants' Mental Health and Wellbeing. *Designs* **2020**, *4*, 26. [CrossRef]
5. Evans, G.W. The built environment and mental health. *J. Urban Health* **2003**, *80*, 536–555. [CrossRef]
6. Goldhagen, S.W. *Welcome to Your World: How the Built Environment Shapes Our Lives*; Harper Collins: New York, NY, USA, 2017.
7. Kopec, D.A. *Environmental Psychology for Design*, 3rd ed.; Bloomsbury Publishing Inc.: London, UK, 2018.

8. BRAC. The State of Cities 2017: Housing in Dhaka. In *BIGD Report*; BRAC Institute of Governance and Development, BRAC University: Dhaka, Bangladesh, 2017.
9. Kamruzzaman, M.; Ogura, N. Apartment housing in Dhaka City: Past, present and characteristic outlook. In Proceedings of the Building Stock Activation, Tokyo, Japan, 5–7 November 2007.
10. Choguill, C.L. Problems in providing low-income urban housing in Bangladesh. *J. Habitat Int.* **1988**, *12*, 29–39. [CrossRef]
11. Khare, H.S. *Barriers Constraining the Low and Middle Income Housing Finance Market in Bangladesh*; World Bank: Washington, DC, USA, 2016.
12. Noguchi, M.; Ma, N.; Woo, C.M.M.; Chau, H.W.; Zhou, J. The usability study of a proposed environmental experience design framework for active ageing. *Buildings* **2018**, *8*, 167. [CrossRef]
13. United Nations. *World Population Ageing 2019: Highlights (ST/ESA/SER. A/430)*; Department of Economic and Social Affairs, Population Division, United Nations: New York, NY, USA, 2019. Available online: https://www.un.org/en/development/desa/population/publications/pdf/ageing/WorldPopulationAgeing2019-Highlights.pdf (accessed on 16 February 2020).
14. Kharas, H. *The Unprecedented Expansion of the Global Middle Class: An Update. Global Economy & Development.* Working Paper 100. Available online: https://www.brookings.edu/wp-content/uploads/2017/02/global_20170228_global-middle-class.pdf (accessed on 6 February 2021).
15. *Under Pressure: The Squeezed Middle Class*; OECD (Organisation for Economic Co-operation and Development): Paris, France, 2019. Available online: https://www.oecd.org/social/under-pressure-the-squeezed-middle-class-689afed1-en.htm (accessed on 6 November 2020).
16. Asian Development Bank. Development Indi, Asian Development Bank. *Economics & Re. Key Indicators for Asia and the Pacific 2011.* Available online: https://www.adb.org/publications/key-indicators-asia-and-pacific-2011 (accessed on 7 February 2021).
17. Beyer, R.C.M. *South Asia Economic Focus, Fall 2019: Making (De) Centralization Work*; The World Bank: Washington, DC, USA, 2019; pp. 1–88.
18. Khatun, F.; Saadat, S.Y. *Youth Employment in Bangladesh*; Springer: Singapore, 2020.
19. *Global Employment Trends for Youth 2020: Technology and the Future of Jobs*; International Labour Organization: Geneva, Switzerland, 2020.
20. Sadeque, C.M.Z. The Housing Affordability Problems of the Middle-Income Groups in Dhaka: A Policy Environment Analysis. Doctoral Dissertation, The University of Hong Kong, Hong Kong, 2013.
21. *Depression and Other Common Mental Disorders: Global Health Estimates*; World Health Organization: Geneva, Switzerland, 2017; pp. 1–24.
22. MoHFW (Ministry of Health and Family Welfare). *National Mental Health Survey, Bangladesh, 2018–2019*; National Institute of Mental Health: Dhaka, Bangladesh, 2019.
23. Shams, S.; Mahruf, M.; Shohel, C.; Ahsan, A. Housing problems for middle and low income people in Bangladesh: Challenges of Dhaka Megacity. *J. Environ. Urban. Asia* **2014**, *5*, 175–184. [CrossRef]
24. Swapan, M.S.H.; Zaman, A.U.; Ahsan, T.; Ahmed, F. Transforming urban dichotomies and challenges of South Asian megacities: Rethinking sustainable growth of Dhaka, Bangladesh. *Urban Sci.* **2017**, *1*, 31. [CrossRef]
25. Sarker, P. Analyzing Urban Sprawl and Sustainable Development in Dhaka, Bangladesh. *J. Econ. Sustain. Dev.* **2020**. [CrossRef]
26. Satu, S.A.; Chiu, R.L. Livability in dense residential neighbourhoods of Dhaka. *Hous. Stud.* **2019**, *34*, 538–559. [CrossRef]
27. *Dhaka Structure Plan 2016–2035*; RAJUK (Rajdhani Unnayan Kartripakkha): Dhaka, Bangladesh, 2015.
28. BBS (Bangladesh Bureau of Statistics). *Household Income and Expenditure Survey 2016–2017*; Ministry of Planning, Government of the People's Republic of Bangladesh: Dhaka, Bangladesh, 2017.
29. Barua, S.; Mridha, A.H.A.M.; Khan, R.H. Housing real estate sector in Bangladesh present status and policies implications. *ASA Univ. Rev.* **2010**, *4*, 240–253.
30. Mridha, M. The effect of age, gender and marital status on residential satisfaction. *Local Environ.* **2020**, *25*, 540–558. [CrossRef]
31. *Towards an Unliveable City*; The Business Standard: Dhaka, Bangladesh, 2019. Available online: https://tbsnews.net/bangladesh/towards-unliveable-city (accessed on 7 February 2021).
32. Larcombe, D.L.; van Etten, E.; Logan, A.; Prescott, S.L.; Horwitz, P. High-Rise Apartments and Urban Mental Health—Historical and Contemporary Views. *Challenges* **2019**, *10*, 34. [CrossRef]
33. McClure, W.R.; Bartuska, T.J.; Young, G.L. *The Built Environment: A Collaborative Inquiry into Design and Planning*; John Wiley & Sons Inc.: Hoboken, NJ, USA, 2011.
34. Bower, I.; Tucker, R.; Enticott, P.G. Impact of built environment design on emotion measured via neurophysiological correlates and subjective indicators: A systematic review. *J. Environ. Psychol.* **2019**, *66*, 101344. [CrossRef]
35. Maslow, A.H. *The Farther Reaches of Human Nature*; Viking Press: New York, NY, USA, 1971; Volume 19711.
36. Mehrabian, A.; Russell, J.A. *An Approach to Environmental Psychology*; The MIT Press: Cambridge, MA, USA, 1974.
37. Kaplan, S. The restorative benefits of nature: Toward an integrative framework. *J. Environ. Psychol.* **1995**, *15*, 169–182. [CrossRef]
38. Ulrich, R.S.; Simons, R.F.; Losito, B.D.; Fiorito, E.; Miles, M.A.; Zelson, M. Stress recovery during exposure to natural and urban environments. *J. Environ. Psychol.* **1991**, *11*, 201–230. [CrossRef]
39. Lawlor, A. *A Home for the Soul: A Guide for Dwelling with Spirit and Imagination*; Clarkson Potter Publishers: New York, NY, USA, 1997.

40. Ergan, S.; Shi, Z.; Yu, X. Towards quantifying human experience in the built environment: A crowdsourcing based experiment to identify influential architectural design features. *J. Build. Eng.* **2018**, *20*, 51–59. [CrossRef]
41. Bluyssen, P.M. *The Indoor Environment Handbook: How to Make Buildings Healthy and Comfortable*; Taylor & Francis: London, UK, 2009.
42. Pallasmaa, J. *The Eyes of the Skin: Architecture and the Senses*; Wiley: Hoboken, NJ, USA, 2005.
43. Chowdhury, S.; Noguchi, M.; Doloi, H. Conceptual Parametric Relationship for Occupants' Domestic Environmental Experience. *Sustainability* **2021**, *13*, 2982. [CrossRef]
44. Sussman, A.; Hollander, J.B. *Cognitive Architecture: Designing for How We Respond to the Built Environment*; Routledge: Abingdon, UK, 2015.
45. Hollander, J.B.; Sussman, A.; Lowitt, P.; Angus, N.; Situ, M. Eye-tracking emulation software: A promising urban design tool. *Archit. Sci. Rev.* **2021**, 1–11. [CrossRef]
46. Cooper, C.J.L. *Designing for Human Behavior*; Dowden-Hutchingson: Stroudsburg, PA, USA, 1974.
47. Mallett, S. Understanding home: A critical review of the literature. *J. Sociol. Rev.* **2004**, *52*, 62–89. [CrossRef]
48. Stokols, D.J.P.R. On the distinction between density and crowding: Some implications for future research. *Psychol. Rev.* **1972**, *79*, 275. [CrossRef]
49. Hayward, D. Psychological concepts of home. *J. HUD Chall.* **1977**, *8*, 10–13.
50. Dovey, K. Home and Homelessness. In *Home Environments. Human Behavior and Environment. Advances in Theory and Research*; Altman, I., Werner, C.M., Eds.; Springer: Boston, MA, USA, 1985; Volume 8.
51. Shirazi, M. *Towards an Articulated Phenomenological Interpretation of Architecture: Phenomenal Phenomenology*; Routledge Research in Architecture: New York, NY, USA, 2013.
52. Graham, L.T.; Gosling, S.D.; Travis, C.K. The psychology of home environments: A call for research on residential space. *Perspect. Psychol. Sci.* **2015**, *10*, 346–356. [CrossRef] [PubMed]
53. Moore, J. Placing home in context. *J. Psychol.* **2000**, *20*, 207–217.
54. Miller, S.; Schlitt, J.K. *Interior Space: Design Concepts for Personal Needs*; Praeger Publishers: Westport, CT, USA, 1985.
55. Blossom, N.H. Human Nature and the Near Environment. In *The Built Environment: A Collaborative Inquiry into Design and Planning*; John Wiley & Sons, Inc.: Hoboken, NJ, USA, 2011.
56. Gifford, R.; Steg, L.; Reser, J.P. *Environmental Psychology*; Wiley Blackwell: Hoboken, NJ, USA, 2011.
57. Rybczynski, W. *Home: A Short History of an Idea*; Penguin Books: London, UK, 1987.
58. Lawrence, D.L.; Low, S.M. The built environment and spatial form. *Annu. Rev. Anthropol.* **1990**, *19*, 453–505. [CrossRef]
59. Vischer, J.C. Towards a user-centred theory of the built environment. *Build. Res. Inf.* **2008**, *36*, 231–240. [CrossRef]
60. Kling, R. *The Organizational Context of User-Centered Software Designs*; MIS Quarterly: Minneapolis, MN, USA, 1977; pp. 41–52.
61. Norman, D.A. *The Psychology of Everyday Things*; Basic Books: New York, NY, USA, 1988.
62. Norman, D.A. *Emotional Design: Why We Love (or Hate) Everyday Things*; Basic Civitas Books: New York, NY, USA, 2004.
63. Ma, N.; Chau, H.-h.; Zhou, J.; Noguchi, M. Structuring the Environmental Experience Design Research Framework through Selected Aged Care Facility Data Analyses in Victoria. *Sustainability* **2017**, *9*, 2172. [CrossRef]
64. Chowdhury, S.; Noguchi, M.; Doloi, H. Research Methods to investigate occupants' domestic environmental experiences for EXD framework. In Proceedings of the 3rd International Conference on Smart Villages and Rural Development (COSVARD 2020), Gowahati, Assam, India, 7–8 December 2020; The University of Melbourne: Melbourne, Australia, 2020.

Entry

Social Housing Customization in Brazil

Luisa Felix Dalla Vecchia * and Nirce Saffer Medvedovski

Faculty of Architecture and Urbanism, Universidade Federal de Pelotas, Pelotas 96010-610, Brazil; nirce.sul@gmail.com
* Correspondence: luisa.vecchia@ufpel.edu.br

Definition: Social housing customization in Brazil refers to the current processes of development and evolution of government-funded neighborhoods for the lowest-income population. The mass production of small housing units that do not satisfy family needs instigates a self-design and self-construction process post-occupancy to customize the units. Ultimately, these changes to the units bring unintended negative consequences for the families and the city. In this context, mass customization is seen as an alternative to address some of the problems related to unit design.

Keywords: housing production; low-cost; design quality; social housing; government programs; post-occupancy; healthy environments; mass customization

Citation: Dalla Vecchia, L.F.; Medvedovski, N.S. Social Housing Customization in Brazil. *Encyclopedia* **2021**, *1*, 589–601. https://doi.org/10.3390/encyclopedia1030049

Academic Editors: Masa Noguchi, Nikos A. Salingaros, Antonio Frattari, Carlos Torres Formoso, Haşim Altan, John Odhiambo Onyango, Jun-Tae Kim, Kheira Anissa Tabet Aoul, Mehdi Amirkhani, Sara Jane Wilkinson and Shaila Bantanur

Received: 14 June 2021
Accepted: 19 July 2021
Published: 21 July 2021

Publisher's Note: MDPI stays neutral with regard to jurisdictional claims in published maps and institutional affiliations.

1. Introduction

This paper shows how low-cost social housing neighborhoods are produced through government-funded programs and how these neighborhoods develop post-occupancy. Furthermore, the paper shows potential paths to maintain adequate environments in these neighborhoods as they evolve. It is based on the revision of literature including research papers, legislation, case reports and post-occupancy studies, as well as results from the authors' previous research. The paper starts with a historic overview to provide the necessary background for the understanding of the social housing processes currently in place. Following the historic background, the paper outlines, in Section 3, the current processes for the production of social housing neighborhoods for the lowest income range, highlighting the main stakeholders and their roles. Section 4 then shows the post-occupancy processes in neighborhoods of house units focusing on the changes the families make to their houses and the problems these changes bring to themselves and the city. In Section 5, the paper describes ways to address these unit changes while maintaining adequate environments as the neighborhoods evolve. In particular, Section 5 shows possibilities for mass customization to contribute to the processes in this context. Concluding remarks are outlined in Section 6.

2. History

The United Nations' Universal Declaration of Human Rights, approved in 1948, recognized housing as a human right and had vast repercussions on social housing initiatives worldwide. However, in Brazil, it resulted in political discourse rather than actions [1]. It was only after the military takeover in 1964 when the National Housing Bank (BNH) was created that large-scale social housing took off [2–4]. Social housing was primarily in the form of mass-produced apartment blocks resembling post-war housing blocks in Europe. However, BNH also subsidized lots with infrastructure and embryo units that were very small unfinished houses meant for further development by the owners. In some cases, only the bathroom was provided, as in the neighborhoods implemented through the Profilurb initiative [5]. These lots and embryo units were meant for people earning up to three times the minimum wage. This bank was the primary source of financing, not only for housing but also for infrastructure and sanitation, until it collapsed in 1986. BNH's long-term legacy goes beyond housing and infrastructure. The idea that local governments are dependent on Federal initiatives is also a legacy from BNH [6]. This idea persists today, embedded in Brazilian legislation, as evidenced by the lack of local programs and recent, laws such as [7,8].

Much of BNH's resources were directed at financing homeownership. Unlike in Europe or North America, homeownership subsidies in Brazil were granted to impoverished formal workers, but still excluded the most vulnerable population. This contributed to the extensive growth of slums (favelas). Attempting to deal with this exclusion, BNH implemented some specific initiatives to remove favelas, transferring the population to new apartment blocks. However, those initiatives ultimately failed as people moved back to the slums, illegally selling their apartments to higher-income families [6].

To start addressing the UN Committee's General Comment no.4 [9] on the right to adequate housing, and the outcome of the 1996 United Nations Conference on Human Settlements, subsequent social housing legislation in Brazil associated the building of new social housing neighborhoods with the provision of services and public transportation. Despite these efforts, the unstable economic and political circumstances of the late 1980s to early 2000s significantly limited the provision of social housing. This was especially significant for poorer municipalities and those less politically aligned with the federal government [1,6]. During this time, the federal government did not have a unified social housing policy or funding avenue. Therefore, states and municipalities had to be creative to implement local initiatives, which focused mainly on urbanizing precarious settlements [3,10]. The housing deficit grew significantly in South America between 1990 and 2000, increasing informal living arrangements [11].

In 1999 the Brazilian government authorized a new housing program based on the French leasing program. In Brazil, the program was called PAR (program of residential leasing) and had some significant differences from its French inspiration. The main difference was the option of ownership at the end of the 15-year lease agreement [1]. This program was initially aimed at families earning three to six times the minimum wage, but was later expanded to families earning up to three times the minimum wage. The cost of the lease was subsidized depending on the income of the family. In line with the international emphasis on housing ownership and internal pressure to overcome the financial deficit in this sector, in 2007 the government passed further legislation to allow early ownership [12].

Following the financialization trend, in Brazil, the government established a new program in 2009, the MCMV (my house my life, translated from Minha Casa Minha Vida) program, which subsidized most of the cost for low-income families to buy homes in a financing process [8]. Its creation was an economic response to deal with the financial crisis that had started in 2007. It was meant to benefit large construction companies, and to boost the economy as much as provide housing for those in need. The MCMV program broke free from the social representation and sustainable urban development aspects of the existing Social Housing National System (SNHIS) [13]. This occurred due to concerns about agility in project approval and the interests of entrepreneurs in the construction industry [14–16]. The MCMV program was implemented based on similar programs previously implemented in other Latin American countries, such as Chile and Mexico [11]. The state's role in these programs is mainly to provide financial support for private companies to build housing for the poor. Even though the government establishes some rules, the companies have autonomy to choose the location, typology, and size of the development. As a result, these developments often remove populations from well-located informal settlements to distant and isolated, but formally owned, housing. Ultimately, this type of program has contributed to creating new problems of urban ghettos in the peripheries [11]. Such top-down schemes based on industrial geometries have been criticized for bringing more benefits to banks and construction companies while producing inhumane, monotonous neighborhoods [17].

Over the past decade, the MCMV program has produced significantly more housing units than any other social housing program in Brazil, having delivered more than one million housing units for the lowest income range between 2009 and 2019 [18]. Furthermore, MCMV's processes and regulations became the model for local programs and a requirement for federally funded programs. Therefore, the processes and policies outlined in this article pertain mainly to the MCMV program. In late 2020, the federal government ended the MCMV program to implement the new Green and Yellow House program (translated from Casa Verde e Amarela), named in reference to the colors of the Brazilian flag, for which legislation was approved in early 2021 [7]. However, this new housing program has kept the MCMV processes for the production of new housing units. In addition, the new program also incorporates avenues for land regularization and improvement initiatives for existing housing. Nevertheless, evaluating the new program beyond theoretical speculation at this point would be premature.

3. Production of Housing Units: Stakeholders, Policies, and Operations

The enterprise sub-section of the MCMV is the main way housing can be financed. It has enabled markedly more housing than other sub-sections. From 2009 to 2019, the MCMV enterprise delivered 1,094,698 housing units for the lowest income range, while all the other urban housing avenues within the program delivered 183,904 units collectively [18]. Importantly, many processes in the MCMV for the lowest income range, including unit design, are similar to those of prior programs for the lowest income range. The units produced, the needs of the intended inhabitants, and the problems that emerge post-occupancy are also consistent with previous programs. Therefore, this section discusses the production of social housing units created under the enterprise sub-section of the MCMV program. However, some specific changes brought about by the Green and Yellow House program are noted.

Historically the lowest income range of social housing programs in Brazil has considered families earning up to three times the minimum wage. More recently the MCMV program changed the cap to BRL 1800, which is less than two times the minimum wage. Currently, the Green and Yellow House program considers families earning up to BRL 2000 [19]. For this population, the programs subsidize most of the costs to build new housing units. According to their individual income, families of the lowest income range pay a small monthly amount to acquire a home in developments built by private companies. Although the program has been successful to some extent in reducing the housing shortfall [20], it has also received much criticism. The role of local housing and planning authorities became limited in this program, with most of the decision power left to the financing bank, Caixa Econômica Federal (CEF), and private companies [21]. Consequently, the lowest income range, which accounts for 70% of the housing demand, only received around 30% of the resources [22].

There are four main stakeholders involved in the design, decisions, approval, and construction of new social housing units: the federal government, the municipal authorities, private developers, and the national bank responsible for financing, CEF. For the lowest income range of such programs, the families that will live in the units have no decision power and are only assigned to a unit once they are built. Several authors indicate that private companies are the main promoting agents in the production of new housing units. It is their role to take the initiative to propose new developments, making the local authorities dependent on the private companies' willingness to build [14,15,21]. These companies prefer cheaper land on the outskirts of cities to increase profits. Similarly, the standardization of designs re-using pre-approved unit designs in several developments has become common practice [21,23]. This standardization can be seen in the examples of newly built house units shown in Figure 1.

Legislation surrounding social housing developments is vast and involves different levels of government. The regulation around how funds can be used for social housing is mainly federal, governed by federal law complemented by standards set by the federal executive power. However, all levels of government have some say regarding where and how things can be built, sometimes with diverging focuses. Given the design of the MCMV program in which the construction companies can propose the developments and location, the federal government must establish some minimum standards for the proposal of new developments. Such standards are very broad to encompass all municipalities' urban planning and building codes. However, this broad regulation is often used to circumvent the more restrictive local legislation. The developers pressure the municipalities to approve new developments that comply only with the federal regulation. Often, this pressure is successful in changing the local regulation for all future developments. Such changes include moving urban perimeters, reducing requirements of green and public spaces, and reducing unit requirements [21,22,24]. These changes in local legislation highlight that the power of choice of location and typology lies with the construction companies [24]. Furthermore, federal legislation prioritizes funding for cities that implement policies like tax exemptions for social housing construction and expedited approval processes [8]. Often, these local policies further diminish the local authority's decision power.

(a)   (b)   (c)

Figure 1. Examples of newly built developments of house units: (**a**) Residencial Veneza in the state of Maranhão—development of detached units built in 2014; (**b**) Residencial Pinheiros in the state of Paraná—development of row-houses built in 2015; (**c**) Residencial Dona Sílvia in the state of Minas Gerais—development of detached units built in 2014. Photos by Ministério do Planejamento [25].

In this context, it is not the role of the municipal authorities to design or propose new developments. The municipality can, however, decide where the development will be by donating the land. While this does happen, in most cases, the developers prefer to acquire the land. Acquiring the land gives the developer a financial advantage to start construction on the development. The developer receives the amount designated for land purchase upfront, but can often negotiate with the previous owner to pay in instalments, leaving a significant amount to start construction. The local authorities approve the final design of the development; however, they cannot deny approval if the project is within the legislation [14,26]. This hinders the municipality's ability to suggest changes and require higher design quality.

The other major role of the municipal authorities is identifying and promoting social actions with the families. The local authority registers the families on an ongoing basis according to the federal criteria for eligibility. It is the process that both the city and the federal government use to determine the demand for social housing in the city. Once a new development is in construction, the municipal authorities rank the eligible families according to national and local criteria to determine the list of families to be sent to CEF for further consideration to receive a unit in that development. It is also the role of the municipal authorities to lead the social work actions with the population of the new development [27,28].

Current legislation guarantees that a minimum of 2.5% of the total amount of federal funds given to any social housing project must be used for social work with the families. However, these funds are often mismanaged, used for other purposes, or not used. When the funds are used according to the legislation, the social work can begin before the families move into the new development, with actions to prepare them to live in the new house and neighborhood. The social work continues for about one year after occupancy. The social work team that works within a given development (a private company) is usually hired specifically to work on that project. The municipal housing social work department oversees and manages this team [28]. One of the services provided by the social work team is providing workshops on skills that will allow the family to generate income, such as fixing mobile phones, makeup, and nail styling. However, such actions reinforce the social stratification of the populations of these neighborhoods. No real effort is made to effectively train them to obtain a higher income, which could lead them out of a low-income situation. Another important action is having on-call days. On these days the families can bring up any issues they are having, and the social work team mobilizes the municipal network to find a solution. Some of the issues the families bring up include finding schools to enroll children, scheduling health exams, etc.

The bank CEF is the primary agent responsible for financial operations. In this regard, there are two main streams of action that are under CEF's responsibility: first, the management of funds for the construction of the development, and second, the financing process for the acquisition of the housing units by the families. Regarding the management of funds for construction, it is the role of this bank to verify if the developers can complete projects under the program. Hence, before proposing the specific project, the construction company must provide a series of documents to CEF to prove their experience, legal standing, and technical and financial capacity [8]. Once this process is completed, the company is enabled to produce a certain number of housing units with a pre-established typology [14]. Later in the process, CEF is responsible for approving the specific project in relation to the requested funding. This approval refers to budget aspects and the compliance of the project to the program's regulation. After the approval of the project, CEF also manages this funding, releasing monthly funds based on inspections and the company's monthly construction completion report. For the lowest income range, CEF acquires all the housing units on behalf of the existing fund used to finance the construction.

When construction of the new development is about 50% completed, CEF receives a list of the selected families from the city to determine their eligibility to finance a housing unit [29]. The list the city provides includes more families than housing units in the development since not all families will remain eligible after CEF's analysis. CEF requests financial information from the families and the documentation that proves their stated financial situation. It is CEF's responsibility to cross-check the financial information provided by the family, including tax and banking information [8]. Establishing a family's eligibility can be difficult given the frequent informal work arrangements of the candidate families. Once the family is approved, CEF determines the amount of the subsidy and charges the monthly payments for the duration of the ten-year contract. The families are only assigned to a specific housing unit when the construction of the development is almost finished, and sometimes after it is finished.

The federal government is responsible for providing most of the funds and outlining, through regulation, the minimum standards required to access the funds. These include requirements in terms of the program's organization, the roles of stakeholders, and the final product itself in urban, architectural, and technical terms. Currently, the product of any housing program that uses federal funding is required to comply with the minimum standards established in this regulation [30]. The federal government also establishes the national demand and goals, deciding how much funding will be allocated for each region, state, and even municipality. However, because the program requires the initiative of private developers, the release of funds for certain regions often does not meet the established goal. Thus, some cities or regions receive extra funding, while others receive much less than the goal [14,15].

For the design of house units, the federal regulation indicates that the housing units must include a living room and kitchen, two bedrooms, a bathroom, and a laundry area. For houses, the minimum area is 36 m^2 if the laundry area is outside or 38 m^2 if the laundry area is inside the house [30]. Since private developers have increased decision power in unit design, they usually have a single unit design that is repeated throughout the development, and even in several different sites, without considering the diverse needs of end-users [21,23,31,32]. It is expected that families will expand their houses after occupancy. However, only recently has this expectation been acknowledged in the regulation of the program. Even so, it is far from being a solution. The legislation includes only one line about allowing the growth of the units, which translates into English as: the housing unit shall be designed to enable its future expansion without loss to the lighting conditions and natural ventilation of the existing rooms [30]. As a result of the lack of guidance, this line of the regulation is often not followed or enforced. Furthermore, when it is followed, the resulting design usually only allows the addition of one room in a specific location [26].

Each stakeholder has the main aspects of their role defined in legislation. However, how they carry out that role can vary significantly from one city to another, and even from one project to another within the same city. Much of the provision of social housing depends on the individual subjective judgements of the many actors involved in the process, and on the political will of government officials and institutions. Thus, this process is susceptible to the capacities, operational standards, and will of those who carry it out.

4. Post-Occupancy Processes and Design Problems

The way a social housing program is implemented can negatively impact the target population. Housing developments for the lowest-income population are usually isolated from commercial areas where most jobs and services are found, thus also increasing the cost of transportation for the families [21,22,24,32]. These segregated neighborhoods are vulnerable, often becoming ghettos of violence and drug trafficking dominated by militias, especially in large cities [11,21]. Such segregation also diminishes the retention rate, with many units illegally sold while families move back to better located informal arrangements [21]. Furthermore, the standardization in unit design, incorporating only the minimum required by the program, results in units that do not satisfy the needs of the many different families that depend on such programs.

Several studies have shown the variety of families that live in these housing units. The units may have only one occupant, more than five family members, or, in many cases, more than one family in the same unit. These studies highlight that the standardization of the housing product is inadequate, considering the diversity of families who live in these developments [4,32,33]. Standardization issues were also present in previous housing programs. They are a consequence of a minimalistic mass-production approach and the lack of interest in the demand during the production process. Families are only selected and assigned to units at the end of the process [31,32,34]. In this context, the families start making changes and expanding their units to satisfy their needs and build spaces for activities to complement their income. Such activities include small shops and services for the local population. Other factors that also contribute to the families' perceived need to make changes to their units include: the lack of safety and privacy, the need to personalize and define the family's territory, the desire for higher quality in surface materials such as floors and walls, and changes in family circumstance (family size, economic, educational, among others) [35–37].

National and local authorities know that the families will make changes and expand their house units. However, the units are not designed and built to facilitate changes, making it difficult and more expensive for families to do so [21]. Moreover, the homeowners design and illegally build most of the changes themselves. This practice often results in inadequate situations, such as the inappropriate discharge of rainwater, encroachment onto the public space, lack of ventilation and natural lighting, among others [31,35,37]. As well as the families, the problems with self-designed and built expansions also affect the municipality and broader society by diminishing the public authorities' capacity to provide safe and healthy environments and services. For example, there have been several reports of flooding inside the houses because the families had built over the water drainage system, which consequently blocked the access for maintenance [26]. Other examples can be seen in Figures 2 and 3. These figures show two different neighborhoods in the city of Pelotas in the south of Brazil, which were built more than thirty years apart. Figure 2 shows an older neighborhood, which originally consisted of duplex units. Currently, access to most of the public lighting and distribution is difficult because the electrical network posts are now inside people's houses or locked gardens. In some areas, the posts had to be moved onto the carriageway. Despite being much more recent, the neighborhood of single-story row-houses in Figure 3 shows the same process of encroachment into the public space in many of the houses.

Figure 2. Example of significant transformation and expansions encroaching into public space in the Guabiroba development. Photos by the authors.

(a)　　　　　　　　　　　　　　　　　　(b)

Figure 3. Examples of expansions encroaching onto the public space in the Anglo development: (**a**) advancing onto the neighborhood's green area, and (**b**) advancing over the sidewalk. Photos by NAUrb-UFPel [38].

Health problems are also among the most significant negative consequences that can emerge from self-designed expansions. The lack of ventilation and natural light often creates problems of humidity inside the unit, aggravating health problems such as respiratory and skin conditions [31,33,36,39,40]. This lack of ventilation, often associated with overcrowded units, has also been an aggravating factor in the spread of COVID-19 in such neighborhoods [39]. Another concern stemming from such expansions is the significant amount of serious accidents occurring due to the use of areas on top of ceiling slabs and inappropriately proportioned staircases [40].

The families make many kinds of changes to their units, such as changes to surface materials, adding objects of significance or painting the façade to personalize and show territoriality, and adding high walls and bars on windows to increase the appearance of safety [35–37]. However, the most relevant kinds of changes that the families make, and that are often associated with negative consequences, are expansions. Expansions usually occur because the dimensions of the initial unit and its spaces are inadequate to their functions and the needs of the users [36,37]. These changes include expanding existing rooms, such as the kitchen and living room, adding rooms such as bedrooms and bathrooms, and adding a room for business [31,33,35–37]. These modifications affect the spatial aspects of the unit, with significantly diverse design solutions depending on the needs of each family. This variety also appears when considering the same kind of change within the same development. Even in cases wherein an expansion plan is provided, it is often not followed [41]. It is also relevant to note that most changes are made soon after the families move in [33,36]. However, although slower, this process continues through time, as illustrated by the greater amount of such changes in older neighborhoods [4].

Although these processes of self-construction and their consequences are well known both in social housing neighborhoods and informal settlements, it was only in 2008 that the government approved legislation to provide professional assistance to low-income families. The federal law Lei No 11.888, from 2008, establishes that low-income families should have free technical assistance in the design and construction of housing [42]. This law indicates that the services for such assistance should be provided by professionals from the areas of architecture, urban planning, and engineering. Among the several objectives of this legislation are: optimizing and qualifying the use of the built space and its surroundings, and formalizing the process of building, renovating or expanding housing before the municipal public authority and other public bodies [42]. The law allows professionals from four different sectors to act through partnerships with the public executive power: public servants, NGOs, autonomous professionals, and professionals enrolled in university residency or outreach initiatives [42,43]. Thus, this law shows great potential to "channel the forces of self-building" [17] (p. 547) by providing some guidance and legal validation in a self-construction context [17]. However, municipalities and other entities have been struggling with the lack of resources to provide such assistance. As a result, there are still very few cases of actions carried out under this law. More recently, some entities, especially CAU (architecture and urbanism professional council), have been mobilized in facilitating and disseminating the enforcement of this law and providing funding for such actions. However, funding for technical assistance is usually destined to informal settlements, since the families in formally established social housing neighborhoods are considered as already adequately housed [26]. This further demonstrates that the construction of social housing units, post-occupancy processes, and the management of informal settlements are all seen as completely unrelated by national and local authorities.

5. Possibilities and Challenges for Design Improvement through Mass Customization

As seen in the previous sections, many of the problems related to unit design are a consequence of ignoring the needs of the many different families in the production process. The small size of the units, contemplating only the minimum required by legislation in standard designs, leads to the immediate transformation of the neighborhoods in a self-design process, often with significant negative consequences for the families and city. The literature shows some paths of investigation and actions to address this problem.

Building more adaptable initial units is an option that has been extensively researched. More adaptable housing units are designed to make it easier for the occupants to change their homes to suit new circumstances [44]. Some studies have developed guidelines for the design of adaptable housing units specifically for this social housing context [35,37]. However, such guidelines are seldom adopted by developers. Some of the aspects of such adaptability strategies that are not attractive to developers include the higher costs of construction, and the greater effort required in the design stage. These aspects reduce profits for the developer. This dynamic is evidenced in the few cases that have used such guidelines in their design. An example is the pilot project in Parauapebas, which incorporated housing units that facilitate change along with other urban quality guidelines. In this case, it was possible because the company Vale guaranteed funding and higher profit margins if the developer followed the guidelines, given that the process required more time and involved a larger number of stakeholders [45]. The added effort in the design stage and the higher costs of building housing units following adaptability guidelines could explain why such guidelines have not been incorporated in the social housing programs' legislation.

Initiatives that operate under the technical assistance law [42] have great potential to improve the quality of environments not only in informal settlements, but also in the development over time of formal social housing neighborhoods. However, there have been few technical assistance actions since the inception of this law in 2008. The costs for a professional to work individually with each family to design their home are high, further limiting the reach of the social assistance law as regards individual renovations. Some groups promote actions to teach the families some design and construction skills, such as the currently active Arquitetura na Periferia organization [46]. Such actions empower the families to self-design and self-build with more quality, while reaching more families with each action. Mass customization, seen as "the mass production of individually customized goods and services" [47] (p. 48), is another approach that shows potential for reaching a large number of families without needing to increase the funding to the same degree.

In a more practical definition, mass customization is seen as "a system that uses information technology, flexible processes, and organizational structures to deliver a wide range of products and services that meet specific needs of individual customers (often defined by a series of options), at a cost near that of mass-produced items", [48] (p. 2). Organizing social housing programs for the lowest income range to provide units designed to satisfy the needs of each family, without needing to increase funding, could significantly increase the quality of environments both within the units and for the neighborhood. However, there are many challenges to the adoption of mass customization in this context.

A significant challenge is engaging the stakeholders with the idea of mass customization for the lowest income range of the social housing programs. Several studies and industry examples explore the concept of mass customization for social housing programs in Brazil. However, such studies and examples are usually limited to the higher income ranges of the social housing programs. The higher income ranges of the programs have a company–customer relationship like that of any other product: the customer chooses the product they want to buy and how much they are willing to pay for it. The lowest income range does not have this market-oriented logic. Only a few studies explore mass customization in social housing programs for the lowest income range, such as [23,26,34,49,50]. These studies consider the benefits that mass customization could bring and some directions for its implementation.

Although these studies agree that it would be beneficial for the families to have a housing unit customized to their needs, mass customization has not yet been applied in any development for the lowest income range. One of the reasons for this is that many of the motives that lead companies to mass customize—such as attracting more customers with customized products, gaining their fidelity, or taking advantage of their willingness to pay a premium—are not available for the lowest income range [26]. In the case of the lowest income range, the families are chosen and assigned to a unit by the local authorities, and the cost per unit is regulated by the program. Thus, despite the benefit it could bring to the families, the developers are not interested in mass customizing the initial unit for the lowest income range [26]. Therefore, the mass customization initiative and management for the lowest income range cannot depend on the developers. They should be in the hands of a stakeholder who has more to gain from unit customization and that also has management capabilities, such as the local authorities. However, currently, the stakeholder with the decision power over the design of the initial housing units is the developer.

Parallel to this are challenges related to the programs' budget and the regulation of the initial housing units. Some studies indicate potential paths to mass customize the initial unit in this context [23,34]. However, legislation currently regulates the number and type of rooms that each unit must have, as well as the types of materials that must be used, which leaves little room for customization [26]. Although the regulation does not set the sizes for the individual rooms, it establishes the furniture that each room must fit and the minimum size of this furniture [30]. All these parameters are expressed in the regulation as the minimum required, thus the housing units could include more rooms or larger rooms, for example. Nonetheless, the program's regulation also favors proposals that have the lowest cost per unit, thus benefiting the largest number of families [51]. It does not include any incentive for incrementing quality in design or being more considerate of user needs. This encourages developers to produce the smallest possible units, accommodating only the minimum required by regulation. City workers from the social housing sector and researchers have strongly criticized this minimalistic approach of the regulation [24,49,52]. However, changing this approach would require significantly higher funding to benefit the same number of families. Therefore, this existing dynamic is a challenge for the implementation of the mass customization of the initial units.

Some challenges could be overcome by considering the post-occupancy processes as an integral part of the housing provision instead of a separate problem. Most studies and cases of mass customization in housing consider the customization of the initial unit, without taking into account the post-occupancy processes. Therefore, they consider that manufacturing must be finished once the family moves in. However, for individually owned houses, the manufacturing process continues long after the family moves in. A shift in perception to match the reality of the users, in which manufacturing continues after occupancy, broadens the scope in which the mass customization agent can operate. This broader scope allows the inclusion of more actors in the value chain of the mass-customized product. Thus, the developer who builds the initial units can be seen as one of the suppliers, while the local authorities, who have closer contact with the families, especially post-occupancy, become the main mass customization agent, and manage the differentiation of the units [49]. Furthermore, considering post-occupancy processes as part of the manufacturing process allows changes to the units over time, as the needs of the family change. Therefore, as concluded in a previous study, it would be more feasible and sustainable over time to apply a mass customization strategy with post-occupancy differentiation of the housing units [26]. Such an approach could also open doors to structure the mass customization processes considering the technical assistance law. Thus, different modes of funding could be considered for the differentiation of the units in a retrofit approach based on mass customization. Furthermore, organizing processes with a mass customization approach could facilitate technical assistance with the individual renovation of units that reaches a large number of families, differently from the current craft processes, in which a specific and different process is put in place for each renovation.

Importantly, expansions affect the spatial aspects of the units, including the size, number, and types of rooms in the house. Therefore, a mass customization strategy in this context must consider dimensional or geometric customization [26]. The lack of confidence, knowledge and will of the users to engage in the design of their own homes has been identified as the main reason for geometric customization not being more widely adopted in housing developments [53–55]. In this social housing context, such cultural limitation is not present since the families have shown they are willing to design their own homes. Thus, a significant challenge in this context is how to guide the families into designing adequate solutions that do not result in negative consequences.

Mass customization toolkits are widely used to match the needs of the user, with custom possibilities available for the product in mass customization strategies. The use of such a configurator, a co-design system, could be useful to engage the families, post-occupancy, in deciding how to expand their units [26]. For housing, digital configurators can include spatial aspects, allowing geometric or dimensional customization [56–58]. For this social housing context, the co-design system would only validate solutions within the parameters determined by the local authorities, thus avoiding problematic situations with the designs [49]. Most importantly, this system would serve as an educational tool, providing a medium through which the families can interact with, visualize, and receive feedback on their solutions before starting construction. Thus, it would not only allow the users to gain better insight into their preferences [59], but would also allow users to gain a better understanding of the design solutions. Such a system could also take advantage of other forms of interaction that are useful in increasing the users' perception of space in social housing design processes, such as physical models, virtual reality and augmented reality [50,60–62].

Mass customization for this context is not seen as a business strategy that brings profit to the agent that provides it. Instead, it is seen as a strategy to optimize the use of resources to improve environments in the city, bringing benefits, through the customized product, not only to the families, but also to other stakeholders and broader society [49]. Thus, mass customization is not seen as a solution separate from other approaches for the improvement of environments and design quality. It is seen as an approach that complements other approaches, such as increased unit adaptability and providing technical assistance.

6. Conclusions

The goal of providing adequate housing for those in need is far from being achieved under the current structure of social housing programs in Brazil. This structure leaves significant decision power in the hands of private companies and benefits quantity regardless of quality. Additionally, it limits the design input of other stakeholders to checking for compliance to minimalistic parameters. Most importantly, it excludes the users from any decision. This process results in segregated neighborhoods, and housing units that do not satisfy the needs of users, as evidenced by extensive renovations post-occupancy. Current research shows that it would be beneficial to apply mass customization in the lowest income range of social housing programs [23,26,34,49]. The literature shows there are feasible avenues and sufficient technologies to apply the concept. It demonstrates the potential of mass customization to significantly improve living environments in these neighborhoods, and to maintain this quality as the neighborhoods evolve. Moreover, this is feasible without significant budget increases. The discussions and studies of mass customization, alongside studies of housing adaptability and architectural technical assistance, should lead to new social housing programs that effectively satisfy the needs of the families and ensure the continued quality of the environments as the neighborhoods evolve.

Author Contributions: Conceptualization, L.F.D.V.; data curation, L.F.D.V. and N.S.M.; writing—original draft preparation, L.F.D.V.; writing—review and editing, L.F.D.V. and N.S.M. Both authors have read and agreed to the published version of the manuscript.

Funding: This research received no external funding.

Conflicts of Interest: The authors declare no conflict of interest.

Entry Link on the Encyclopedia Platform: https://encyclopedia.pub/13306.

References

1. Chiarelli, L.M.A. Habitação Social em Pelotas (1987–2010) Influência das Políticas Públicas na Promoção de Conjuntos Habitacionais. Ph.D. Thesis, Pontifícia Universidade Católica do Rio Grande do Sul, Porto Alegre, Brazil, 2014.
2. Bonduki, N. La Nueva Política Nacional de Vivienda en Brasil: Desafíos y Limitaciones. *Rev. Ing.* **2012**, *35*, 88–94.
3. Bonduki, N. Política habitacional e inclusão social no Brasil: Revisão histórica e novas perspectivas no governo Lula. *Rev. Eletron. Arquit.* **2008**, *1*, 70–104.
4. Medvedovski, N.S. A Vida Sem Condomínio: Configuração e Serviços Públicos Urbanos em Conjuntos Habitacionais de Interesse Social. Ph.D. Thesis, Universidade de São Paulo, São Paulo, Brazil, 1998.
5. Melo, M.A. Políticas Públicas e Habitação Popular: Continuidade e Ruptura, 1979–1988. *Revista de Urbanismo e Arquitetura* **1989**, *2*, 37–59.
6. Cardoso, A.L. Política Habitacional no Brasil: Balanço e perspectivas. *Rev. Propos.* **2003**, *95*, 6–17.

7. Lei No. 14.118. 2021. Available online: https://www.in.gov.br/en/web/dou/-/lei-n-14.118-de-12-de-janeiro-de-2021- (accessed on 2 July 2021).
8. Lei No. 11.977. 2009. Available online: http://www.planalto.gov.br/ccivil_03/_Ato2007-2010/2009/Lei/L11977.htm (accessed on 1 February 2020).
9. UN E/1992/23. *The Committee's General Comment No. 4 on the Right to Adequate Housing*; Office of the High Commissioner for Human Rights: New York, NY, USA, 1992. Available online: http://repository.un.org (accessed on 13 November 2017).
10. Cardoso, A.L.; Aragão, T.A. Do Fim do BNH ao Programa Minha Casa Minha Vida: 25 Anos da Política Habitacional no Brasil. In *O Programa Minha Casa Minha Vida e Seus Efeitos Territoriais*; Cardoso, A.L., Ed.; Letra Capital: Rio de Janeiro, Brazil, 2013; pp. 17–65.
11. Rolnik, R. Late Neoliberalism: The Financialization of Homeownership and Housing Rights. *Int. J. Urban Reg. Res.* **2013**, *37*, 1058–1066. [CrossRef]
12. Lei No. 11.474. 2007. Available online: http://www.planalto.gov.br/ccivil_03/_ato2007-2010/2007/lei/l11474.htm (accessed on 2 July 2021).
13. Lei No. 11.124. 2005. Available online: http://www.planalto.gov.br/ccivil_03/_Ato2004-2006/2005/Lei/L11124.htm (accessed on 1 February 2021).
14. Cardoso, A.L.; Mello, I.d.Q.; Jaenisch, S.T. A Implementação do Programa Minha Casa Minha Vida na Região Metropolitana do Rio de Janeiro: Agentes, Processos e Contradições. In *Minha Casa e a Cidade? Avaliação Do Programa Minha Casa Minha Vida Em Seis Estados Brasileiros*; Amore, C.S., Shimbo, L.Z., Rufino, M.B.C., Eds.; Letra Capital: Rio de Janeiro, Brazil, 2015; pp. 73–102.
15. Ferreira, G.; Calmon, P.; Fernandes, A.; Araújo, S. Política habitacional no Brasil: Uma análise das coalizões de defesa do Sistema Nacional de Habitação de Interesse Social versus o Programa Minha Casa, Minha Vida. *Rev. Bras. Gestão Urbana* **2019**, *11*. [CrossRef]
16. Amore, C.S. Minha Casa Minha Vida' para Iniciantes. In *Minha Casa e a Cidade? Avaliação Do Programa Minha Casa Minha Vida Em Seis Estados Brasileiros*; Amore, C.S., Shimbo, L.Z., Rufino, M.B.C., Eds.; Letra Capital: Rio de Janeiro, Brazil, 2015; pp. 11–27.
17. Salingaros, N. Spontaneous Cities: Lessons to Improve Planning for Housing. *Land* **2021**, *10*, 535. [CrossRef]
18. Ministério do Desenvolvimento Regional. Sistema de Gerenciamento da Habitação. 2019. Available online: http://sishab.cidades.gov.br/ (accessed on 16 February 2020).
19. Casa Verde e Amarela. 2020. Available online: https://www.gov.br/mdr/pt-br/assuntos/habitacao/casa-verde-e-amarela/ (accessed on 17 March 2021).
20. Menezes, D.B. Provisão de habitação de interesse social nos municípios gaúchos: Resultados de programas federais entre 2007 e 2016. *Indicadores Econômicos FEE* **2017**, *44*, 97–111.
21. Rufino, M.B.C. Um olhar sobre a produção do PMCMV a partir de eixos analíticos. In *Minha Casa e a Cidade? Avaliação Do Programa Minha Casa Minha Vida Em Seis Estados Brasileiros*; Amore, C.S., Shimbo, L.Z., Rufino, M.B.C., Eds.; Letra Capital: Rio de Janeiro, Brazil, 2015; pp. 51–70.
22. Pinto, J.V. Contribuições para estudo do 'Programa Minha Casa, Minha Vida' para uma Cidade de Porte Médio, Pelotas-RS: Caracterização das Empresas Construtoras e Incorporadoras Privadas e Inserção Urbana. Master's Thesis, Universidade Federal de Pelotas, Pelotas, Brazil, 2016.
23. Taube, J. Reflexões Sobre a Customização em Massa no Processo de Provisão de Habitações de Interesse Social: Estudo de caso na COHAB de Londrina-PR. Master's Thesis, Universidade Estadual de Londrina, Londrina, Brazil, 2015.
24. Ribeiro, C.J.; Kruger, N.R.M.; Oliveira, T.C. A Cidade e a Moradia: O caso de Pelotas. *PIXO Rev. Arquit. Cid. Contemp.* **2017**, *1*. [CrossRef]
25. Ministério do Planejamento. Programa de Aceleração do Crescimento. Available online: https://www.flickr.com/photos/pacgov/ (accessed on 25 May 2021).
26. Vecchia, L.; Kolarevic, B. Mass Customization for Social Housing in Evolving Neighborhoods in Brazil. *Sustainability* **2020**, *12*, 9027. [CrossRef]
27. Ministério das Cidades. Portaria No. 21. 2014. Available online: http://www.cidades.gov.br (accessed on 14 November 2019).
28. Ministério das Cidades. Portaria No. 464. 2018. Available online: http://www.cidades.gov.br (accessed on 14 November 2019).
29. Ministério das Cidades. Portaria No. 595. 2013. Available online: http://www.cidades.gov.br (accessed on 14 November 2019).
30. Ministério das Cidades. Portaria No. 660. 2018. Available online: http://www.cidades.gov.br (accessed on 14 November 2019).
31. Palermo, C. Avaliação da qualidade no projeto de HIS: Uma parceria com a Cohab/SC. In *Qualidade Ambiental Na Habitação: Avaliação Pós-Ocupação*; Villa, S.B., Ornstein, S., Eds.; Oficina de Textos: São Paulo, Brazil, 2013.
32. Rolnik, R.; Pereira, A.L.d.S.; Lopes, A.P.d.O.; Moreira, F.A.; Borrelli, J.F.d.S.; Vannuchi, L.V.B.; Royer, L.d.O.; Rossi, L.G.A.; Iacovini, R.F.G.; Nisida, V.C. Inserção Urbana no PMCMV e a Efetivação do Direito à Moradia Adequada:Uma avaliação de sete empreendimentos no estado de São Paulo. In *Minha Casa e a Cidade? Avaliação Do Programa Minha Casa Minha Vida Em Seis Estados Brasileiros*; Amore, C.S., Shimbo, L.Z., Rufino, M.B.C., Eds.; Letra Capital: Rio de Janeiro, Brazil, 2015; pp. 391–416.
33. Jorge, L.O.; Medvedovski, N.S.; Santos, S.; Junges, P.; da Silva, F.N. A transformação espontânea das unidades habitacionais do loteamento Anglo em Pelotas/ RS: Reflexões sobre a urgência do conceito de Habitação Social Evolutiva. *Cad. Proarq.* **2017**, *29*, 122–153. Available online: http://cadernos.proarq.fau.ufrj.br/en/paginas/edicao/29 (accessed on 23 October 2019).
34. Taube, J.; Hirota, E.H. Customização em massa no processo de provisão de Habitações de Interesse Social: Um estudo de caso. *Ambient. Constr.* **2017**, *17*, 253–268. [CrossRef]

35. Brandão, D.Q. Disposições técnicas e diretrizes para projeto de habitações sociais evolutivas. *Ambient. Constr.* **2011**, *11*, 73–96. [CrossRef]
36. Marroquim, F.M.G.; Barbirato, G.M. Flexibilidade Espacial em Projetos de Habitações de Interesse Social. In *Colóquio de Pesquisas Em Habitação*; EAUFMG: Belo Horizonte, Brazil, 2007. Available online: http://www.mom.arq.ufmg.br/mom/coloquiomom/comunicacoes/marroquim.pdf (accessed on 15 March 2018).
37. Digiacomo, M.C. Estratégias de Projeto para a Habilitação Social Flexível. Masters' Thesis, Universidade Federal de Santa Catarina, Santa Catarina, Brazil, 2004.
38. NAURB-UFPel. Research Center in Architecture and Urbanism—Núcleo de Pesquisa em Arquitetura e Urbanismo. Available online: https://wp.ufpel.edu.br/naurb/ (accessed on 11 August 2019).
39. Parlato, S.; dos Santos, L.H.; Medvedovski, N. Novos Desafios da Extensão Universitária em Tempos de Covid: Assistência Técnica Em Assentamentos Precários. *PIXO Rev. Arquit. Cid. Contemp.* **2020**, *5*. [CrossRef]
40. Estevão, M.; Medvedovski, N.S. Entrevista com Mariana Estevão: A Prática da Arquitetura e Urbanismo com a Promoção à Saúde da População Brasileira. *Expr. Ext.* **2017**, *22*, 9–12. [CrossRef]
41. Larcher, J.V.M. Diretrizes Visando a Melhoria de Projetos e Soluções Construtivas na Expansão de Habitações de Interesse Social. Masters' Thesis, Universidade Federal do Paraná, Curitiba, Brazil, 2005.
42. Lei No. 11.888. 2008. Available online: http://www.planalto.gov.br/ccivil_03/_ato2007-2010/2008/lei/l11888.htm (accessed on 24 March 2018).
43. Medvedovski, N.S.; Santos, L.A.; Santiago, G.B. Assistência Técnica para Habitação de Interesse Social (Athis): O Ciclo Caat e Suas Contribuições. *Expr. Ext.* **2020**, *25*, 85–98. [CrossRef]
44. Friedman, A. *The Adaptable House: Designing Homes for Change*; McGraw-Hill: New York, NY, USA, 2002.
45. Eskes, N.; Vieira, A. Rethinking Minha Casa, Minha Vida: The Resurgence of Public Space. *Arch. Des.* **2016**, *86*, 54–59. [CrossRef]
46. Arquitetura na Periferia. Available online: https://arquiteturanaperiferia.org.br/ (accessed on 30 June 2021).
47. Pine, B.J. *Mass Customization: The New Frontier in Business Competition*; Harvard Business School Press: Boston, MA, USA, 1993.
48. Da Silveira, G.J.; Borenstein, D.; Fogliatto, F. Mass customization: Literature review and research directions. *Int. J. Prod. Econ.* **2001**, *72*, 1–13. [CrossRef]
49. Dalla Vecchia, L.F. The use of Mass Customization to Improve Environments in Social Housing Neighbourhoods in Brazil. Ph.D. Thesis, University of Calgary, Calgary, AB, Canada, 2021. Available online: https://prism.ucalgary.ca/handle/1880/112960 (accessed on 23 March 2021).
50. Azuma, M.H. Customização em Massa de Projeto de Habitação de Interesse Social por Meio de Modelos Físicos Paramétricos. Ph.D. Thesis, Universidade de São Paulo, São Carlos, Brazil, 2016.
51. Caixa Habitação Urbana—Minha Casa Minha Vida. 2019. Available online: http://www.caixa.gov.br/voce/habitacao/minha-casa-minha-vida/urbana/Paginas/default.aspx (accessed on 16 May 2019).
52. Nascimento, D.M.; Costa, H.S.; Mendonca, J.G.; Lopes, M.S.D.; Lamounirer, R.d.F.; Salomao, T.M.N.; Soares, A.C.B. Programa Minha Casa Minha Vida: Desafios e Avanços na Região Metropolitana de Belo Horizonte. In *Minha Casa e a Cidade? Avaliação Do Programa Minha Casa Minha Vida Em Seis Estados Brasileiros*; Amore, C.S., Shimbo, L.Z., Rufino, M.B.C., Eds.; Letra Capital: Rio de Janeiro, Brazil, 2015; pp. 195–228.
53. Kolarevic, B. Metadesigning Customizable Houses. In *Mass Customization and Design Democratization*; Kolarevic, B., Duarte, J.P., Eds.; Routledge: New York, NY, USA, 2019; pp. 117–128.
54. Kolarevic, B. From Mass Customisation to Design 'Democratisation'. *Arch. Des.* **2015**, *85*, 48–53. [CrossRef]
55. Kolarevic, B.; Duarte, J.P. From massive to mass customization and design democratization. In *Mass Customization and Design Democratization*; Kolarevic, B., Duarte, J.P., Eds.; Routledge: New York, NY, USA, 2019; pp. 1–12.
56. Khalili-Araghi, S.; Kolarevic, B. Development of a framework for dimensional customization system: A novel method for customer participation. *J. Build. Eng.* **2016**, *5*, 231–238. [CrossRef]
57. Khalili-Araghi, S.; Kolarevic, B. Variability and validity: Flexibility of a dimensional customization system. *Autom. Constr.* **2020**, *109*, 102970. [CrossRef]
58. Lo, T.T.; Schnabel, M.A.; Gao, Y. ModRule: A User-Centric Mass Housing Design Platform. In *Computer-Aided Architectural Design Futures. The Next City—New Technologies and the Future of the Built Environment. CAAD Futures 2015*; Celani, G., Sperling, D., Franco, J., Eds.; Communications in Computer and Information Science; Springer: Berlin/Heidelberg, Germany, 2015; Volume 527, pp. 236–254. [CrossRef]
59. Franke, N.; Hader, C. Mass or Only "Niche Customization"? Why We Should Interpret Configuration Toolkits as Learning Instruments. *J. Prod. Innov. Manag.* **2014**, *31*, 1214–1234. [CrossRef]
60. Souza, M.P.; Imai, C.; Azuma, M.H. Contribuições e limitações de modelos físicos e de realidade virtual na análise de projetos de HIS por usuários leigos. *Gestão Tecnol. Proj.* **2018**, *13*, 21–38. [CrossRef]
61. Imai, C. O Processo Projetual e a Percepção dos Usuários: O Uso de Modelos Tridimensionais Físicos na Elaboração de Projetos de Habitação Social. *Ambient. Construído* **2009**, *9*, 105–118.
62. Cuperschmid, A.R.M.; Ruschel, R.C.; Goes, A.M. Augmented Reality: Recognition of multiple models simultaneously. In *The Next City—New Technologies and the Future of the Built Environment, Proceedings of the 16th International Conference CAAD Futures 2015, Sao Paulo, Brazil, 8–10 July 2015*; Electronic Proceedings: São Paulo, Brazil, 2015; pp. 135–154.

 *encyclopedia*

Entry

Age-Friendly Built Environment

Hing-Wah Chau * and Elmira Jamei

Built Environment Discipline Group, Institute for Sustainable Industries and Liveable Cities,
College of Engineering and Science, Victoria University, Footscray, VIC 3155, Australia; elmira.jamei@vu.edu.au
* Correspondence: Hing-Wah.Chau@vu.edu.au

Definition: Age-friendly built environments have been promoted by the World Health Organisation (WHO, Geneva, Switzerland) under the Global Age-friendly Cities (AFC) movement in which three domains are related to the built environment. These are: housing, transportation, outdoor spaces and public buildings. The aim is to foster active ageing by optimising opportunities for older adults to maximise their independent living ability and participate in their communities to enhance their quality of life and wellbeing. An age-friendly built environment is inclusive, accessible, respects individual needs and addresses the wide range of capacities across the course of life. Age-friendly housing promotes ageing in familiar surroundings and maintains social connections at the neighbourhood and community levels. Both age-friendly housing and buildings provide barrier-free provisions to minimise the needs for subsequent adaptations. Age-friendly public and outdoor spaces encourage older adults to spend time outside and engage with others against isolation and loneliness. Age-friendly public transport enables older adults to get around and enhances their mobility. For achieving an age-friendly living environment, a holistic approach is required to enable independent living, inclusion and active participation of older adults in society. The eight domains of the AFC movement are not mutually exclusive but overlap and support with one another.

Keywords: age-friendly; active ageing; ageing in place; walkability; bikeability; accessibility

Citation: Chau, H.-W.; Jamei, E. Age-Friendly Built Environment. *Encyclopedia* **2021**, *1*, 781–791. https://doi.org/10.3390/encyclopedia1030060

Academic Editor: Antonio Frattari

Received: 6 June 2021
Accepted: 5 August 2021
Published: 10 August 2021

Publisher's Note: MDPI stays neutral with regard to jurisdictional claims in published maps and institutional affiliations.

1. Introduction

The world population is growing older according to the *World Population Prospects 2019* published by the United Nations. Persons aged 65 or above outnumbered children under five years old globally for the first time in history in 2018. One in eleven people in the world (9%) were aged 65 or above in 2019, which is projected to increase to one in six people (16%) by 2050. This figure is even projected to be one to four (25%) in Europe and North America by 2050. Lower birth rates and higher life expectancies are transforming the age pyramid shape, especially for developed countries. Another significant trend is urbanisation, which is projected to increase from 53% of people living in urban areas in 2020 to 70% in 2050 [1]. In response to the twin factors of population ageing and urbanisation, the World Health Organisation (WHO, Geneva, Switzerland) published *Global Age-friendly Cities: A Guide* (AFC Guide) in 2007 to enhance the age-friendliness of urban environments [2]. The demographic change towards a much older population structure has considerable consequences for the built environment, social welfare and community services. There is a framework of eight domains to enhance the age-friendliness of cities which can be classified into three categories: (1) the built environment to cover housing, outdoor spaces, public buildings and transportation; (2) social aspects to cover respect and social inclusion, employment, social and civic participation; and (3) service provisions to cover community support and health services, communication and information. The aim of this encyclopedic entry is to unfold the underlying idea of the AFC movement, highlight the importance of creating age-friendly built environments and provide recommendations.

The Global AFC movement can be traced back to the resolution 33/52 of the United Nations General Assembly on 14 December 1978 in which worldwide attention was called to the problems of ageing [3]. Subsequently, the First World Assembly on Ageing was held in Vienna in 1982, resulting in the *Vienna International Plan of Action on Ageing* with 62 recommendations for formulating relevant policies and programmes on ageing [4]. After 20 years, the Second World Assembly on Ageing was held in Madrid in 2002 to review the implementation of the previous *Vienna International Plan* and to release the *Madrid International Plan of Action on Ageing* which focused on three priority areas: older persons and development, advancing health and wellbeing into old age, and ensuring enabling and supportive environments [5].

As a contribution to the Second World Assembly on Ageing, the WHO published the *Active Ageing: A Policy Framework* in 2002 to formulate action plans that promote health and active aging [6]. The underlying idea of the AFC movement is the policy framework for active ageing. Active ageing aims to extend healthy life expectancy and enhance quality of life as people age. The word 'active' does not merely refer to physically active but also refers to the continual participation in social, economic, civil, cultural and spiritual affairs according to personal needs, desires and capabilities. Active ageing arouses people's awareness of their social, physical and mental wellbeing and highlights the importance of maximising autonomy, mobility and engagement. On one hand, maintaining independence to make personal decisions in relation to daily living based on own preferences is a key goal; on the other hand, maintaining social interaction including intergenerational relationships for older adults is crucial as ageing takes place within the context of others, including friends, neighbours and family members. Since the decline in personal abilities and skills vary from a life course perspective, the supportive and enabling living environments promoted by the age-friendly city movement thus do not only cater for older adults but also address the needs of a diverse group of people.

The WHO Global Network of Age-friendly Cities and Communities (GNAFCC) was then launched in 2010. It is a broad-scale effort under rapid expansion and now involves a total of 1114 cities and communities in 44 countries [7]. The Global Network serves as a platform for providing connection and support among different communities and cities worldwide to inspire change and to find appropriate innovative and evidence-based solutions. The vision of the Global Network is for every city and community to strive to become increasingly age-friendly [8].

For achieving an age-friendly living environment, a holistic approach is required to enable independent living, inclusion and active participation of older adults in society. The eight domains of the AFC Guide are not mutually exclusive but overlap and support with one another. For example, respect and social inclusion are reflected in the barrier-free access to outdoor spaces and public buildings. Availability of communication and information allow people to connect for social and civic participation.

2. Applications and Influences

The built environment refers to human-made physical spaces for living, working and recreation. It ranges in scale from buildings, public and open spaces to neighbourhoods and communities, as well as streets and transportation systems. The built environment has a significant impact on the wellbeing and quality of life of older adults.

2.1. Age-Friendly Housing and Buildings

Older adults tend to spend considerably more time at home compared to other age groups [9]. Homes provide older adults with familiar physical setting and emotional affinity in terms of personal experiences and memories [10]. Although ageing may involve deterioration in mobility, visual or hearing abilities and affect cognition and mental capability, it is still desirable for older adults to continue to live in their own homes as long as possible under the notion of 'ageing in place' [11]. The concept of ageing in place is broad, involving personal attachment to place [12]. Attachment to place enables older adults to preserve a sense of identity and independence [13]. Ageing in place does not necessarily imply the continual stay in the same dwelling throughout the later life, but also includes the scenario of downsizing or moving to another home with better accessibility to maximise the use of remaining abilities for independent or assisted living [14].

There are a range of housing options for older adults depending on their degree of mobility and levels of impairment, from ordinary houses, apartment units and retirement housing for independent living, extra care housing with required personal support for assisted living, to residential aged care facilities with nursing care for institutional living [15]. In this continuum of housing options, one end is the long-term family home available in the ordinary housing market, whereas another end is the institutional setting of residential aged care facilities for those who are vulnerable with high levels of dependency. Supporting older adults to age in place at their home settings can delay or obviate the need to move to institutional aged care facilities [16].

Housing design with due consideration of the changing needs across the life course caters for ageing in place. The rationale is to support accessibility and facilitate the ease of movement for a wide range of abilities of residents to enable them to live independently, safely and comfortably to minimise the need for subsequent alternation and retrofitting. Age-friendly housing not only provides benefits to residents for ageing well, but also has public value to reduce the resources spent on institutional care facilities [17]. People have spent more time at home than ever before especially during the coronavirus lockdown periods. Under the impact of the COVID-19 pandemic, residential aged care facilities have experienced significant challenges to control infection and minimise disease transmission by enforcing no outside visitor policies and requiring residents to remain in their rooms. However, these measures have had ongoing detrimental effects on the overall health and wellbeing of residents [18]. This highlights the importance of enabling older adults to remain in their own homes as long as possible through age-friendly housing design and technology-supported smart home provisions.

There are different approaches towards increased accessibility and adaptability of housing and buildings such as universal design, inclusive design, design for all and accessible design [19–21]. Despite the lack of consistent terminology, the rationale behind accessibility is quite similar, highlighting the importance of non-discrimination, social inclusion and equity [22]. Design guidelines and accreditation schemes of age-friendly housing have been established in some developed countries, such as Lifetime Homes Design Guide in the UK, Livable Housing Design Guidelines in Australia and the newly developed Homes4Life in Europe.

The idea of the Lifetime Homes Standard originated from the Helen Hamlyn foundation and the Habinteg Housing Association in the late 1980s. It is based on five overarching principles: inclusivity, accessibility, adaptability, sustainability and good value with 16 design criteria to address the changing needs of individuals and families at different stages of life [23]. The Lifetime Homes Design Guide heavily influenced the development of accessibility requirements of the Building Regulations in the UK. Referring to Part M of the Building Regulations, M4(1) is mandatory to set the minimum standards of accessibility for visitable dwellings, while M4(2) and M4(3) are optional for accessible and adaptable dwellings and wheelchair user dwellings, respectively, in which the optional requirement of M4(2) is broadly equivalent to the Lifetime Homes Design Guide. The minimum baseline standard of M4(1) has been criticised to be too low to meet the actual needs [24,25]. A consultation paper was published by the UK government in 2020 to consider how to raise accessibility standards of new homes [26].

The Livable Housing Design Guidelines was developed as an outcome of the National Dialogue on Universal Housing Design which was formed in 2009 consisting of representatives from the government, key stakeholder groups from the ageing, disability and community support sectors and the residential building and property industry. The voluntary guidelines describe 15 liveable design elements which provide guidance on the expected performance to achieve silver, gold or platinum level accreditation. The first seven elements cover the core requirements of the basic silver level accreditation, whereas the platinum level requires the compliance of all 15 elements [27]. According to the strategic plan in 2010, the original target was set to require all new houses to comply with silver level by 2020 [28]. However, such voluntary provision was hard to achieve the aspirational target [29,30]. Only a low proportion of 5% of new housing could comply with the guidelines by 2020 [31]. Australia is a signatory to the United Nations Convention on the Rights of Persons with Disabilities (UNCRPD) and has obligations about universal design in housing [32]. Considering the failure to meet the 2020 target, the United Nations Committee (New York, NY, USA) supported the recommendation for a mandated approach through legislation [33]. A majority of building ministers in Australia finally agreed to include minimum accessibility provisions for residential housing and apartments in the National Construction Code 2022 based on the silver level of the Livable Housing Design Guidelines [34].

The Homes4Life project titled 'certified smart and integrated living environments for ageing well' is a new European certification scheme which started in December 2018 after receiving funding from the European Union's Horizon 2020 research and innovation programme (grant agreement no. 826295). The project is run by a multidisciplinary group from five countries (Spain, France, The Netherlands, Italy and Belgium) [35]. By adopting a holistic approach, the evaluation framework of Homes4Life covers six aspects: physical, outdoor access, economic, social, personal and management domains for both new and existing residential buildings. Affordability, privacy and dignity, identity and emotional connectivity are taken into consideration. Compared with the Lifetime Homes Design Guide in the UK and the Livable Housing Design Guidelines in Australia, Home4Life puts a specific focus on smart home technologies. For encouraging digitally enriched dwellings to support home-based independent living, the category of smart readiness is introduced in the Homes4Life certification scheme, covering wireless and wired connectivity, network infrastructure and interoperability, IT infrastructure and application programming interfaces, as well as digital security and data protection [36]. Such comprehensive evaluation framework aims to provide a more responsive age-friendly housing for older adults equipped with digital services (such as telecommunication, telehealth and telecare) to maintain social connectedness, health and wellbeing across the life course, together with home-based monitoring technologies with motion sensors, real-time monitoring and reliable alerts to mitigate risks and ensure their safety and security at home [37–39].

It is important for buildings, especially public facilities, to be easily accessible and conveniently used regardless of the age, ability or status of end users. Features that are considered necessary for age-friendly buildings include ramps, elevators, handrails, slip-resistant flooring, seating, accessible public toilets, signage and priority parking bays. Barrier-free buildings address the needs of older adults and empower them to participate in society with inclusion and equity.

2.2. Age-Friendly Public Spaces and Neighbourhoods/Communities

Age-friendly housing and buildings are associated with accessible public and outdoor spaces as well as effective social support services and facilities within neighbourhoods to enhance the quality of life and wellbeing of older adults. It is common for older adults to feel lonely and depressed, especially for those who are living alone without a spouse or a partner [40,41]. If public spaces are age-friendly for people to feel safe and comfortable, their willingness to spend time outside will increase. Popular public spaces, such as cafes, parks, shopping malls, libraries, markets and community centres, are regarded as third places, which are different from the first place (home) or the second place (work/school) [42]. Third places foster voluntary social interaction and provide opportunities for older adults to engage in their local communities to form weak and strong social ties with others against isolation and loneliness [43].

There is a close relationship between the built environment, physical activity and health [44,45]. Regular physical activity contributes to beneficial health effects against overweight or obesity [45]. Since walking is a key outdoor physical activity of older adults, so the built environment is commonly assessed by its walkability [46]. A neighbourhood walkability depends on the land use mix, street connectivity and residential density [47]. An appropriate functional mix integrating residential, commercial and communal uses attracts people to go outside on various schedules. Street connectivity correlates with the street pattern and block size and is an indicator of how accessible a neighbourhood is. Residential density refers to the dense concentration of people, which is regarded as an important condition for 'flourishing city life' and 'a visibly lively public street life' [48]. Older adults living in mixed-use, better connected and more compact neighbourhoods are more likely to be active as the walking distances between destinations and homes are shortened [49]. Other factors that affect the walkability of the built environment include pedestrian safety from vehicular traffic with lower speed limits and more road crossings, connectivity of pedestrian facilities, public safety from crime and violence, trees and vegetation for shading and amenity, adequate lighting for illumination, distinctive signage for wayfinding, user-friendly street furniture and inclusive urban design with barrier-free access [50–52].

The notion of ageing in place extends from age-friendly housing to neighbourhoods and communities. Ageing in place encourages older adults with some degree of independent living ability to remain in their familiar surroundings for longer to maintain their routines, habits and ongoing social connections with others [53]. It is beneficial for older adults to stay in familiar neighbourhoods and communities in their later life, which provides them with a sense of security and belonging. Secure neighbourhoods encourage older adults to spend time outside in confidence to participate in various activities, which enhance cross-generational social encounters. The strategy of lifetime homes and lifetime neighbourhoods was launched in the UK in 2008 to promote a range of affordable and accessible housing choices, to enable residents to connect with other people at the community level and to empower them to develop lifetime neighbourhoods [54]. The six key components of lifetime neighbourhoods are: housing, the built and natural environments, services and amenities, social networks/wellbeing, accessibility and resident empowerment [55]. Similarly, the *Lifelong Communities Handbook: Creating Opportunities for Lifelong Living* was published in 2009 in the US [56]. The seven principles of lifelong communities are: diversity of dwelling types, neighbourhood retail and services, pedestrian access and transit, connectivity, social interaction, healthy living and consideration for existing residents. An enabling built environment with associated social infrastructure supports health and wellbeing over an individual's lifetime [57].

2.3. Age-Friendly Transportation

Walkable age-friendly neighbourhoods and communities provide a safe and well-connected pedestrian network with various transport modes. The availability and accessibility of safe, frequent and reliable public transport facilitates older adults to get around and enhance their mobility. Service quality and travel safety perceptions affect the perceived accessibility of public transport [58]. The age-friendly design of train stations, tram and bus stops is equipped with seating benches, shelters and adequate lighting. The provisions of ramps, elevators and escalators to platforms and low-floor boarding onto buses offer ease of access. Priority seats on public transport encourage passengers to offer seats to older adults and other users in need. The affordability of public transport is also a factor to be considered. In some countries, older adults are entitled to concession fares or even free access to public transport such as buses, trains and trams. This allows them to access to health-related services and communal facilities. Travelling by public transport provides opportunities for social encounters and enables the enjoyment of 'the spectacle of others' interactions and the sense of life going on' [59]. Mobility enhancement through the use of public transport leads to an increase in the quality of life of older adults and a reduction in social exclusion [60].

In addition to walkability, the built environment can be assessed by its bikeability or level of bicycle-friendliness [61]. Cycling is an environmentally sustainable transportation mode to reduce vehicle use and improve air quality. It is also a form of physical activity that contributes to public health as part of an active ageing agenda [62]. Factors affecting the cycling participation of older adults include segregation from vehicular traffic, safety, other cyclists' behaviour, convenience and availability of the relevant infrastructure [63]. Electric bicycles (also called e-bikes) are increasingly popular and stimulate cycling participation. E-bikes are battery-driven and enable users to ride longer distances in comfort with reduced effort [64]. From the planning perspective, it is important to design and provide well-connected bike tracks (or cycling trails) to stimulate transportation cycling among older adults.

For those older adults living in suburbs and remote areas with a lack of public transport, car ownership is associated with their mobility and wellbeing closely. The continuing decline in cognitive, sensory and physical abilities of older adults may cause slow reaction times, poor eyesight, hearing impairment, attention difficulty and memory loss, resulting in driving risks and other safety concerns [65]. Driving cessation is a difficult and challenging issue for older adults, especially for those who have been used to driving. The transition to a non-driving status leads to a sense of loss of independence and the feeling of depression [66,67]. The emergence of fully-autonomous or driverless vehicles may be a solution to offer stress-free door-to-door mobility and increase travel opportunities for older adults, but the popularity of this emerging transport technology depends on its accessibility, reliability and affordability [68,69].

3. Conclusions and Prospects

Facing population ageing and urbanisation, there is a pressing need to create age-friendly built environments for people to have fair and just opportunities to continue their lifestyles and participate in their communities to enhance their quality of life and wellbeing regardless of their age or ability. Ageing in place is promoted for older adults to stay in their familiar living environment to maintain social engagement at the neighbourhood and community levels. It is beneficial for them to live at home as long as possible to minimise the need of staying in institutional setting of aged care facilities. For achieving ageing in place, accessible, adaptable and affordable housing options should be available and smart technologies are highly preferrable to be integrated in housing design to support independent living, assisted living and various digital services. Although there are voluntary design guidelines of age-friendly housing and buildings, there is a lack of incentives for private developers to put additional resources to address the needs of the ageing population. Therefore, the existing voluntary design guidelines are recommended to be replaced by regulatory approaches through legislation to benefit people of all ages and abilities. Government and local councils should be more proactive in enhancing the age-friendliness of built environment as exemplars.

Built environments are often designed and planned by professional people with limited diversity, but not all voices are heard, especially those from marginalised groups. Co-design is a possible way to engage with older adults as collaborators and partners in which their contribution and participation are valued and acknowledged. Despite the challenge of embedding co-design practice into design processes and procedures involving potential time and cost implications, it is worthwhile to take into due consideration of various views from different stakeholders to ensure that their needs are addressed in the design outcome. In addition to planning and design, the participation of older adults in other decision-making processes should be widened to empower them to share their valuable knowledge and experience in enhancing the overall age-friendliness of their neighbourhoods and communities.

The age-friendly built environment is inclusive and accessible through tangible physical infrastructure, building design, public and outdoor open space arrangement and transportation provisions. Other domains of AFC Guide cover intangible aspects, including respect and social inclusion, community support and communication. Both tangible and intangible aspects are complimentary with one another. It is of paramount importance to take a holistic comprehensive approach to optimise both physical and social environments with supportive service provisions to accommodate the growing ageing population.

The idea of age-friendliness has successfully attracted many cities around the world to join the GNAFCC, which seems to be a distinctive branding strategy for cities. Compared with high-sounding commitment, it is crucial to implement age-friendly initiatives and programs in a sustainable manner at the national and state government levels, as well as local communities and neighbourhoods, to address the changing needs across the life course.

4. Lists of Design Standards and Guidelines

4.1. Age-Friendly Housing and Buildings

- Accessibility of Housing: A Handbook of Inclusive Affordable Housing Solutions for Persons with Disabilities and Older Persons [70];
- ADA Standards for Accessible Design;
- ANSI A117.1 Accessible and Usable Buildings and Facilities;
- AS 1428.1 Design for Access and Mobility–General Requirements for Access–New Building Work;
- AS 1428.2 Design for Access and Mobility–Enhanced and Additional Requirements–Buildings and Facilities;
- AS 4299 Adaptable Housing;
- BS 4467 Guide to Dimensions in Designing for Elderly People;
- BS 8300-2 Design of an Accessible and Inclusive Built Environment–Part 2: Buildings–Code of Practice;
- BS 9266 Design of Accessible and Adaptable General Needs Housing–Code of Practice;
- Design Principles for Extra Care Housing [15]
- Designing for Inclusion and Independence: An Explanatory Guide to support the Briefing and Design of Accessible Housing [71];
- Homes4Life: Certification for Ageing in Place [35];
- Housing Enabler Screening Tool [72];
- Housing for Life [73];
- ICC A117.1 Standard for Accessible and Usable Buildings and Facilities;
- ISO/IEC Guide 71 Guide for Addressing Accessibility in Standards;
- ISO TR 22,411 Ergonomics Data for Use in the Application of ISO/IEC Guide 71
- Lifetime Homes Design Guide [23];
- Livable Housing Design Guidelines [27];
- Lifemark Design Standards Handbook [74];
- FlexHousing: Building Adaptable Housing [75];
- Older Persons Housing Design: A European Good Practice Guide [76];
- Universal Design for Houses [77];
- Universal Housing Design Guidelines [78];
- Wheelchair Housing Design Guide [79].

4.2. Age-Friendly Public Spaces and Neighbourhoods/Communities

- BIP 2228 Inclusive Urban Design: A Guide to Creating Accessible Public Spaces;
- BS 8300-1 Design of an Accessible and Inclusive Built Environment–Part 1: External Environment–Code of Practice;
- DD ENV 14383-2 Prevention of Crime–Urban Planning and Design–Part 2: Urban Planning;
- Enabling Inclusive Cities: Tool Kit for Inclusive Urban Development [80];
- Good Practices of Accessible Urban Development [81];
- Lifelong Communities Handbook: Creating Opportunities for Lifelong Living [56];
- Lifetime Neighbourhoods [55];
- Livable Communities: An Evaluation Guide [82];
- Neighbourhoods for Life: A Checklist of Recommendations for Designing Dementia-friendly Outdoor Environments [83];
- The Inclusive Imperative: Towards Disability-inclusive and Accessible Urban Development: Key Recommendations for an Inclusive Urban Agenda [84].

4.3. Age-Friendly Transportation

- Connecting Transportation & Health: A Guide to Communication & Collaboration. [85];
- Guidelines: Equivalent Access under the Disability Standards for Accessible Public Transport 2002 (Cth) [86];
- Improving Transport Accessibility for All: Guide to Good Practice [87];
- Improving Accessibility to Transport for People with Limited Mobility (PLM): A Practical Guidance Note [88];
- Inclusive Mobility: Making Transport Accessible for Passengers and Pedestrians [89];
- The Whole Journey: A Guide for Thinking Beyond Compliance to Create Accessible Public Transport Journeys [90];
- Transit Universal Design Guidelines: Principles and Best Practices for Implementing Universal Design in Transit [91].

Author Contributions: Conceptualization, H.-W.C.; writing—original draft preparation, H.-W.C. and E.J.; writing—review and editing, H.-W.C. and E.J. Both authors have read and agreed to the published version of the manuscript.

Funding: This research received no external funding.

Acknowledgments: The authors would like to express their thanks to anonymous reviewers for their constructive comments, the Lead Collection Editor, Masa Noguchi and other collection editors for the topical collection 'Encyclopedia of ZEMCH Research and Development'.

Conflicts of Interest: The authors declare no conflict of interest.

Entry Link on the Encyclopedia Platform: https://encyclopedia.pub/14047.

References

1. United Nations. *World Population Prospects 2019*; UN Department of Economic and Social Affairs Population Division: New York, NY, USA, 2019.
2. WHO. *Global Age-Friendly Cities: A Guide*; World Health Organization: Geneva, Switzerland, 2007.
3. United Nations. Resolution 33/52: World Assembly on the Elderly (14 December 1978). Available online: https://undocs.org/en/A/RES/33/52 (accessed on 22 July 2021).
4. United Nations. Report of the World Assembly on Aging: Vienna, 26 July to 6 August 1982. Available online: https://www.un.org/esa/socdev/ageing/documents/Resources/VIPEE-English.pdf (accessed on 24 July 2021).
5. United Nations. Report of the Second World Assembly on Ageing: Madrid, 8–12 April 2002. Available online: https://www.un.org/en/events/pastevents/ageing_assembly2.shtml (accessed on 24 July 2021).
6. WHO. *Active Ageing: A Policy Framework*; World Health Organization: Geneva, Switzerland, 2002.
7. About the Global Network for Age-Friendly Cities and Communities. Available online: https://extranet.who.int/agefriendlyworld/who-network/ (accessed on 26 March 2021).
8. WHO. *The Global Network for Age-Friendly Cities and Communities: Looking Back over the Last Decade, Looking Forward to the Next*; World Health Organization: Geneva, Switzerland, 2018.
9. Oswald, F.; Hieber, A.; Wahl, H.-W.; Mollenkopf, H. Ageing and person-environment fit in different urban neighbourhoods. *Eur. J. Ageing* **2005**, *2*, 88–97. [CrossRef]
10. Stones, D.; Gullifer, J. 'At home it's just so much easier to be yourself': Older adults' perceptions of ageing in place. *Ageing Soc.* **2016**, *36*, 449–481. [CrossRef]
11. Wiles, J.L.; Leibing, A.; Guberman, N.; Reeve, J.; Allen, R.E. The meaning of "aging in place" to older people. *Gerontologist* **2012**, *52*, 357–366. [CrossRef]
12. Pani-Harreman, K.E.; Bours, G.J.; Zander, I.; Kempen, G.I.; van Duren, J.M. Definitions, key themes and aspects of 'ageing in place': A scoping review. *Ageing Soc.* **2020**, *41*, 2026–2059. [CrossRef]
13. Peace, S.; Holland, C.; Kellaher, L. 'Option recognition' in later life: Variations in aging in place. *Ageing Soc.* **2011**, *31*, 734–757. [CrossRef]
14. Boldy, D.; Grenade, L.; Lewin, G.; Karol, E.; Burton, E. Older people's decisions regarding 'ageing in place': A Western Australian case study. *Australas. J. Ageing* **2011**, *30*, 136–142. [CrossRef]
15. Housing LIN. *Design Principles for Extra Care Housing*; Housing Learning and Improvement Network: London, UK, 2020.
16. Grimmer, K.; Kay, D.; Foot, J.; Pastakia, K. Consumer views about aging-in-place. *Clin. Interv. Aging* **2015**, *10*, 1803–1811. [CrossRef] [PubMed]

17. Sinclair, S.; de Silva, A.; Kopanidis, F. *Exploring the Economic Value Embedded in Housing Built to Universal Design Principles: Bridging the Gap between Public Placemaking and Private Residential Housing*; Centre for Urban Research, RMIT University: Melbourne, Australia, 2020.
18. Dawson, A.; Berta, W.B.; Morton-Chang, F.; Palmer, L.; Quirke, M. Long term care and the coronavirus pandemic: A new role for environmental design in a changing context. In *World Alzheimer Report 2020—Design, Dignity, Dementia: Dementia-Related Design and the Built Environment*; Fleming, R., Zeisel, J., Bennett, K., Eds.; Alzheimer's Disease International: London, UK, 2020; Volume 1, pp. 238–245.
19. Carr, K.; Weir, P.L.; Azar, D.; Azar, N.R. Universal design: A step toward successful aging. *J. Aging Res.* **2013**, *2013*, 324624. [CrossRef]
20. Hall, T. Inclusive design and elder housing solutions for the future. *NAELA J.* **2015**, *11*, 61–72.
21. Houston, A.; Mitchell, W.; Ryan, K.; Hullah, N.; Hitchmough, P.; Dunne, T.; Dunne, J.; Edwards, B.; Marshall, M.; Christie, J. Accessible design and dementia: A neglected space in the equality debate. *Dementia* **2020**, *19*, 83–94. [CrossRef]
22. Persson, H.; Åhman, H.; Yngling, A.A.; Gulliksen, J. Universal design, inclusive design, accessible design, design for all: Different concepts—One goal? On the concept of accessibility—Historical, methodological and philosophical aspects. *Univers. Access Inf. Soc.* **2015**, *14*, 505–526. [CrossRef]
23. Goodman, C. *Lifetime Homes Design Guide*; IHS BRE Press: Bracknell, UK, 2011.
24. House of Commons Women and Equalities Committee. *Building for Equality: Disability and the Built Environment*; House of Commons Women and Equalities Committee: London, UK, 2017.
25. RIBA. *A Home for the Ages: Planning for the Future with Age-Friendly Design*; RIBA: London, UK, 2019.
26. Ministry of Housing, Communities and Local Government. *Raising Accessibility Standards for New Homes: A Consultation Paper*; Ministry of Housing: London, UK, 2020.
27. Livable Housing Australia. *Livable Housing Design Guidelines*, 4th ed.; Livable Housing Australia: Forest Lodge, Australia, 2017.
28. Department of Social Services, Australian Government. *National Dialogue on Universal Housing Design—Strategic Plan*; Department of Social Services, Australian Government: Canberra, Australia, 2010.
29. Bringolf, J. *Barriers to Universal Design in Housing*; University of Western Sydney: Sydney, Australia, 2011.
30. Ward, M.; Franz, J.; Adkins, B. Livable Housing Design: The voluntary provision of inclusive housing in Australia. *J. Soc. Incl.* **2014**, *5*, 43–60. [CrossRef]
31. Australian Civil Society. *Disability Rights Now 2019: Australian Civil Society Shadow Report to the United Nations Committee on the Rights of Persons with Disabilities: UN CRPD Review 2019*; Australian Civil Society: Canberra, Australia, 2019.
32. Ward, M. Universal Design in Housing: Reporting on Australia's Obligations to the UNCRPD. In Proceedings of the 4th Australian Universal Design Conference, Melbourne, Australia, 17–18 May 2021.
33. UN Committee on the Rights of Persons with Disabilities. *Concluding Observations on the Combined Second and Third Periodic Reports of Australia*; UN Committee on the Rights of Persons with Disabilities: New York, NY, USA, 2019.
34. Department of Industry, Science, Energy and Resources, Australian Government. Building Ministers' Meeting: Communiqué April 2021. Available online: https://www.industry.gov.au/news/building-ministers-meeting-communique-april-2021 (accessed on 4 June 2021).
35. Homes4Life. Homes4Life: Certification for Ageing in Place. Available online: http://www.homes4life.eu/ (accessed on 30 May 2021).
36. Homes4Life. *D4.4—Technical Reference Framework—Final Version*; Homes4Life: Dubai, United Arab Emirates, 2021.
37. Decorme, R.; Urra, S.; Nicolas, O.; Dantas, C.; Hermann, A.; Peñaloza, G.H.; García, F.Á.; Ollevier, A.; Vassiliou, M.C.; van Staalduinen, W. Sustainable Housing Supporting Health and Well-Being. *Proceedings* **2020**, *65*, 12. [CrossRef]
38. van Kasteren, Y.; Bradford, D.; Zhang, Q.; Karunanithi, M.; Ding, H. Understanding smart home sensor data for ageing in place through everyday household routines: A mixed method case study. *JMIR mHealth uHealth* **2017**, *5*, e52. [CrossRef]
39. Carnemolla, P. Ageing in place and the internet of things—How smart home technologies, the built environment and caregiving intersect. *Vis. Eng.* **2018**, *6*, 7. [CrossRef]
40. Savikko, N.; Routasalo, P.; Tilvis, R.S.; Strandberg, T.E.; Pitkälä, K.H. Predictors and subjective causes of loneliness in an aged population. *Arch. Gerontol. Geriatr.* **2005**, *41*, 223–233. [CrossRef] [PubMed]
41. Sundström, G.; Fransson, E.; Malmberg, B.; Davey, A. Loneliness among older Europeans. *Eur. J. Ageing* **2009**, *6*, 267–275. [CrossRef] [PubMed]
42. Oldenburg, R. *The Great Good Place: Cafés, Coffee Shops, Community Centers, Beauty Parlors, General Stores, Bars, Hangouts, and How They Get You through the Day*; Paragon House: New York, NY, USA, 1989.
43. Alidoust, S.; Bosman, C.; Holden, G. Planning for healthy ageing: How the use of third places contributes to the social health of older populations. *Ageing Soc.* **2018**, *39*, 1459–1484. [CrossRef]
44. Sallis, J.F. Measuring physical activity environments: A brief history. *Am. J. Prev. Med.* **2009**, *36*, S86–S92. [CrossRef]
45. Frank, L.D.; Andresen, M.A.; Schmid, T.L. Obesity relationships with community design, physical activity, and time spent in cars. *Am. J. Prev. Med.* **2004**, *27*, 87–96. [CrossRef]
46. Borst, H.C.; de Vries, S.I.; Graham, J.M.; van Dongen, J.E.; Bakker, I.; Miedema, H.M. Influence of environmental street characteristics on walking route choice of elderly people. *J. Environ. Psychol.* **2009**, *29*, 477–484. [CrossRef]

47. Van Holle, V.; Van Cauwenberg, J.; Van Dyck, D.; Deforche, B.; Van de Weghe, N.; De Bourdeaudhuij, I. Relationship between neighborhood walkability and older adults' physical activity: Results from the Belgian Environmental Physical Activity Study in Seniors (BEPAS Seniors). *Int. J. Behav. Nutr. Phys. Act.* **2014**, *11*, 110. [CrossRef]
48. Jacobs, J. *The Death and Life of Great American Cities*; Vintage Books: New York, NY, USA, 1961.
49. Frank, L.D.; Schmid, T.L.; Sallis, J.F.; Chapman, J.; Saelens, B.E. Linking objectively measured physical activity with objectively measured urban form: Findings from SMARTRAQ. *Am. J. Prev. Med.* **2005**, *28*, 117–125. [CrossRef]
50. Jacobs, A.B. *Great Streets*; MIT Press: Cambridge, MA, USA, 1993.
51. Crowe, T.D. *Crime Prevention through Environmental Design*; Elsevier: Amsterdam, The Netherlands, 2013.
52. Davoudian, N. *Urban Lighting for People: Evidence-Based Lighting Design for the Built Environment*; RIBA Publishing: London, UK, 2019.
53. Houben, P. Changing housing for elderly people and co-ordination issues in Europe. *Hous. Stud.* **2001**, *16*, 651–673. [CrossRef]
54. DCLG. *Lifetime Homes, Lifetime Neighbourhoods: A National Strategy for Housing in an Ageing Society*; Department for Communities and Local Government: London, UK, 2008.
55. Bevan, M.; Croucher, K. *Lifetime Neighbourhoods*; Department for Communities and Local Government: London, UK, 2011.
56. Atlanta Regional Commission. *Lifelong Communities Handbook: Creating Opportunities for Lifelong Living*; Atlanta Regional Commission: Atlanta, Georgia, 2009.
57. Keyes, L.; Rader, C.; Berger, C. Creating communities: Atlanta's lifelong community initiative. *Phys. Occup. Ther. Geriatr.* **2011**, *29*, 59–74. [CrossRef]
58. Friman, M.; Lättman, K.; Olsson, L.E. Public transport quality, safety, and perceived accessibility. *Sustainability* **2020**, *12*, 3563. [CrossRef]
59. Green, J.; Jones, A.; Roberts, H. More than A to B: The role of free bus travel for the mobility and wellbeing of older citizens in London. *Ageing Soc.* **2014**, *34*, 472–494. [CrossRef]
60. Rye, T.; Mykura, W. Concessionary bus fares for older people in Scotland-are they achieving their objectives? *J. Transp. Geogr.* **2009**, *17*, 451–456. [CrossRef]
61. Krenn, P.J.; Oja, P.; Titze, S. Development of a bikeability index to assess the bicycle-friendliness of urban environments. *Open J. Civ. Eng.* **2015**, *5*, 451. [CrossRef]
62. Black, P.; Street, E. The power of perceptions: Exploring the role of urban design in cycling behaviours and healthy ageing. *Transp. Res. Procedia* **2014**, *4*, 68–79. [CrossRef]
63. Winters, M.; Sims-Gould, J.; Franke, T.; McKay, H. "I grew up on a bike": Cycling and older adults. *J. Transp. Health* **2015**, *2*, 58–67. [CrossRef]
64. Van Cauwenberg, J.; De Bourdeaudhuij, I.; Clarys, P.; De Geus, B.; Deforche, B. Older adults' environmental preferences for transportation cycling. *J. Transp. Health* **2019**, *13*, 185–199. [CrossRef]
65. Anstey, K.J.; Wood, J.; Lord, S.; Walker, J.G. Cognitive, sensory and physical factors enabling driving safety in older adults. *Clin. Psychol. Rev.* **2005**, *25*, 45–65. [CrossRef]
66. Adler, G.; Rottunda, S. Older adults' perspectives on driving cessation. *J. Aging Stud.* **2006**, *20*, 227–235. [CrossRef]
67. Deka, D. The effect of mobility loss and car ownership on the feeling of depression, happiness, and loneliness. *J. Transp. Health* **2017**, *4*, 99–107. [CrossRef]
68. Kovacs, F.S.; McLeod, S.; Curtis, C. Aged mobility in the era of transportation disruption: Will autonomous vehicles address impediments to the mobility of ageing populations? *Travel Behav. Soc.* **2020**, *20*, 122–132. [CrossRef]
69. Zandieh, R.; Acheampong, R.A. Mobility and healthy ageing in the city: Exploring opportunities and challenges of autonomous vehicles for older adults' outdoor mobility. *Cities* **2021**, *112*, 103135. [CrossRef]
70. UN Habitat. *Accessibility of Housing: A Handbook of Inclusive Affordable Housing Solutions for Persons with Disabilities and Older Persons*; UN-Habitat: Nairobi, Kenya, 2014.
71. Summer Housing. Designing for Inclusioin and Independence: An Explanatory Guide to Support the Briefing and Design of Accessible Housing. Available online: https://summerhousing.org.au/designing-for-inclusion-and-independence/ (accessed on 29 July 2021).
72. Iwarsson, S.; Slaug, B. The Housing Enabler Screening Tool. Available online: http://www.enabler.nu/ (accessed on 30 May 2021).
73. Master Buildrs Association of the ACT. *Housing for Life*; Master Buildrs Association of the ACT: Lyneham, Australia, 2001.
74. Lifetime Design Limited. *Lifemark Design Standards Handbook*; Lifetime Design Limited: Auckland, New Zealand, 2016.
75. Canada Mortgage and Housing Corporation. *FlexHousingTM: Building Adaptable Housing*; Canada Mortgage and Housing Corporation: Ottawa, ON, Canada, 2000.
76. Emilia-Romagna Territorial Economic Development S.p.A. (ERVET); Blekinge Institute of Technology—Karlskrona University; Brighton and Hove City Council; FAMCP Federazione Aragonese di Municipalità; Györ City Council. *Older Persons' Housing Design: A European Good Practice Guide*; Wel-hops Network: Bologna, Italy, 2007.
77. BRANZ. Universal Design for Houses. Available online: http://www.buildmagazine.org.nz/assets/PDF/Build_168_Universal_Design_Supplement.pdf (accessed on 29 July 2021).
78. Landcom. *Universal Housing Design Guidelines*; Landcom: Parramatta, Australia, 2008.
79. Habinteg Housing Association. *Wheelchair Housing Design Guide*, 3rd ed.; RIBA Publishing: London, UK, 2018.
80. Asian Development Bank. *Enabling Inclusive Cities: Tool Kit for Inclusive Urban Development*; Asian Development Bank: Manila, Philippines, 2017.

81. United Nations. *Good Practices of Accessible Urban Development*; United Nations: New York, NY, USA, 2016.
82. AARP. *Livable Communities: An Evaluation Guide*; American Assocation of Retired Persons: Washington, DC, USA, 2005.
83. Mitchell, L.; Burton, E.; Raman, S. *Neighbourhoods for Life: A Checklist of Recommendations for Designing Dementia-Friendly Outdoor Environments*; Housing Corporation: London, UK, 2004.
84. Disability Inclusive and Accessible Urban Development Network. *The Inclusion Imperative: Towards Disability-Inclusive and Accessible Urban Development—Key Recommendations for an Inclusive Urban Agenda*; Disability Inclusive and Accessible Urban Development Network: New York, NY, USA, 2016.
85. Steedly, A.; Townsend, T.; Huston, B.; Lane, L.B.; Danley, C. Connecting Transportation & Health: A Guide to Communication & Collaboration. Available online: http://onlinepubs.trb.org/onlinepubs/nchrp/docs/NCHRP25-25Task105/NCHRP25-25 Task105Guidebook.pdf (accessed on 29 July 2021).
86. Australian Human Rights Commission. *Guidelines: Equivalent Access under the Disability Standards for Accessible Public Transport 2002 (Cth)*; Australian Human Rights Commission: Sydney, Australia, 2020.
87. European Conference of Ministers of Transport. *Improving Transport Accessibility for All: Guide to Good Practice*; OECD Publications Service: Paris, France, 2006.
88. World Bank. *Improving Accessibility to Transport for People with Limited Mobility (PLM): A Practical Guidance Note*; World Bank: Washington, DC, USA, 2013.
89. Department for Transport, U.G. Inclusive Mobility: Making Transport Accessible for Passengers and Pedestrians. Available online: https://www.gov.uk/government/publications/inclusive-mobility (accessed on 6 June 2021).
90. Department of Infrastructure and Regional Development, A.G. *The Whole Journey: A Guide for Thinking beyond Compliance to Create Accessible Public Transport Journeys*; Department of Infrastructure and Regional Development, A.G.: Canberra, Australia, 2017.
91. American Public Transportation Association. *Transit Universal Design Guidelines: Principles and Best Practices for Implementing Universal Design in Transit*; American Public Transportation Association: Washington, DC, USA, 2020.

Entry

Green Building Rating Systems (GBRSs)

Lia Marchi [1], Ernesto Antonini [1] and Stefano Politi [2,*]

[1]   Department of Architecture, University of Bologna, 40136 Bologna, Italy; lia.marchi3@unibo.it (L.M.);
      ernesto.antonini@unibo.it (E.A.)
[2]   Politecnica, 41126 Modena, Italy
*   Correspondence: stefano.politi2@unibo.it

Definition: Green Building Rating Systems (GBRSs) are typically third-party, voluntary, and market driven standards that measure buildings' sustainability level by multi-criteria assessment, and encourage the adoption of environmentally, socially and economically sustainable practices in design, construction and operation of buildings (or neighborhoods). GBRSs aim at guiding and assessing the project throughout all its life cycle, thus limiting the negative impact on the environment, as well as on the building occupants' health and well-being, and even reducing operational costs. Hundreds of GBRSs are now available worldwide, varying in approaches, application processes, and evaluation metrics. BREEAM, CASBEE, Green Star and LEED are among the most applied worldwide. Despite some differences, they all adhere to the same general evaluation structure: project performances ares measured using a set of relevant indicators, grouped per topics such as water management, energy use, materials, site qualities. Each assessed requirement is assigned a score/judgment, the total of which determines the level of sustainability achieved. In addition to regular updates, a current trend is to improve the effectiveness of protocols, making them more comprehensive and accurate, while keeping them easy to use.

Keywords: BREEAM; CASBEE; Green Star; LEED; multi-criteria assessment; sustainable building

Citation: Marchi, L.; Antonini, E.; Politi, S. Green Building Rating Systems (GBRSs). *Encyclopedia* **2021**, *1*, 998–1009. https://doi.org/10.3390/encyclopedia1040076

Academic Editors: Masa Noguchi, Antonio Frattari, Carlos Torres Formoso, Haşim Altan, John Odhiambo Onyango, Jun-Tae Kim, Kheira Anissa Tabet Aoul, Mehdi Amirkhani, Sara Jane Wilkinson and Shaila Bantanur

Received: 27 August 2021
Accepted: 22 September 2021
Published: 26 September 2021

Publisher's Note: MDPI stays neutral with regard to jurisdictional claims in published maps and institutional affiliations.

1. Introduction

Growing attention to global environmental and societal challenges requires the construction sector to be more sustainable, because of its major impact on these challenges. Beyond regulations and policy enforcements, a voluntary effort is required of all the stakeholders to design, construct, run and manage buildings assuming a holistic approach to sustainability. This requires that the effect of construction features on the triple bottom line (planet, people, profit), as well as possible mitigation actions, are clearly understood. Accordingly, sustainability assessment has been recognized as a crucial mean to this end [1], and Green Building Rating Systems (GBRSs) have emerged as a valuable tool to assess and guide the whole construction process to be greener.

1.1. Terms and Definition

In a nutshell, GBRSs can be defined as third-party, voluntary, and market driven standards that measure buildings' sustainability by multi-criteria assessment, and encourage the adoption of environmentally, socially and economically sustainable practices in design, construction and operation of buildings (or neighborhoods).

Since the notion combines two different elements—Green Building and Rating System—examining them separately may help to provide a suitable definition for the whole. First, Green Building is a multifaceted notion that refers to a broad variety of issues. The World Green Building Council (World GBC) defines it as "a building that, in its design, construction or operation, reduces or eliminates negative impacts, and can create positive impacts, on our climate and natural environment. Green buildings preserve precious natural resources and improve our quality of life" [2]. Recurrent topics addressed by a green project are efficient use of energy, water, and other natural resources; pollution reduction and waste management; good indoor environmental standards; sustainable use of materials along their life cycle; occupants health and well-being; design featured flexibility and adaptability to a changing environment; and quality of open spaces. Since its appearance in the early 1990s [3], the notion of Green Building has mainly addressed physical and functional features of an eco-architecture, encompassing only few elements of the broader view of sustainability and its multiple social, economic and institutional issues [4]. However, as the sustainability goals have been increasingly enlarged, a wider approach to Green Building has also emerged during the last decades. In fact, the recent ISO 15392:2019 Regulation (*Sustainability in buildings and civil engineering works*) set that sustainable construction works should consider sustainable development in terms of its three primary aspects (economic, environmental and social), while meeting the requirements for technical and functional performance [5].

A Rating System (RS) is a tool for classifying objects based on how well they comply with one or more relevant requirements, which are those that affect the object's performance whose level the system is intended to appraise. A RS evaluating the level of sustainability of a building must take into account several requirements, detecting the level of performance for each of them respect to a common baseline, which might be regulatory thresholds or a comparison benchmark with other buildings. In other words, a RS "rates or rewards relative levels of building performance or their compliance with specific environmental goals and requirements" [6].

Therefore, the vast range of green buildings' performances (i.e., different topics such as energy consumption, water use, indoor environmental quality, location) are evaluated by means of a specific baseline. Since only an overall assessment of the building's sustainability can make the system useful and effective, a GBRS must combine "apples and oranges" [7] into a score that expresses how much that project respects the environment, the building's occupants, and the local community (*multi-criteria*). Thus, the measures of these performances are weighted, using a balancing process specific to each scheme, and are combined into a single grade/judgment that shortly communicate the building's overall level of sustainability. As a result, the building is "classified" or "rated" by the organization who manages the GBRS (*third-party*).

Both in literature and practice, many names are used to designate these tools that assess the impact a construction has on both its local surrounding and the broader environment. Among the most common include, Green Building Rating Systems, Sustainable Building Rating Systems, Sustainable Rating Tools, Green Building Assessment systems, certifications, and protocols. Each of them has a few distinctions, as well as variations in approaches, methodology and applications that are found within the same family of tools. Therefore, it is difficult to give a unified definition of these tools. However, the World GBC has provided a description with the aim of being comprehensive: a GBRS is a tool "used to assess and recognize buildings which meet certain green requirements or standards" [8]. This leads to making the rating tools, whose adoption is often *voluntary*, a means to recognize and reward organizations or individuals who build and operate greener buildings (*market-driven*), thereby encouraging them to push the boundaries of sustainability forward.

1.2. Background Condition to Development of GBRS

The need to deal with growing global challenges is the context in which these tools have been developed. As the built environment is one of the main contributors to socio-ecological issues, above all of which is climate change, the adoption of environmentally- and socially-friendly approaches in this sector has been recognized as crucial for decades.

This stems primarily from the significant impact the built environment makes in terms of energy and resource consumption, as well as emissions. In fact, the building sector accounts globally for 31% of the entire final energy use, 54% of the final electricity demand, and 23% of the global energy-related CO_2 emissions, one third of which come from the direct consumption of fossil fuels [9]. Even in Europe, where a great effort has been put on reducing construction-related impact, buildings are still responsible for around 40% of the EU energy consumption and for 36% of greenhouse gas (GHG) total emissions [10]. The building industry also consumes more than 40% of the raw materials of the global economy every year, the majority of which are non-renewable, and generates over 35% of global waste [11]. Water use is also relevant, as over 17% of fresh water is used globally by the construction and operation of buildings [12]. In addition, the quality of the built environment affects building occupants' and local communities' health and well-being (e.g., by indoor VOC emissions, thermo-hygrometric discomfort, and lack of open spaces).

Although the energy demand of the building sector is higher than industry and transport, the construction has the highest potential of emission reduction, mainly due to the flexibility of its demand [13]. According to Berardi, large room for improvement may especially come from adoption of more sustainable building practices. This leads to the concept of environmentally-friendly building, which has gradually taken place worldwide [4].

As a result, regulations, codes, strategies, and tools have been developed in the last decades to push a rapid and effective transition in design and construction processes, with a particular stress on energy efficiency.

At a global scale, several goals of the 2030 UN Agenda for Sustainable Development addresses the issue [14], including but not limited to: SDGs n. 11 (Sustainable cities and communities), n. 9 (Industry, innovation and infrastructures), and n. 7 (Clean and affordable energy). In Europe, this path is mainly regulated by the EU Directives on Energy Performance of Buildings (EPBD III) 2018/844/EC [15] and Energy Efficiency (EED) 2018/2002/EU [16] that attempts to improve energy efficiency in both new and refurbished buildings across the Union. However, less than 40% of energy consumption and less than 50% of CO_2 emissions from buildings are currently subject to mandatory performance policies [17]. Hence, beyond regulations, a great voluntary effort is needed to drive the change toward a more sustainable living environment.

Therefore, the construction industry has gradually taken important steps, especially boosting the market to improve the building environmental performances by considering the whole building life cycle, from raw material extraction to the potential reuse of the outputs coming from the dismissal. In this particular circumstance, sustainability building assessment systems have spread as a means to detect and mitigate the impacts.

1.3. History: First Appearance and Developments

Although the assessment of the building's environmental features such as air quality and indoor comfort had begun earlier, it was only with the introduction of multi-criteria assessment schemes (e.g., GBRS) that the problem was addressed as a whole, instead of one topic at a time, using separate indicators and criteria [1]. In this context, GBRSs appeared as a more comprehensive, user-friendly, and informative way of spreading sustainable construction, as compared to the more accurate but time-consuming and extremely technical approaches such as Life Cycle Assessment (LCA) and comparable frameworks that use a set of complex indicators.

The first multi-criteria tool for this scope was developed by the Building Research Establishment (BRE Group) in the United Kingdom, as even prior to today the current concept of green building had entered international agendas. The British Building Research Establishment Environmental Assessment Method (BREEAM) was launched in 1993 for new building sustainable design [18]. Many other GBRSs have been developed following this example, though they have adopted different assessment schemes, metrics and indicators [19].

The Leadership in Energy and Environmental Design (LEED) was launched by the United States Green Building Council (USGBC) in 1998 [20]. The Japanese Comprehensive Assessment System for Building Environmental Efficiency (CASBEE) made its appearance in 2001, three years after LEED [21].

The need to make these national instruments global soon emerged, pushing the Natural Resources Canada (NRC) to lead the development of SBMethod, at the end of 1990s. As a common protocol on whose scheme national and even regional adaptations could be made, SBMethod resulted in several applications, such as Verde (Spain), SBTool PT (Portugal), SBTool CZ (Czech Republic) and SBTool IT (Italy), this latest re-named ITACA protocol in 2011 [22].

The success of sustainability assessment tools had been so extraordinary that over 600 schemes were available worldwide in 2004, within just one decade from the first one that had been launched [13,23]. Although the reason for this abundance lies mainly in the need of each country to adapt the instrument to its peculiarities (e.g., climate and environmental priorities, type of building stock, culture, and local codes), this led each rating to adopt different parameters and metrics as well, making any cross comparison and transnational collaboration difficult [23]. Therefore, two important standards were developed in order to harmonize the many available assessment tools: ISO 21931–1, 2010 *Framework for Methods of Assessment of the Environmental Performance of Construction Works* [24] and ISO 15643–1, 2010 *Sustainability Assessment of Buildings* [25].

Furthermore, many other RSs have been developed and more are being developed by both public and private organizations. At present, the World GBC alone enlists 58 rating tools administered by its national Councils [8]: beyond the already mentioned BREEAM (UK), LEED (US), and CASBEE (JP), quite popular are also DGNB System (D-), Green Star (AUS), HQE (F), Green Globes (US and Canada), and GBTool (South Africa).

At present, thousands of buildings have been certified by GBRSs worldwide. Table 1 shows key facts about the diffusion and application of the four most predominant, according to [12].

Table 1. Facts of most diffused GBRSs worldwide according to Say and Wood (2008).

GBRS	N. of Certified Buildings	Countries	Data Source
BREEAM (UK)—since 1993	594,011 (2021)	89	[18]
LEED (US)—since 1998	79,418 (2021)	all (mainly US)	[20]
CASBEE (JP)—since 2001	500 (2016)	JP	[21]
GREEN STAR (AU)—since 2003	2827 (2020)	AU + NZ + SA	[26]

2. Methods: Among Similarities and Differences

2.1. General Aim and Scope

The main objective of GBRSs is to encourage all stakeholders in the construction industry to move toward a more sustainable, healthy, and just living environment. This objective hence aims to elevate the ambition of governments, corporations, owners, and practitioners.

The specific goal is to provide a structured procedure by which the sustainability level of a building or a neighborhood can be assessed. However, despite being conceived for assessment purposes, GBRSs also serve as design support tools [27] and are powerful means for project management, as they offer a framed protocol to map important synergies among the building elements and functions.

Moreover, these methods include a comprehensive analysis of the site and its surroundings that highlight strengths and weaknesses on which to focus the envisaged enhancement strategies. By doing so, GBRSs are good instruments to balance the many facets of a project, seeking for the "best compromise", on a case-by-case basis

Since green building clients often do not have clearly in mind what that entails, and at the same time architects are not always able to clearly motivate sustainable design strategies they envisage, a further scope is to improve communication [28] by making the building's performance evident to the market. This is a crucial purpose for the rating processes, being that they are typically voluntary.

Accordingly, it derives that the most recurrent motivations for applying a rating tool are: *educational and marketing* purposes, as a GBRS tool can provide owners and design teams with an effective means of ascertaining and communicating the ecological performance of the building; *guidance effects* in identifying the green standards to be followed and in selecting environmentally-friendly products, materials and strategies that the project can benefit from; and *means* of supporting stakeholders in becoming more aware of the value of green building and its long-term benefits.

In addition to environmental benefits, achieving a certification is also proven to offer a series of economic gains/savings, especially those related to the lowering of building operating costs [28]. For example, the LEED Concepts Guide reports several studies that prove maintenance costs in green buildings are lower than traditional buildings (around 12% less), occupants satisfaction is higher (from 2 to 27%), while the average additional cost is 2%, despite the general public's perception that it is higher. Furthermore, the USGBC reported that certified green office buildings are usually rented for 2% more than similar buildings nearby [29].

2.2. Comparison in Approaches

Given the large number of GBRSs available today, as well as the wide range of methods, terms, models, and indicators they use, there is a consistent body of literature that analyzes and compares them [1,4,12,13,19,23,28,30–37].

There are features and approaches common to all GBRSs representing the foundations of this type of tool, e.g., adopting a multi-criteria structure and presenting the final result as an overall score. Many other differences can be found in each RS, which should be analyzed in detail to discuss specific features. However, a good picture of this variability—but not exhaustive—can be given by comparing the four most widespread GBRSs at global level [12], as an opportunity to discuss different approaches. Hence, based on the most frequently addressed issues in the above-mentioned publications, a comparison among BREEAM, LEED, Green Star and CASBEE is made in Table 2, by highlighting: i) available building adaptation schemes; ii) core evaluation categories; iii) assessment procedures; and iv) certification levels. This led to pinpoint both the main similarities (that are also basic elements of other GBRSs) and crucial differences. As a result, also strengths and weaknesses of each can be identified and discussed.

Table 2. Comparison of most diffused GBRSs worldwide.

	BREEAM	LEED	CASBEE	Green Star
(i) Building adaptations	New Construction In-Use Refurbish. and Fit-Out Communities	New Construction Exist. Buildings Operations and Maintenance Comm. Interiors Core and Shell Schools Retail Healthcare Homes Neighbor. Develop.	Pre-design New Construction Existing Building and Renovation	Communities Buildings Design and As Built Interiors Performances
(ii) Categories	Management Health and Well-being Energy Transport Water Material Waste Land Use and Ecology Pollution Innovation	Integrative Process Location and Transportation Sustainable Site Water Efficiency Energy and Atmosphere Material and Resources Indoor Env. Quality Regional Priority Innovation	Indoor Environment Quality of Service On-site Environment Energy Resource and Materials Off-site Environment	Management Indoor Environment Quality Energy Transport Water Material Land Use and Ecology Emissions Innovation

Table 2. *Cont.*

	BREEAM	LEED	CASBEE	Green Star
(iii) Assessment method	Pre-weighted categories	Additive credits	BEE ranking chart	Pre-weighted categories
(iv) Certification levels	Pass $\geq$ 30 Good $\geq$ 45 Very Good $\geq$ 55 Excellent $\geq$ 70 Outstanding $\geq$ 85	Certified 40–49 Silver 50–59 Gold 60–79 Platinum $\geq$ 80	Poor: BEE < 0.5 Fairy Poor: BEE 0.5–1.0 Good: BEE 1–1.5 Very Good: BEE 1.5–3 or BEE $\geq$ 3 and Q < 50 Excellent: BEE $\geq$ 3 and Q $\geq$ 50	Min. Practice (1 star) Average Practice (2) Good Practice (3) Best Practice (4) Austr. Excellence (5) World Leader. (6)
Data source	[18]	[20]	[21]	[26]

2.2.1. Building Adaptations and Diffusion

As CASBEE and Green Star are specifically designed to their own geo-political context, their diffusion is significantly lower than the other two (as shown in Table 1), which instead offer flexibility in adapting to different areas, offering international equivalent codes as reference standards.

Since sustainable performances may depend not only on geographical features but also on the building and intervention type, many GBRSs have scheme adaptations [6]. All of the main four have been updated and new version with building adaptations released over the years, so that today they are available for many types of buildings, both existing and new. Specific schemes include, for example, single-family houses, multi-family rise, commercial buildings, schools, offices, and warehouses, as well as campuses and entire neighborhoods. Some credits and indicators may vary for new construction, renovation, interior fit-out or existing building operation.

Given that CASBEE is the system where government plays a predominant role, the extent to which a project can be applied is broad up to the city level [4], while the others reach only the neighborhood scale.

2.2.2. Categories

Topics in different tools are referred to as Credit Category, Evaluation Area, Topic, Theme, etc., and gather a set of Criteria, Credits, Requirements, and Core Indicators that measure specific performances, by Indicators [36].

Despite differences in names and details, almost all GBRSs address the same core topics, confirming that the green building industry shares the same concerns at a global level.

The main assessment categories of BREEAM target low impact design and carbon emissions reduction, durability and resilience, adaption to climate change, ecological value and biodiversity protection. Similarly, LEED covers a comprehensive set of sustainability goals as a means of coping with climate change, enhancing occupants' well-being, and protecting water resources, as well as promoting biodiversity, regenerative material cycles, green economy, community justice, and quality of life. Accordingly, the other two GBRSs use very analogous categories.

What usually emerges from comparisons in the literature is that energy is the most important category for all, followed by materials, and health and well-being. Transport, land use and ecology are usually weighted less. Regarding sustainability pillars, the comparison performed by Doan shows all the ratings address well the environmental sphere; the social sphere is given less but fair importance, while very few credits address the economic sustainability of the project [4].

Except for CASBEE, the schemes include an Innovation category that demonstrates the willingness of being updated and even preceding new steps and approaches from the construction industry.

In addition, particularly worth mentioning is the Integrative Process (IP) by LEED, the general principle of which is common to many other rating schemes [29]. In this topic, the tool assesses and promotes the project team to collaborate tightly, looking for synergies among systems and components, in order to achieve high levels of building performance, human comfort, environmental benefits and cost-effectiveness. The many professionals involved work closely and simultaneously from the beginning rather than separately as in more conventional design processes.

2.2.3. Assessment Method

The multi-criteria structure of these tools allows each design team to select their proper method (i.e., credits) to reach the desired target (i.e., certification level). Therefore, for example, one can decide whether to gain LEED points by eight credits in the Energy and Atmosphere category or by five credits scattered through all categories. Some GBRSs, however, have set a number of mandatory credits in order to avoid only particular topics being addressed (i.e., prerequisites in LEED and BREEAM).

Each credit has in general a maximum number of points allocated over the entire assessment, and the overall assessment of sustainability is determined by adding up the results. The weighting process adopted by BREEAM and Green Star consist of pre-weighting categories before adding up selected credits, whilst the "additive credits" method of LEED allows a simple sum up of all selected credits. CASBEE adopts a totally different and more complex approach: instead of calculating single credits and summing up, all the measurements are divided in internal and external loads and visualized on a graph.

Despite that the final score can be quantitative or qualitative (e.g., points out of the total, qualitative judgment, graphs or diagrams), the LEED points system is considered all in all the more user-friendly method to calculating the final result [4].

2.2.4. Certification Levels

According to the score achieved in the assessment, a project earns a certification level or grade. BREEAM rates from Acceptable (existing buildings only) to Pass (>30), Good, Very Good, Excellent, up to Outstanding (>85), corresponding to an increasing number of stars reported on the certificate. LEED uses four levels: Certified (>40), Silver, Gold, and Platinum (>80). Green Star assigns an increasing number of stars from Minimum Practice (1 star) to World Leadership (6).

Being based on the concept of closed systems, CASBEE differs the most, as it only focuses the assessment on two elements: building performance and environmental load. The rating is expressed as an eco-efficiency gauge (BEE) given on a graph with environmental loads (L) on one axis and quality (Q) on the other; more sustainable buildings have the lowest environmental loads and highest quality.

Each rating system adopt its own levels of certification, which are hardly comparable each other. However, what can be observed is that all associate a qualitative or symbolic description to the numerical score or index. This demonstrates that the communication of results to the market and non-technicians is of utmost importance for GBRSs.

The process to earning the certification varies for each GBRS. BREEAM has five stages, LEED has six, and Green Star has four. Despite the number, a registration phase is usually required by paying fees, followed by the assessment of project performance and submission of forms, which are then verified by a third-party organization who is usually in charge of issuing the certificate.

In some cases, as in LEED and BREEAM, certification can be broken down into multiple stages, such as the Design and Construction stages. Goals are specified throughout the design process, and techniques and materials are chosen accordingly; nonetheless, it is the contractor's responsibility to purchase materials and ensure compliance with designated credits, as well as construction waste management.

3. Evolutions, Research Directions and Development of GBRSs

3.1. Limits and Improvements of Available Tools

This brief overview of the tools highlighted that there are some limitations to GBRSs. Beyond differences and weaknesses specific to LEED, BREEAM, CASBEE or Green Star, three main issues are usually criticized in the literature for almost every GBRSs on the market: a) regionality and different languages; b) lack in considering of social and economic issues; and c) affordability and risk of green washing.

3.1.1. Regionality and Different Languages

Most GBRSs have been initially designed to assess buildings in a specific country or region. Some of them have then decided to adapt and become more flexible to spread all over the world, by including, for example, equivalent international codes and regulations for single credits calculation or even an entire evaluation category (i.e., LEED Regional Priority). However, each has continued to adopt their own metrics, procedures, and levels: in other words, their own language. As a result, a LEED Platinum is hardly comparable with a BREEAM Outstanding (both are top levels) and the stakeholders become confused, especially the non-technicians (e.g., owners, tenants, estate managers).

On the one hand, this has pushed some Councils to seek common metrics, as in the case of BREEAM, LEED, and Green Star, which in 2009 announced to collaborating in a common measurement of CO_2 emissions equivalents from new homes and buildings [38]. On the other hand, although these and other rating systems are spending efforts to be comparable, some authors point out that a "one size fits all" approach to ranking buildings across the world would be questionable [12], because geographical, cultural, social and economic differences impact sustainability level and it is not possible to have the "winner" or the world's most sustainable building.

Therefore, some effort has been put at the very least in making the general features more coherent. To this purpose, in 2015, the World GBC published the *Quality Assurance Guide for Green Building Rating Tools*, a step-by-step guidance for operators of forthcoming, emerging, and established rating tools to ensure that their development and implementation are reliable, transparent, and of high quality [39].

3.1.2. Lack of Social and Economic Issues

Because the public is becoming more aware of negative externalities, GBRSs are increasingly required to address the triple bottom line of sustainability by introducing categories or specific credits to assess the social and economic impact of constructions, a lack of which has frequently been criticized [1].

However, the vast majority of these systems still focus on environmental issues, as the relative weights of related topics demonstrate. Taking for example LEED—as the most diffused system—social and economic topics are mainly related to accomplishing two of the less relevant impact categories: building a greener economy, and enhancing social equity, environmental justice, and community quality of life. These are both given 5% weight compared to reverse contribution to global climate change, which is given 35% [40].

To take the pace with new challenges and global targets (e.g., Sustainable Development by SDGs) many GBRSs have evolved during the years by including new topics and evaluation categories, by including and balancing within them both environmental and social issues, as well the economic aspects. Even though many criticize the weight assigned to this topic is not enough yet, an excellent example is given by Germany, whose rating tool DGNB—that is taking exceptional ground in Europe—assigns proper credits for the evaluation of social and economic aspects [13].

3.1.3. Affordability and Green Washing

The extraordinary success that GBRSs have had in recent times raised the risk of green washing, in that the cost of certifications are sometimes prohibitive, and only big clients can afford it. In reality, the real cost is not much in the registration fee itself, rather in the consultants fees to prepare and submit the project documents to certification. As a result, instead of being a powerful tool for spreading sustainable buildings, it may become a marketing tool for the elite. This also raises another issue, which is to focus on gaining more rewarding credits rather than developing effective case-by-case solutions: the phenomenon is also called 'LEED brain' [23] or 'point-chasing' [12].

3.2. Integration of Existing GBRSs

While several Green Building Councils and other organizations around the world are updating and improving their rating tools to fill the gap and eliminate the detected flaws, other important elements are being integrated to enhance and expand their applications.

The first to mention is certainly the integration of LCA methods and indicators within GBRSs. Since they aim at driving the building industry toward more sustainable design, construction, operation, maintenance, renovation and demolition practices, the assessment protocols have more and more adopted a life cycle perspective [6]. This means not only the design or construction stages are evaluated, but also some projections are made of previous and following phases. Many GBRSs have integrated a cradle-to-grave approach, some others have even included the disposal and reuse practices within the scoring, as well the recycling potentials of building components and materials (focusing on the cradle-to-cradle assessment stage). Several have now included LCA indicators and EPD certification requirements within their credits [34,41], albeit adopting different approaches and imputing different relevance to the outputs [42]. Although these types of credits are generally attributed a lower relative weight than energy-related credits [35], the involvement of these topics within the RSs suggests a growing attention to construction materials and products lifecycle which, thanks, for example, to EPDs, can positively stimulate the construction market to be greener [43].

On the same wave length, another attempt of integration that deserves mention is *Level(s)*, the framework for assessing and reporting on the sustainability performance of buildings recently released by the European Union [36,44]. In line with the call to uniform GBRSs' languages, the EU proposes a common set of simple indicators that relates the various building energy performances to the EU priorities as set by the Circular Economy Action Plan (CEAP) in 2020. As a result, many international sustainability certification tools are aligning to Level(s), ensuring common EU policy objectives are being integrated. Additionally, as an open-source tool, Level(s) allows an easy integration within GBRSs by means of modules to fill and attach.

3.3. Transformations/Mutations of GBRS

Beyond continuous upgrading and refinements of existing schemes, the most recent developments in the field are giving birth to other families of instruments. These share the same structure and approach but vary in scope.

A group of tools shifts from sustainability to resilience, as considered a more suitable framework to the changing paradigm of global challenges. Among the most known are the Resilience-based Earthquake Design initiative (REDi) by Arups Advanced Technology and Research team [45] and Resilience Action List (RELi) by the USGBC [46]. These combine resilience-based innovative design criteria with integrative design processes for assessing and guiding the design of neighborhoods, buildings, homes and infrastructures according to acute events (e.g., earthquakes, flooding) or chronic critical status of the ecosystem. In parallel, the USGBC also attempts to integrate principles of resilience into the most common international RSs, such as LEED [47], by adding specific pilot credits to existing schemes.

Another interesting branch is instead focusing on measuring, certifying and monitoring those aspects of buildings that impact occupants' health and well-being (e.g., air, water, light, and comfort). This is the approach of the WELL rating system, by International WELL Building Institute (IWBI), or Fitwel, by US Centers for Disease Control (CDC) and Prevention, and US General Services Administration.

4. Conclusions and Prospects

GBRSs have been primarily developed to assess the sustainability level of constructions. However, their actual contribution goes much further. For nearly three decades these tools have played a critical role in encouraging institutions, practitioners, constructors, owners, and managers to transform the built environment in a sustainable manner.

Since BREEAM was launched on the market, GBRSs have had extraordinary success and hundreds of schemes are accounted today, varying in approaches, methods, and diffusion. The open, consensus-based process that most of them adopt to engage stakeholders, and keep continuously updating, are certainly important factors of this achievement.

With their aim of shaping the construction market to become greener, rating tools have also been precursors for important international codes such as International green Construction Code (IgCC) [48].

In addition, these tools highly contribute to raise the awareness of owners, tenants, practitioners, and constructors on the issue. As proven by indicators, sustainable buildings have significant environmental benefits, as well as economic and social values (e.g., reduced operational costs and improved indoor air quality of working place). This is mainly due to the structured methodology that steers stakeholders and provides them information and proof of the benefits of their decisions, as well as encourages new and more effective ways of collaboration among professionals (e.g., Integrative Process) and the harnessing of value and opportunities not yet included within the system (e.g., system thinking and life cycle thinking).

In spite of specific criticisms and need for improvements, GBRSs are demonstrating to be valuable and flexible tools to support the transition of the built environment toward a more health, smart and just future, under the large umbrella of Agenda 2030 for Sustainable Development. The open challenge is now to strengthen and expand their role in leading the design and construction process toward truly holistic sustainability.

Author Contributions: Conceptualization, L.M. and E.A.; investigation, L.M.; resources, L.M. and S.P.; writing—original draft preparation, L.M.; writing—review and editing, E.A. and S.P.; visualization, L.M. All authors have read and agreed to the published version of the manuscript.

Funding: This research received no external funding.

Institutional Review Board Statement: Not applicable.

Informed Consent Statement: Not applicable.

Entry Link on the Encyclopedia Platform: https://encyclopedia.pub/15671.

References

1. Ding, G.K.C. Sustainable construction-The role of environmental assessment tools. *J. Environ. Manag.* **2008**, *86*, 451–464. [CrossRef] [PubMed]
2. World Green Building Council. What is Green Building? Available online: https://www.worldgbc.org/what-green-building (accessed on 12 July 2021).
3. Korkmaz, K.; Erten, D.; Syal, M.; Potbhare, V. A Review of Green Building Movement Time lines in Developed and Developing Countries to Build an International Adoption Framework. In Proceedings of the Fifth International Conference on Construction in the 21st Century: Collaboration and Integration in Engineering, Management and Technology, Istanbul, Turkey, 20–22 May 2009.
4. Doan, D.T.; Ghaffarianhoseini, A.; Naismith, N.; Zhang, T.; Ghaffarianhoseini, A.; Tookey, J. A Critical Comparison of Green Building Rating Systems. *Build. Environ.* **2017**, *123*, 243–260. [CrossRef]
5. ISO/TC 59/SC 17. *ISO 15392:2019—Sustainability in Buildings and Civil Engineering Works—General Principles*; ISO: Geneva, Switzerland, 2019.
6. Vierra, S. Green Building Standards And Certification Systems. Available online: https://www.wbdg.org/resources/green-building-standards-and-certification-systems (accessed on 25 July 2021).
7. Jesinghaus, J. *On the Art of Aggregating Apples and Oranges*; Fondazione Eni Enrico Mattei: Milano, Italy, 2000.
8. World Green Building Council Rating tools. Available online: https://www.worldgbc.org/rating-tools (accessed on 27 July 2021).

9.	IEA. World Energy Outlook. Paris. 2017. Available online: https://www.iea.org/reports/world-energy-outlook-2017 (accessed on 24 June 2021).

10.	European Commission. Energy Performance of Buildings Directive. Available online: https://ec.europa.eu/energy/topics/energy-efficiency/energy-efficient-buildings/energy-performance-buildings-directive_en (accessed on 7 April 2021).

11.	OECD. *Global Material Resources Outlook to 2060: Economic Drivers and Environmental Consequences*; OECD Publishing: Paris, France, 2019.

12.	Say, C.; Wood, A. Sustainable rating systems around the world. *CTBUH J.* **2008**, *2*, 18–29.

13.	Berardi, U. Sustainability Assessment in the Construction Sector: Rating Systems and Rated Buildings. *Sustain. Dev.* **2012**, *20*, 411–424. [CrossRef]

14.	United Nations. Transforming Our World: The 2030 Agenda for Sustainable Development (A/RES/70/1). 2015. Available online: https://documents-dds-ny.un.org/doc/UNDOC/GEN/N15/291/89/PDF/N1529189.pdf?OpenElement (accessed on 12 December 2020).

15.	European Commission. *Energy Performance of Buildings Directive (EPBD III) 2018/844/EC*; European Commission: Brussels, Belgium, 2018.

16.	European Union. *Energy Efficiency Directive (EED) 2018/2002/EU*; European Union: Brussels, Belgium, 2018.

17.	GlobalABC/IEA/UNEP. GlobalABC Roadmap for Buildings and Construction: Towards a Zero-Emission, Efficient and Resilient Buildings and Construction Sector. Paris. 2020. Available online: www.globalabc.org (accessed on 10 August 2021).

18.	BREEAM. Available online: https://www.breeam.com/ (accessed on 20 July 2021).

19.	Saunders, T. A Discussion Document Comparing International Environmental Assessment Methods for Buildings. 2008. Available online: http://www.dgbc.nl/images/uploads/rapport_vergelijking.pdf (accessed on 10 August 2021).

20.	LEED. Available online: https://www.usgbc.org/leed (accessed on 21 July 2021).

21.	CASBEE. Available online: https://www.ibec.or.jp/CASBEE (accessed on 21 July 2021).

22.	iiSBE Italia. Available online: http://iisbeitalia.org/ (accessed on 21 July 2021).

23.	Reed, R.G.; Bilos, A.; Wilkinson, S.; Schulte, K. International comparison of sustainable rating tools. *JOSRE* **2009**, *1*, 1–15.

24.	ISO 21931–1: 2010. *Sustainability in Building Construction—Framework for Methods of Assessment of the Environmental Performance of Construction Works—Part 1: Buildings*; ISO: Geneva, Switzerland, 2010.

25.	ISO 15643–1: 2010. *Sustainability of Construction Works—Sustainability Assessment of Buildings—General Framework*; ISO: Geneva, Switzerland, 2010.

26.	Green Star. Available online: https://new.gbca.org.au/rate/rating-system/ (accessed on 21 July 2021).

27.	Marchi, L. Rating System as design tool to manage complexity. In *Producing Project*; Lauria, M., Mussinelli, E., Tucci, F., Eds.; e-book; Maggioli Editore: Santarcangelo di Romagna, Italy, 2020; pp. 141–146. ISBN 9788891643087.

28.	Reeder, L. *Guide To Green Building Rating Systems. Understanding LEED, Green Globes, ENERGY STAR, the National Green Building Standard, and More*; John Wiley & Sons: Hoboken, NJ, USA, 2010; ISBN 9780470401941.

29.	U.S. GBC. *LEED Core Concepts Guide*; U.S. Green Building Council: Washington, DC, USA, 2014; ISBN 9781932444346.

30.	Sant, R.; Borg, R.P. A review of green building rating tools and their application in Malta. In Proceedings of the CESB 2016—Central Europe Towards Sustainable Building: Innovations for Sustainable Future, Prague, Czech Republic, 22–24 June 2016.

31.	Fowler, K.M.; Rauch, E.M. Sustainable Building Rating Systems Summary. 2006. Available online: https://www.wbdg.org/FFC/GSA/sustainable_bldg_rating_systems.pdf (accessed on 24 June 2021).

32.	Liu, Z.; Wang, Q.; Gan, V.J.L.; Peh, L. Envelope thermal performance analysis based on building information model (BIM) cloud platform—Proposed green mark collaboration environment. *Energies* **2020**, *13*, 586. [CrossRef]

33.	Lu, W.; Chi, B.; Bao, Z.; Zetkulic, A. Evaluating the effects of green building on construction waste management: A comparative study of three green building rating systems. *Build. Environ.* **2019**, *155*, 247–256. [CrossRef]

34.	Sartori, T.; Drogemuller, R.; Omrani, S.; Lamari, F. A schematic framework for Life Cycle Assessment (LCA) and Green Building Rating System (GBRS). *J. Build. Eng.* **2021**, *38*, 102180. [CrossRef]

35.	Politi, S.; Antonini, E. An expeditious method for comparing sustainable rating systems for residential buildings. *Energy Procedia* **2017**, *111*, 41–50. [CrossRef]

36.	Cordero, A.S.; Melgar, S.G.; Márquez, J.M.A. Green building rating systems and the new framework level(s): A critical review of sustainability certification within Europe. *Energies* **2019**, *13*, 1–25.

37.	A Comparison of the World's Various Green Rating Systems. Available online: https://www.fmlink.com/articles/a-comparison-of-the-worlds-various-green-rating-systems/ (accessed on 27 July 2021).

38.	BREEAM and LEED to Collaborate on Global Standard. Available online: https://www.building.co.uk/breeam-and-leed-to-collaborate-on-global-standard/3135221.article (accessed on 10 July 2020).

39.	Towell, B.H. Quality Assurance Guide for Green Building Rating Tools. WGBC. 2015. Available online: http://www.worldgbc.org/sites/default/files/WorldGBC_QA_Guide_for_Green_Building_Rating_Tools.pdf (accessed on 24 June 2021).

40.	U.S. GBC. *LEED v4. Impact Category and Point Allocation Development Process*; U.S. Green Building Council: Washington, DC, USA, 2013.

41.	Campioli, A.; Dalla Valle, A.; Ganassali, S.; Giorgi, S. Progettare il ciclo di vita della materia: Nuove tendenze in prospettiva ambientale. *Techne* **2018**, *16*, 86–95.

42.   Politi, S.; Antonini, E.; Wilkinson, S.J. Overview of Building LCA from the Sustainability Rating Tools Perspective. In Proceedings of the ZEMCH 2018 International Conference, Melbourne, Australia, 29 January–1 February 2018; pp. 237–250.
43.   Božiček, D.; Kunič, R.; Košir, M. Interpreting environmental impacts in building design: Application of a comparative assertion method in the context of the EPD scheme for building products. *J. Clean. Prod.* **2021**, *279*, 123399. [CrossRef]
44.   Level(s). Available online: https://ec.europa.eu/environment/levels_it (accessed on 21 July 2021).
45.   Almufti, I.; Willford, M. *REDi Rating System: Resilience-based Earthquake Design Initiative for the Next Generation of Buildings*; Arup Co.: London, UK, October 2013; pp. 1–68.
46.   Wholey, F. Building resilience: A Framework for Assessing and Communicating the Costs and Benefits of Resilient Design Strategies. *Res. J.* **2015**, *7*, 7–18.
47.   Wilson, A. LEED Pilot Credits on Resilient Design Adopted! 2015. Available online: www.resilientdesign.org (accessed on 12 September 2019).
48.   ANSI/ASHRAE/ICC/USGBC/IES. *International Green Construction Code (189.1-2017)—Standard for the Design of High-Performance Green Buildings Except Low-Rise Residential Buildings*; ANSI: New York, NY, USA; ASHRAE: Atlanta, GA, USA; ICC: Dubai, United Arab Emirates; USGBC: Washington, DC, USA; IESUSA: Houston, TX, USA, 2018.

 *encyclopedia*

Entry

Degrowth Perspective for Sustainability in Built Environments

Iana Nesterova

Department of Geography, Umeå University, 901 87 Umeå, Sweden; iana.nesterova@umu.se

Definition: Degrowth, as a social movement, a political project, and an academic paradigm, aims to find ways that can lead to harmonious co-existence between humanity and nature, between humans and non-humans, and within humanity, including oneself. Seen through the lens of degrowth, everything becomes subject to reflection, critique, re-evaluation, and re-imagining. This concerns environments created by humans in a long process of interaction with nature, i.e., built environments. Built environments are always in becoming. This entry contemplates the implications of degrowth for intentionally directing this becoming towards genuine sustainability.

Keywords: degrowth; post-growth; built environment; sustainability transformations

Citation: Nesterova, I. Degrowth Perspective for Sustainability in Built Environments. *Encyclopedia* **2022**, *2*, 466–472. https://doi.org/10.3390/encyclopedia2010029

Academic Editors: Masa Noguchi, Antonio Frattari, Carlos Torres Formoso, Haşim Altan, John Odhiambo Onyango, Jun-Tae Kim, Kheira Anissa Tabet Aoul, Mehdi Amirkhani, Sara Jane Wilkinson, Shaila Bantanur and Raffaele Barretta

Received: 9 January 2022
Accepted: 7 February 2022
Published: 9 February 2022

Publisher's Note: MDPI stays neutral with regard to jurisdictional claims in published maps and institutional affiliations.

1. Introduction

Degrowth does not have a single definition, nor is degrowth a single discipline. It is at once a social movement, a political project, and an academic paradigm [1]. What unites these pursuits is a desire for sustainable, harmonious co-existence between humanity and nature, between humans and non-humans, and within humanity, including oneself [2,3]. The name of the concept, degrowth, reflects the roots of the concept, i.e., a critique of prevailing economic growth orientation which disregards the negative effects of pursuing economic growth, such as ecological and social degradation [2,4]. Despite its name, degrowth is not opposed to growth, but the growth that it advocates, supports and encourages is mainly non-material, such as growth in care, solidarity, empathy, creativity, generosity, and connectedness. Neither is material growth fully alien to degrowth thought. Selective growth (such as in renewable energy provision, in organic agriculture and permaculture) is welcomed, as are improvements in material conditions of those whose genuine needs are not met. While degrowth originated in the 1970s with the rise in prominence of a variety of ecological movements, the ideas which degrowth scholars explore originated much earlier and have longer intellectual heritage and histories. For instance, degrowth calls for simpler living, deviation from consumerism, and viewing non-humans as one's neighbours, calls that could be readily found in the 19th century (see, e.g., [5]). Likewise, in the 19th century, the "hunger for wealth, which reduces the planet to a garden", and the need to live harmoniously in and with nature, were noticed [6] (p. 161). Going even further back in time, harmonious co-existence with nature, as well as with other humans, were prominent thoughts in ancient China [7].

As an academic paradigm, degrowth is very broad. While influential scientists within this paradigm can be identified [1], including perhaps most notably Nicholas Georgescu-Roegen [8,9] and Serge Latouche [10], currently, multiple scholars from a wide variety of disciplines are working with the concept of degrowth which contributes to the diversity within this academic paradigm. They include sociologists (see, e.g., [11]), political economists (see, e.g., [12]), and geographers (see, e.g., [13]). In recent years, adventurous inter-disciplinarity has become even more evident in the broad and growing field of degrowth. For instance, degrowth has been included in dialogues with philosophies such as existentialism [3] and critical realism [14], and with diverse economies thinking [15]. Degrowth thinking has been applied to multiple phenomena, such as business and organisation [3,13,15–17], housing [18], tourism [19], and technology [20]. Many other strands within academia share similar pursuits with degrowth, exemplified in a desire for a good and harmonious life for all, including humans, nature and non-human beings. Such strands include, for instance, diverse economies [21], deep ecology [22] and technological scepticism and pessimism [23], which are often found in dialogues with degrowth. Outside academia, a variety of movements and initiatives can be seen as degrowth compatible, whether they explicitly identify themselves as such or not [24]. Implicitly degrowth compatible movements such as voluntary simplicity, the tiny house movement and the zero-waste movement share many commonalities with degrowth but do not necessarily identify themselves as part of degrowth, though some practitioners might. As a political project, degrowth is likewise diverse. In terms of political ideologies, some see degrowth as an eco–socialist project [25], while others advocate for anarchism [26] and call for bottom-up transformations. Such transformations entail, first and foremost, changes in individuals' values and worldviews, and progress in moral agency rather than (high) technology [3,27]. Yet others do not propose a single political ideology but rather see the current political systems as diverse. They propose a variety of pathways of how degrowth can be achieved, which includes both top-down and bottom-up strategies [12]. The journey of degrowth as a social movement, a political project and an academic paradigm is unfolding, and the collection of knowledge is becoming ever deeper, more diverse, and broader. Since degrowth seeks answers to the question of how a harmonious co-existence can be achieved, nothing remains unquestioned. Every social domain, phenomenon, norm, institution can be looked at from a critical degrowth perspective. Likewise, the deep and broad knowledge that degrowth has accumulated so far can be used for proposing better alternatives that ensure a good life for all humans and non-humans far into the future. This includes built environments.

Built environments are environments that humans have created in a long process of interaction with nature, claiming nature's own spaces and transforming them into their own. Such human spaces can be viewed as "cocoons that humans have woven in order to feel at home in nature" [28] (p. 13). Such cocoons span across the face of the earth, affecting even those areas where human presence is not immediately visible. Untouched natural environments that, for instance, von Humboldt [29] describes, largely remain in the past. Even where no human activity is visible to the naked eye, the effects are felt due to human imposed activities resulting in climate change, ocean acidification, pollution, and other detrimental effects.

Built environments are diverse, ranging from simple dwellings to large and complex cities [28]. They depend on cultures, geographies, and topographies, and are always in development. They are constantly unfolding and changing. Thus, the past and past participle form of the verb, as well as the adjective, "built" gives a false sense of completion, stability, and finality. In critical realist terms, reality consists of social and physical realities [30]. Built environments are where social and physical realities meet, empower, and impose constraints onto each other. A question arises: what could be implications of degrowth thinking for sustainability in built environments?

2. Degrowth in Built Environments

While it has become normalised to see humans and nature as opposites, ontologically this is not the case [31]. There is no strict separation between humans and nature. Humans are embodied personalities, and one's body is of nature. In its turn, nature is impacted by humans, our pursuits and creative thought manifested in action and transformation. Nature shapes humans' worldviews [22,28] and has always provided inspiration for humans, including in its own transformation [32]. Built environments provide excellent examples of inseparability between humans and nature. For instance, a house is made from materials derived from nature, assembled together (and most often collectively) by humans guided by their creativity and skill. Looking closer at the house, one may notice ladybirds on a balcony and birds residing under the roof, building a nest. The house is home not only for humans, but also for non-humans. Due to this inseparability, it is perhaps best to see ourselves not as tenants in a particular house or a town, but rather as "tenants of the earth" [33] (p. 7). This is precisely what degrowth scholars advocate and encourage humans to do. Beyond seeing ourselves as tenants of the earth, it is essential to behave as such and in practice strive for sustainable, harmonious co-existence within humanity and with nature. The implications of this thinking for built environments are profound, though, as we will see, not always precise at this stage of degrowth's journey as an academic paradigm.

From a degrowth perspective, the central question when contemplating built environments is the question of harmonious co-existence. It is perhaps best captured by Thoreau's [5] chapter title in Walden, "Where I lived, and what I lived for", but this time applied to societal level as well as the level of a human individual. Degrowth does not provide a concrete answer to the question of what a human individual or societies should live for. However, it does make suggestions regarding what is destructive and may be misleading. While it calls for a deviation from materially excessive, hedonistic lifestyles, it invites humans to focus on justice, an increase in local self-sufficiency, independence from global economic forces and from commercialisation [34–36]. It encourages humans to make a step from overproduction and overconsumption to sufficient production and consumption, from material wealth to wellbeing [2]. In this invitation lie hints and more precise suggestions of where to live, how to approach our built environments, how to make them more sustainable.

It is suggested that in a degrowth society, community needs to replace commerce [37]. Thus, the life of human settlements will not be structured around commercial pursuits, commercial infrastructure, daily migration for work, shopping, and other market centred activities. It will be centred around community and fully exploring the possibilities of being in the world with others [38]. The focus of degrowth living is primarily on locations, i.e., the town, the suburb, the neighbourhood where most needs will be met [36]. Trainer [36] paints a picture of what a degrowth built environment may look like and entail in practice. It would include backyard and communal gardens and workshops, and multiple small, privately and collectively owned businesses instead of large businesses. Travelling on foot or by bicycle would replace travelling by private cars [36], thus creating a need for bicycle compatible infrastructure. Railway services would remain and, in most cases, replace air travel, which naturally leaves us with questions regarding existing airports and associated with them infrastructure. Trainer [36] suggests that derelict sites would be repurposed for productive activities by communities, and streets would become permaculture commons. Existing retail infrastructure could be repurposed to house repair stores or become meeting places, libraries, and places of barter [36]. Other suggestions comparable to Trainer's [36] include ecovillages [39] and intentional communities, which challenge the norm of what built environment in industrialised countries should look like and be centred around. They actively deviate from standardisation that is common in capitalist settings [4]. While such visions may appear utopian and radically different from what our environments currently look like, such alternatives exist already, though in the niches and pockets of current societies [21]. In a degrowth society, such initiatives would come to prominence and claim spaces. Such claiming of space should be seen as a process rather than a sudden shift. Oftentimes, existing alternatives and modes of being have one foot in a capitalist economy and the other in a non-capitalist economy [40], making the process of transformation and space claiming challenging and arduous, though not impossible [41].

What characterises degrowth inspired thought in relation to built environments is diversity, plurality, possibility of change and alternative modes of living, relating and beings. Diversity, for instance, can be seen in the forms of housing for degrowth (tiny house, yurts, using natural, reclaimed, repurposed materials), as well as in living arrangements (e.g., co-living) and ownership possibilities (common rather than private). While in the current capitalist society the emphasis is put on private property and the isolated individual, degrowth emphasises togetherness, transcending the situation of ecological and social degradation in cooperation with others.

Being with others includes fellow humans, but also non-human beings [3]. Thus, the notion of the built environment itself from a degrowth perspective needs to be infused with a different and much expanded meaning. It needs to recognise labour beyond human labour [42] in building, such as the labour of nature itself in providing materials for built environments and keeping these environments pleasant, healthy, and liveable. It also remains important to acknowledge the built environments of other species, such as beehives, termite mounds, rabbits' burrows, and ant nests, where ants live in symbiosis with fungus and establishing fungus gardens [42]. Human beings are not the only beings who transform their environments, and harmonious co-existence with others presupposes respect for another being's built environment, dwelling and space. Moreover, non-human beings such as birds, foxes, bees, and hedgehogs live in proximity to and within built human spaces such as cities, gardens, and parks. Since degrowth emphasises the wellbeing of others beyond humans, transformations of human-built environments need to be attentive to the diversity of beings within human-built spaces, ensuring their safety and allowing them to flourish.

Apart from directing our imagination towards more harmonious built environments, degrowth offers useful suggestions which can assist transformation in practice. For instance, creating new spaces should focus on nurturing healthy relationships with nature, others, and the self. Tuan [28] notes how different places interact with humans' senses differently. While a skyscraper is a pathological space, "the experience of the interior of a cathedral involves sight, sound, touch, and smell. Each sense reinforces the other so that together they clarify the structure and substance of the entire building, revealing its essential character" [28] (p. 11). One's experience in a forest can be even more profound in terms of connection and self-transcendence [5,32]. New buildings can be inspired by non-commercial spaces and by nature and use natural, renewable, biodegradable materials. Wherever possible, repurposing existing buildings and infrastructure should take place [36]. Reduction in resource use is central to degrowth, thus making use of the existing stock should be considered, and improvements (insulation, increased energy efficiency, water reuse systems, safety) made where possible. Degrowth scholars often question the pursuit of constant technological innovation [10,20], thus less technologically intensive solutions can be sought. Such solutions can be sought outside Western science: degrowth scholars emphasise plurality of knowledge, such as lay and indigenous knowledge [43]. Another proposal is removing advertisements from human spaces which take attention and direct it towards consumption and having rather than being [44].

While such a wide diversity of considerations may seem overwhelming, it is possible to structure them around a limited number of domains or planes of social being. For instance, Bhaskar [45] suggests that social being, and any social phenomenon, exist and unfold on four interconnected planes. They include (1) material transactions with nature, (2) social interactions, relationships between people, (3) social structure, and (4) embodied personality. When considering transformations of built environments or any aspect thereof, one may consider, and become attentive to all of these planes at once. For instance, when planning a housing development, in terms of material transactions with nature, one may consider the geography of the location and the materials to be used, whether they are biodegradable, durable and will serve a long time, can be sourced locally or reused. In terms of the second plane, one may consider the possibilities of communal activities which would facilitate healthy human interactions, nurturing relationships and cooperation, such as establishing and looking after a community garden nearby. The third plane, that of social structure, is abstract and unites a plurality of social structures, such as norms, institutions, as well as buildings themselves. On this plane, one may consider a wide variety of aspects such as affordability of housing, accessibility and quality of useful infrastructure, and challenging the norms (e.g., deviation from standardised architecture, seeking creative architectural solutions, changing attitudes towards non-humans). The plane of embodied personality is local and intimate and concerns oneself. While degrowth tends to overlook this dimension, the self is essential to consider, since this is where creative, transformative change resides [46]. On this plane, one may consider whether a house may reflect the personality and the needs of a human being, if it can become a place and not only a location to a person, if it accounts for the complexity of human nature (such as the drives towards and also away from other humans).

Recently, the scientific pursuits of degrowth scholars were referred to as a science of deep transformations [14]. Deep transformations concern not only scientific pursuits, but also, and perhaps most importantly, the nature of transformations in the real world. From this perspective, a built environment needs to be deeply transformed and become deeply transformative for human beings; it needs to produce spaces for sustainable, harmonious, peaceful co-existence, for humans and non-humans, for life (e.g., homes) and also death (e.g., cemeteries).

3. Constraints

While spaces are socially created [47] and thus, like all social structures, are only relatively enduring [41], there are constraints to the unfolding of degrowth compatible built environments. Despite the growth in degrowth as a movement, a political project and an academic paradigm, degrowth still does not enjoy wide support from the general public [48]. While degrowth indeed pursues human flourishing, its emphasis on nature and non-humans, whose flourishing degrowth likewise defends, may invoke scepticism from the electorate [49]. Thus, attempts to reclaim spaces and repurpose them, e.g., replacing car parks with gardens, may not be welcomed by the public. A change in values and world-views is required for such proposals to be met with a positive attitude [3]. A wide range of constraints is related to capitalism as a system, including state capitalism. Capitalism is a powerful hurdle on the path of transformation [4,15,50]. This is because this system is orientated towards valorisation of capital [50]. It is evident, for instance, in the strife of businesses to make a profit [16,51] and more generally, in expecting monetary return on one's investments. Capitalism is centred around private property and often its accumulation and rent, thus ownership rights. Ownership rights need to be modified for degrowth built environments to be possible, so choices can be made not on the grounds of what is economically best for the owner (e.g., to build a shopping centre, a standardised block of flats, an airport, or a mansion), but what is best for the sustainable thriving of human and non-human life (e.g., an affordable housing development, an organic agriculture project, or a forest). However, with ownership rights come responsibilities. These include responsibilities for health, safety, and maintenance of built environments. If ownership rights are transferred to communities, questions concerning these aspects and how decisions can be made democratically will need to be considered.

4. Conclusions

This entry attempted to capture the implications of degrowth for built environments. First and foremost, an important implication is ontological, i.e., being mindful, as scholars, practitioners and citizens of the ontology of human societies. This is exemplified in the acute realisation of our inter-connectedness with nature. Every human act of transformation in our built environments and beyond entails the use of natural resources and energy. Due to the ongoing ecological degradation, such transformation needs to reduce as much as possible.

Furthermore, the design, building and repurposing of sites, from a degrowth perspective, would be aimed at much simpler, place-sensitive and more local and communal living. Thus, principles such as simplicity, localisation and community can be used to guide urban and beyond-urban planning. The question of localisation, local built environment and local living is of immediate relevance. At the time of writing this manuscript, the world is facing the COVID-19 pandemic. As one of its implications, many humans found their lives constrained to their local areas. While neither this crisis itself, nor the sudden localisation of living can be described in terms of degrowth, the localisation of living showed the importance of the local area, local community, local infrastructure and local nature. While some found themselves in attractive locations in proximity to natural settings and useful infrastructure, others found themselves in areas where the traces of natural environments are scarce, the skyline is human-made, and where tall buildings conceal the horizon. It is challenging to know what role cities may play in a degrowth society. However, degrowth may provide inspiration for better localised living. Creation of accessible green spaces, encouraging community life and solidarity, proliferation of local small businesses, creating opportunities for outdoor exercise, spiritual practices and nature connection are but some elements of the direction in which degrowth inspires us to go. This requires a profound change in values, since pursuing this direction entails deviation from the opposite: appropriation of green spaces, atomised existence, car-centred infrastructure, shopping centres and other manifestations of modern built environments. Since degrowth aims at a good life for all, what remains essential to consider is accessibility to healthy and harmonious built environments for everyone, not only those who can currently afford it.

A useful way to apply degrowth to built environments in order to make them more sustainable is to think in terms of four planes of social being at once, which captures all aspects of social ontology, including humans' material transactions with nature, social relationships, social structures, and the self. While degrowth provides a wealth of ideas and encouragement to think freely, creatively, adventurously and to embrace alternatives, there are constraints to a degrowth compatible built environment. Apart from the capitalist system itself, which has valorisation of capital and not sustaining life at its core, there is a variety of its manifestations which act as constraining structures. They include, for instance, profit seeking and ownership rights. Despite these constraints, alternatives thrive currently and are plentiful, and the built environment will without a doubt also be part of degrowth society. Opportunities for contemplating what it will look like, and bringing harmonious and nurturing alternatives to existence, remain immense.

Funding: This research received no external funding.

Data Availability Statement: Not applicable.

Conflicts of Interest: The author declares no conflict of interest.

Entry Link on the Encyclopedia Platform: https://encyclopedia.pub/20570.

References

1. Buch-Hansen, H. Modvækst som paradigme, politisk projekt og bevægelse. *Nyt Fokus Fra Okon. Vækst Til Bæredygtig Udvikl.* **2021**, *17*, 6–9. Available online: http://www.nytfokus.nu/nummer-17/modvaekst-som-paradigme-politisk-projekt-og-bevaegelse/ (accessed on 5 January 2022).
2. Bonnedahl, K.J.; Heikkurinen, P. The case for strong sustainability. In *Strongly Sustainable Societies: Organising Human Activities on a Hot and Full Earth*; Bonnedahl, K.J., Heikkurinen, P., Eds.; Routledge: London, UK, 2019; pp. 1–20.
3. Nesterova, I. Addressing the obscurity of change in values in degrowth business. *J. Clean. Prod.* **2021**, *315*, 128152. [CrossRef]
4. Koch, M. *Capitalism and Climate Change: Theoretical Discussion, Historical Development and Policy Responses*; Palgrave Macmillan: New York, NY, USA, 2012.
5. Thoreau, H.D. *Walden*; Pan Macmillan: New York, NY, USA, 2016.
6. Emerson, R.W. *Essays: Second Series*; Floating Press: Auckland, New Zealand, 2009.
7. Mote, F.W. *Intellectual Foundations of China*, 2nd ed.; McGraw-Hill: New York, NY, USA, 1989.
8. Georgescu-Roegen, N. *The Entropy Law and the Economics Process*; Harvard University Press: Cambridge, MA, USA, 1971.
9. Georgescu-Roegen, N. Energy and economic myths. *South. Econ. J.* **1975**, *41*, 347–381. [CrossRef]
10. Latouche, S. *Farewell to Growth*; Polity Press: Cambridge, UK, 2009.
11. Koch, M. State-civil society relations in Gramsci, Poulantzas and Bourdieu: Strategic implications for the degrowth movement. *Ecol. Econ.* **2022**, *193*, 107275. [CrossRef]
12. Buch-Hansen, H. Capitalist diversity and de-growth trajectories to steady-state economies. *Ecol. Econ.* **2014**, *106*, 167–173. [CrossRef]
13. Schmid, B. Structured diversity: A practice theory approach to post-growth organisations. *Manag. Rev.* **2018**, *29*, 281–310. [CrossRef]
14. Buch-Hansen, H.; Nesterova, I. Towards a science of deep transformations: Initiating a dialogue between degrowth and critical realism. *Ecol. Econ.* **2021**, *190*, 107188. [CrossRef]
15. Schmid, B. *Making Transformative Geographies: Lessons from Stuttgart's Community Economy*; Transcript: Bielefeld, Germany, 2020.

16. Nesterova, I. Degrowth business framework: Implications for sustainable development. *J. Clean. Prod.* **2020**, *262*, 121382. [CrossRef]
17. Nesterova, I. Small firms as agents of sustainable change. *Futures* **2021**, *127*, 102705. [CrossRef]
18. Nelson, A.; Schneider, F. (Eds.) *Housing for Degrowth: Principles, Models, Challenges and Opportunities*; Routledge: Abingdon, UK, 2019.
19. Hall, C.M.; Lundmark, L.; Zhang, J.J. (Eds.) *Degrowth and Tourism: New Perspectives on Tourism Entrepreneurship, Destinations and Policy*; Routledge: Abingdon, UK, 2021.
20. Heikkurinen, P. Degrowth by means of technology? A treatise for an ethos of releasement. *J. Clean. Prod.* **2018**, *197*, 1654–1665. [CrossRef]
21. Gibson-Graham, J.K.; Dombroski, K. Introduction to the handbook of diverse economies: Inventory as ethical intervention. In *The Handbook of Diverse Economies*; Gibson-Graham, J.K., Dombroski, K., Eds.; Edward Elgar: Cheltenham, UK, 2020; pp. 1–24.
22. Næss, A. *Ecology of Wisdom*; Penguin Classics: London, UK, 2016.
23. Heikkurinen, P.; Ruuska, T. (Eds.) *Sustainability beyond Technology: Philosophy, Critique, and Implications for Human Organization*; Oxford University Press: Oxford, UK, 2021.
24. Burkhart, C.; Schmelzer, M.; Treu, N. (Eds.) *Degrowth in Movement(s): Exploring Pathways for Transformation*; Zero Books: Winchester, UK, 2020.
25. Swift, R.S. *O.S.: Alternatives To Capitalism*; New Internationalist Publications: Oxford, UK, 2014.
26. Trainer, T. The degrowth movement from the perspective of the Simpler Way. *Capital. Nat. Soc.* **2015**, *26*, 58–75. [CrossRef]
27. Nesterova, I. Small, local, and low-tech firms as agents of sustainable change. In *Sustainability beyond Technology: Philosophy, Critique, and Implications for Human Organization*; Heikkurinen, P., Ruuska, T., Eds.; Oxford University Press: Oxford, UK, 2021; pp. 230–253.
28. Tuan, Y.-F. *Topophilia: A Study of Environmental Perception, Attitudes, and Values*; Columbia University Press: New York, NY, USA, 1974.
29. Von Humboldt, A. *Personal Narrative of a Journey to the Equinoctial Regions of the New Continent: Abridged Edition*; Penguin: London, UK, 1995.
30. Buch-Hansen, H.; Nielsen, P. *Critical Realism: Basics and Beyond*; Red Globe Press: London, UK, 2020.
31. Latour, B. *We Have Never Been Modern*; Harvard University Press: Cambridge, MA, USA, 2013.
32. Emerson, R.W. *Essays: First Series*; Floating Press: Auckland, New Zealand, 2009.
33. Tuan, Y.-F. *Space and Place: The Perspective of Experience*; University of Minnesota Press: Minneapolis, MN, USA, 2001.
34. Max-Neef, M. The world on a collision course and the need for a new economy. In *Co-Operatives in a Post-Growth Era*; Novkovic, S., Webb, T., Eds.; Zed Books: London, UK, 2014; pp. 15–38.
35. Trainer, F.E. Environmental significance of development theory. *Ecol. Econ.* **1990**, *2*, 277–286. [CrossRef]
36. Trainer, T. De-growth: Do you realise what it means? *Futures* **2012**, *44*, 590–599. [CrossRef]
37. Klitgaard, K. Heterodox political economy and the degrowth perspective. *Sustainability* **2013**, *5*, 276–297. [CrossRef]
38. Heidegger, M. *Being and Time*; Blackwell: Oxford, UK, 2001.
39. Kliemann, C. Ecovillages: Living degrowth as a community. In *Degrowth in Movement(s): Exploring Pathways for Transformation*; Burkhart, C., Schmelzer, M., Treu, N., Eds.; Zero Books: Winchester, UK, 2020; pp. 172–186.
40. Boonstra, W.J.; Joosse, S. The social dynamics of degrowth. *Environ. Values* **2013**, *22*, 171–189. [CrossRef]
41. Collier, A. *Critical Realism: An Introduction to Roy Bhaskar's Philosophy*; Verso: London, UK, 1994.
42. Barron, E.; Hess, J. Non-human 'labour': The work of Earth Others. In *The Handbook of Diverse Economies*; Gibson-Graham, J.K., Dombroski, K., Eds.; Edward Elgar: Cheltenham, UK, 2020; pp. 163–169.
43. Spash, C.L. New foundations for ecological economics. *Ecol. Econ.* **2012**, *77*, 36–47. [CrossRef]
44. Fromm, E. *To Have or to Be?* Bloomsbury: London, UK, 2013.
45. Bhaskar, R. *Dialectic. The Pulse of Freedom*; Verso: London, UK, 1993.
46. Bhaskar, R. *From East to West: Odyssey of a Soul*; Routledge: London, UK, 2000.
47. Lefebvre, H. *The Production of Space*; Basil Blackwell: Oxford, UK, 1991.
48. Buch-Hansen, H. The prerequisites for a degrowth paradigm shift: Insights from critical political economy. *Ecol. Econ.* **2018**, *146*, 157–163. [CrossRef]
49. Latour, B. To modernize or to ecologize? That's the question. In *Remaking Reality: Nature at the Millennium*; Castree, N., Willems-Braun, B., Eds.; Routledge: New York, NY, USA, 1998; pp. 221–242.
50. Gorz, A. *Capitalism, Socialism, Ecology*; Verso: London, UK, 2012.
51. Nesterova, I. Small Business Transition towards Degrowth. Ph.D. Thesis, University of Derby, Derby, UK, 2020.

 *encyclopedia*

Entry

Wooden Additional Floor in Finland

Anu Soikkeli [1], Hüseyin Emre Ilgın [2,*] and Markku Karjalainen [2]

[1] School of Architecture, University of Oulu, FI-90014 Oulu, Finland; anu.soikkeli@oulu.fi
[2] School of Architecture, Faculty of Built Environment, Tampere University, P.O. Box 600, FI-33014 Tampere, Finland; markku.karjalainen@tuni.fi
[*] Correspondence: emre.ilgin@tuni.fi

Definition: One of the most effective ways to cover real estate development and renovation processes by improving functionality and energy efficiency is wooden additional floor construction. This entry maps out, organizes, and collates scattered information on the current state of the art and the benefits of this practice including its different stages, focusing on the case of Finland. The entry presents this topic in an accessible and understandable discourse for non-technical readers. By highlighting the benefits and opportunities of this sustainable application, the entry will contribute to increasing the awareness of wooden additional floor construction, which has many advantages, and therefore to gain more widespread use in Finland and other countries.

Keywords: wood; additional floor construction; sustainability; Finland

Citation: Soikkeli, A.; Ilgın, H.E.; Karjalainen, M. Wooden Additional Floor in Finland. *Encyclopedia* **2022**, *2*, 578–592. https://doi.org/10.3390/encyclopedia2010038

Academic Editors: Masa Noguchi, Antonio Frattari, Carlos Torres Formoso, Haşim Altan, John Odhiambo Onyango, Jun-Tae Kim, Kheira Anissa Tabet Aoul, Mehdi Amirkhani, Sara Jane Wilkinson, Shaila Bantanur and Raffaele Barretta

Received: 3 February 2022
Accepted: 3 March 2022
Published: 7 March 2022

Publisher's Note: MDPI stays neutral with regard to jurisdictional claims in published maps and institutional affiliations.

1. Introduction

Among the targets of the European Union's 2050 Energy Roadmap are the decarbonization of increasing energy resources, making more use of renewable energy, and improving energy efficiency [1,2]. In line with these sustainable goals, Finnish building regulations were revised and developed, allowing new construction methods to be more energy-efficient [3–5]. Most of Finland's building stock consists of buildings that were erected before the 1990s, have very low energy efficiency, and are mostly in the renovation era [6,7]. More than 30% of residences, which constitutes a substantial part of the building inventory in Finland, are apartment buildings constructed in the Finnish suburbs in the 1960s and 1970s and need refurbishment [8–10].

The thermal insulation of these Finnish suburban apartments was poor, and at that time, no regulations or targets were set for this in Finland, and as a result, these buildings needed a serious energy upgrade [11]. Energy upgrade strategies, especially for the old housing inventory, should be adopted as an important approach to increase energy efficiency because, with the erection of new buildings, the rate of renewal of the building stock has not even reached two percent per year [12]. Overall, besides the lack of equipment and poor technical conditions, among the most important problems of Finnish suburban apartments is poor energy efficiency [13,14].

As in many other countries, it takes a lot of investment and government subsidies to renovate a building by increasing its energy efficiency in Finland [15–18]. Additionally, it is difficult to find a contractor who can undertake or is willing to undertake suburban apartment renovations, and it is often necessary to seek and hire more than one contractor for a project in Finland [19]. Real estate and housing companies play a key role in renovating old apartments in Finland [20]. There are more than 60,000 flats where almost half of Finland's population resides [6]. Real estate and housing companies that require financing instruments to refurbish and improve their properties often play an important role in the maintenance and modernization of apartments [21]. In addition, renovation projects involve excessive work and strong coordination of residents as well as building managers and housing companies [22]. In practice, building renovations are a slow, expensive, dirty, and destructive process [23]. This is mainly because refurbishment projects in Finland often use operational models created for new construction [24].

It is worth mentioning here a local Finnish challenge stems from the ownership of buildings [25]. Single people have their flats and a piece of land below. They co-manage housing companies that must make a joint decision to fund their new investments, where redevelopments are often financed by a bank loan. This amount is then directly attributed to the occupant's share of the total renovation costs. Parking space is another challenging issue if parking lots need to be constructed to replace old parking lots and offer additional parking spaces.

In Finland, one of the most effective ways to cover real estate development and renovation processes by improving functionality and energy efficiency is the construction of an additional floor [24] (Figure 1). When the building height and the number of building stories increase or the roof form changes, the terms additional floor, roof, or elevation construction are used [26]. Furthermore, additional floor construction provides numerous opportunities and benefits such as increased owner income, short-term income to housing companies by selling the building rights or areas of additional floors, as well as a relatively lower carbon footprint, increased gross floor area, and improved building appearance [26–28] (see Section 3).

(a)(b)

Figure 1. Wooden additional floor project examples: (**a**) (Image courtesy of Aino Hirvilammi, Eetu Salminen, and Joel Lehtola); (**b**) (Image courtesy of Risto Piirainen and Silja Suitiala).

Important issues requiring special attention in additional floor projects in the Finnish context are as follows [29]: (i) Economic feasibility: This consideration, which has become more important with the sale of building rights to the outside party that built the additional floor, poses a significant problem if the commercial return of the additional floors is not properly estimated, which means the targeted profit for the project will not be achieved; (ii) change in the city plan or deviation from the city plan: This issue, which directly affects the right to build on the land, also has an impact on the amount of property tax; (iii) finding a suitable contractor: As the cost of maintaining the property increases over time, it is very important to find a contractor who will build the additional floors as soon as possible; (iv) obtaining expert opinion: The presence of an expert is especially important for identifying and then minimizing risks. In some municipalities in Finland, the procedure for making changes to the city plan is suspended unless the relevant expert is included in the project.

The intensification of Finnish urban environments is an adopted Eurocentric goal in tackling climate change, driven by the needs of continued urbanization and the environmental impact of low-density urban structures. In this sense, European building retrofit and urban renewal applications have shown that the expansion of building volumes, such as the construction of additional floors, has significant potential [25].

Technical features (e.g., structural and architectural issues) of reinforced concrete apartment blocks built between the 1960s and 1980s in Finland generally allow the construction of additional floors, often designed as lightweight structures [29]. They are suitable for these implementations with both their structural capacity and flat roof. The current Finnish fire code also gives the green light to the construction of additional floors [30]. The Finnish fire regulations make it possible to construct the top floor of a class P1 building with a timber-framed additional floor without automatic extinguishing equipment if the building has no more than seven stories. Two additional stories require a sprinkler system on the topmost old floor and additional floors (Figure 2). There are three main fire classes, P0, P1, and P2, used for apartments in Finland [30]. While the P1 fire class represents the structural frame in which non-combustible materials such as concrete are used, wooden load-bearing systems are classified in the P2 category. On the other hand, the P0 category is based on the calculation method and is used when deviating from standard table values.

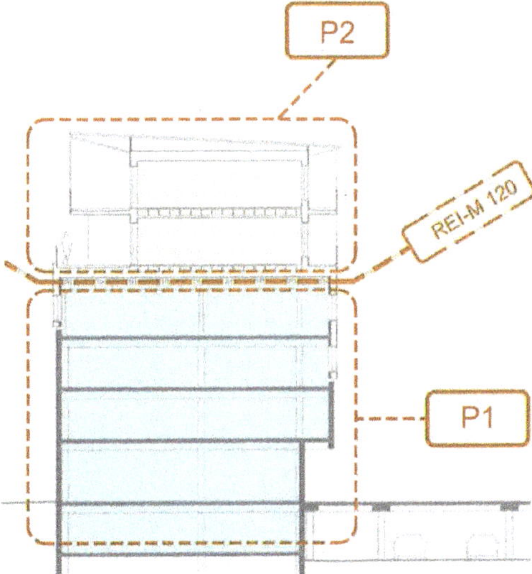

Figure 2. Two-story-high wooden additional floor (Image courtesy of Petri Pettersson).

Material selection is important for refurbishment in terms of sustainability. Considering sustainable construction concerns, the renovation materials must be environmentally friendly, long-lasting, renewable, reusable, and their production must require the least amount of energy and produce a minimum amount of greenhouse gas emissions [31–34]. Studies in the literature have indicated that timber has numerous benefits over traditional building materials such as brick, concrete, and steel, and with its environmentally friendly properties in particular, timber is a suitable material for renovation [35–41].

In this context, wooden structures are considered lower carbon structures and represent lower embodied energy consumption compared to non-wood structures [42–47]. In addition, buildings using concrete and steel structural systems embody and consume 20% and 12% more energy, respectively, compared to buildings with wooden structures, so structural material selection plays an essential role in the amount of embodied carbon [48]. Furthermore, both in production and on-site construction, concrete and steel structures utilize 50% and 7% more resources compared to wooden structures, generating 16% and 6% more solid waste, respectively [49]. Overall, the construction of wooden buildings is in line with the sustainability goals of the European Union [50], where timber as a building material is considered to lower carbon emissions in the building construction sector and is a method of transitioning to a sustainable bio-economy.

The construction phase of timber buildings can deliver considerable savings with over 50% faster assembly times compared to traditional construction materials [51]. Timber construction offers light and prefabricated alternatives with various size and thermal insulation options to respond to special demands [52–54]. The prefabrication process ensures that facade elements such as doors and windows are integrated into the prefabricated units (Figure 3) [55]. In addition to being used as a construction material, after completing its service life, timber can be reused as a raw material for other buildings, or it can be burned instead of fossil fuels as a last resort [56–58].

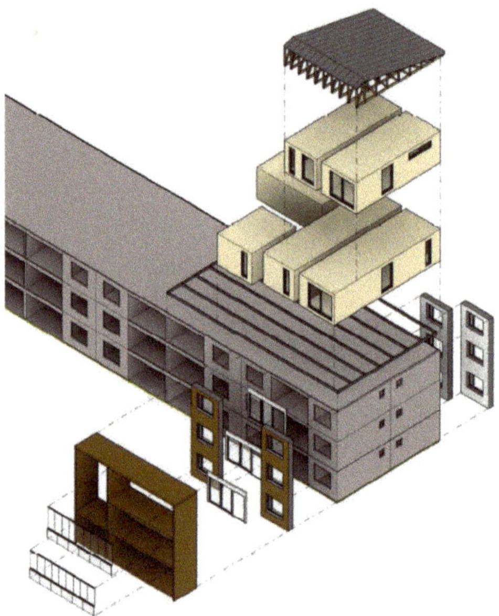

Figure 3. Additional floor project with wooden prefabricated units (Image courtesy of Simo Rasmussen).

The attitudes of residents towards new construction methods (e.g., a wooden additional floor) have an important role in the spread of these practices [59]. Moreover, the positive attitude of residents is a critical aspect in the effective execution of extensive refurbishment [60]. In this sense, the survey by Karjalainen et al. [25] showed that participants generally assessed the construction of wooden additional floors positively and thought that it would contribute to the attractiveness of the residential area.

The combustibility of timber may limit its use as a construction material in Finland, as in many countries, due to constraints on building regulations [61–63]. Various studies have been carried out recently on the fire behavior of wooden buildings around the world, aiming to provide fundamental data on the safe use of wood (e.g., [64]). As a result of extensive testing, new fire design concepts and models were developed, and existing advanced knowledge in the fire design area of wooden structures together with technical precautions, especially well-equipped fire services and sprinkler systems, ensure the safe use of wood in a wide range of applications as seen in the building code relaxations introduced in recent years [65]. In this sense, fire safety engineering and performance-based design offer benefits and challenges for the use of timber in buildings, where the performance-based approach is primarily based on the use of fire engineering principles, calculations, and modeling instruments (e.g., structural models, thermal models) to meet building regulations, considering fire modeling, full-scale structural fire experiments, and experience from fire accidents in timber structures [66–69]. Additionally, the following considerations stand out in terms of the implementation of fire safety design in wooden structures [70–73]: Manual firefighting, sprinklers, encapsulation, fire retardants, fire performance and fall-off times of protective systems, the fire performance of connections between structural timber elements, details to prevent the internal spread of fire, external fire spread in the same building, and quality assurance. Furthermore, timber and steel structures have some similarities and differences in terms of fire safety measures [65,74]. Some fire regulations, such as those in Canada, encourage full encapsulation of timber frames to ensure equivalent fire safety to the non-combustible steel frame structure. In terms of performance-based design, performance-based formulations of requirements for timber structures can be considered to provide a fire-safety equivalent to regulatory steel structures. Regarding structural modeling, wooden structures are usually easier to model than steel structures because the wood has poor thermal conductivity and does not undergo considerable thermal expansion. In the manual fire extinguishing strategy, the fire risk will be greatly reduced if immediate action is taken to contain the fire, and this reduction in fire load is adjusted for steel frames. This method is also permissible for timber structures. Moreover, in terms of external fire spread in the same building, timber facades can also be used as fire-resistant facade cladding in steel structures.

Issues with wooden structures, especially sound insulation and moisture, require special insulation and protection techniques. To obtain good air-borne sound insulation, the partitioning wall and intermediate floor structures should be built in layers and the layers should be separated from each other so that the sound does not pass through the structure [75,76]. On the other hand, humidity issues lead to both reduced durability and mold growth, which can affect indoor air quality and have adverse health consequences [77]. The best strategy for providing a moisture-resistant structure is to ensure that the wood is not exposed to water or high relative humidity for extended periods. Neglecting moisture safety can mean a high risk of damage, with extensive costs and consequent time delays for research, decontamination, or material replacement [78].

Wood-based composite materials and wood frame-based hybrid structures are among the important topics in today's wood construction literature. In general, owing to the destruction of forest resources and recently developed technologies for wood-based composite materials in particular, engineered wood products have gradually replaced traditional materials for residential construction [79]. These materials are produced from similar materials based on wood products, e.g., timber or lumber processed into boards, or wood chips [80], and the residential and commercial building construction industry is among the areas where wood-based composites are most in-demand [81,82]. On the other hand, the idea of hybrid structures that combine multiple materials, such as timber, along with steel and/or concrete, is gaining increasing acceptance in the engineering community [83]. Moreover, hybridizing timber with other structural materials is one of the most popular approaches for designing high-rise timber buildings [84–87] as in the case of Brock Commons Tallwood House (Vancouver, BC, Canada, 2017) [88].

The three critical components of timber frame construction are the floor, the roof, and the load-bearing wall, which have significant effects on occupants' comfort. The wood floor, the most common system component, is in frequent physical contact with building inhabitants [89]. The dynamic movement of people or objects caused by defects or deficiencies in the structural performance of the floor can cause occupant discomfort. Movements, e.g., walking, running, jumping, can create structural vibrations on the wooden floor, which adversely affect the efficiency of work and quality of life [90]. However, environmental excitation and impact excitation vibration tests as well as comfort analyses of timber floors offer solutions to these undesirable situations [91]. In addition, particularly nowadays, when standard structures are supported by contemporary technologies such as wooden floors combined with underfloor heating, it is necessary to meet technical guidelines and specifications during the operation of the floor as a whole [92]. Moreover, in line with the 'smart building' concept, wood, namely wood flooring, is used as an ideal material to be applied in triboelectric nanogenerators for large-scale applications in smart houses [93]. This ensures that mechanical energy (for example, the movements of residents) is directly converted into useful electricity [94–97].

Although there are numerous research studies on different construction solutions with the use of engineered timber products with related technical features (e.g., [98–109]), several studies have focused on the use of wood as a building material from the viewpoint of construction professionals (e.g., [110–119]) and consumers or users (e.g., [120–122]). On the other hand, to date, there has been a limited number of studies on wooden additional floor applications, especially in the housing construction industry.

This entry maps out, organizes, and collates scattered information on the current state of the art, as well as benefits and challenges of wooden additional floor projects with their different stages, focusing on the case of Finland, and presents it in an accessible and understandable discourse for non-technical readers. This entry also provides a methodical literature analysis on international peer-reviewed studies and research projects. By highlighting the advantages and opportunities of these sustainable practices, the entry will contribute to an increase in the awareness of wooden additional floor construction, which has many advantages and therefore help to gain more widespread use in Finland and other countries.

In this entry, timber or wood refers to engineered timber products [123,124], e.g., cross-laminated timber ((CLT) is a wood panel product made from gluing together layers of solid-sawn lumber), laminated veneer lumber ((LVL) is produced from veneer and is designed for structural framing where high strength and rigidity are required), and glue-laminated timber (glulam) ((GL) consists of layers of dimensional lumber glued together with durable, moisture-resistant structural adhesives).

The remainder of this entry is composed as follows: First, a literature survey is provided. This was followed by a section on the benefits, challenges, and drawbacks of wooden additional floor construction. Finally, the conclusions and prospects of the research are presented.

2. Literature Survey

As mentioned above, there are a limited number of studies on wooden additional floor construction. Among them, Karjalainen et al. [29] analyzed the different stages of wooden additional floors in old apartments from the standpoint of housing and real estate firms in Finland via interviews with involved professionals. Their result indicated (a) meticulous scrutiny of commercial conditions was critical to a profitable investment; (b) a change in the city plan, finding a suitable contractor, and involving an expert were among the highlights; (c) for the feasibility phase, the importance of parking space, load-bearing capacity, and compatibility with building regulations was emphasized; (d) during the project planning phase, attention was drawn to the importance of current building regulations and building rights concerning taxes and fees; (e) during the implementation planning process, city plan changes and different tender conditions came to the fore; and (f) effective information sharing between parties was a critical parameter for the successful completion of the construction phase.

Similarly, highlights of the study by Karjalainen et al. [125] were (a) the feasibility study emphasized the property's condition and potential targets for improvement, as well as relevant codes and regulations in force; (b) construction cost, profit, and sales of building rights were among the issues that were frequently discussed during the project planning phase; (c) building rights, changes in the city plan, and conditions of the company managing the property were reported as significant issues during the implementation planning phase, and (d) regular and frequent updating of building occupants on the progress of the construction process was the backbone of the construction phase.

On the other hand, Karjalainen et al. [26] conducted a survey and discussed the residents' approach to wood facade renovation and additional floor construction from Finnish residents' standpoint. Their study highlighted the following regarding additional floor construction: (a) Residents' attitudes were mostly positive; (b) younger and more educated people approached these practices much more positively; (c) respondents generally thought that additional floor construction will increase the attractiveness of residential areas; and (d) apartment owners positively appreciated the housing association's decision for additional floors to finance the facade refurbishment.

Soikkeli and Sorri [19], Soikkeli et al. [23], and Soikkeli [24] presented a research project (Finnish national research project, User- and Business-oriented Suburb Renovation Concept [KLIKK]) targeted to develop an industrial-scale, viable, and effective method for refurbishing and implementing additional floors to old apartment buildings. The technical solutions that form part of the concept were actively taking advantage of the opportunity offered by the new Finnish fire codes for the use of timber structures during renovations.

Cronhjort et al. [25] discussed obstacles and benefits for the utilization of wood-based approaches for infill development and building extensions in Finland by examining architecture and engineering master's theses. Their findings highlighted (a) there was a potential for infill development in the Finnish municipalities; (b) the applicability of vertical extensions that increase the functionality and quality of life was also demonstrated; (c) the added value of using timber-based prefabricated solutions in each case was shown; and (d) environmental, financial, and social advantages of infill construction, which is one of the Finnish national targets in provincial and land use, were evident in the case studies.

In addition, there are several studies on the vertical extension of buildings conducted in other countries. Prominent among these studies, Leskovar et al. [126] aimed to verify the effect of a structural lightweight wood–glass upgrade module and building a vertical extension on the energy efficiency of the selected renovated building concerning the relationship between the volume sizes of the existing building and different types of upgrade modules in Slovenia. The results showed great potential to reduce the energy consumption of existing inefficient buildings and simultaneously addressed the problem of urban sprawl by enabling the concentration of usable floor space in urban centers. Sundling et al. [127] evaluated and compared four refurbishment approaches to find an economically feasible model using life cycle profit analysis and life cycle impact assessment, based on a study of six similar buildings constructed in Gothenburg (Sweden) in 1971. The findings showed that vertical extension promotes the energy-efficient renovation of buildings, and the combination of low energy and vertical extension has the highest return on investment and lowest environmental impact. Artes et al. [128] focused on the regeneration of city centers by identifying high proportions of the buildable area remaining on roofs (approximately 2500 buildings making up 800,000 square meters of buildable area in Barcelona, Spain) using two-dimensional panels and three-dimensional partitioning made of wood and steel. Their construction methods made it possible to undertake the upgrading of existing buildings and to provide, in at least seven cases, new high-quality homes that improved the quality of life for the community. Dind et al. [129] introduced the Workspace project aiming to develop a new prefabricated timber structure system tailored to the vertical extension of existing office buildings through a pilot project carried out in Lausanne (Switzerland). The results allowed the technical and architectural suitability of the system to be validated, particularly in terms of its typological flexibility and economic feasibility, with prefabrication, transportation, and assembly of large elements and a building in operation, in a dense urban context.

3. Wooden Additional Floor Construction

Additional floors (Figure 4) increase the building height and the number of building stories or change the shape of the roof, as one of the most effective ways to cover real estate development and renovation processes by increasing energy efficiency and functionality [25]. The benefits, challenges, and drawbacks of additional floor construction can be summarized as follows [23–28,130–135]:

(1) As an efficient and environmentally friendly construction method, it provides beneficial development of the building stock and increases the income of property owners.

(2) Renovation and upgrading of old building stock were more beneficial in terms of the carbon footprint than new construction and demolition. For example, Hu-uhka et al. [27] reported that renovation and upgrading works with additional floors have a 20% lower negative impact on the environment in terms of carbon footprint compared to new construction and demolition.

(3) Additional floor construction provides short-term income to housing companies by selling the building rights or areas of additional floors. Revenues from additional floors could be used to cover the renovation cost of the old building.

(4) Additional floors considerably increase the total floor area without significantly affecting the total energy consumption of the upgraded building.

(5) As energy-efficient passive structures, additional floors can substantially improve the energy efficiency of existing buildings, especially if the upper floors have not been renovated for a long time.

(6) Additional floors improve the appearance of the building and can have a substantial impact on the architectural impression.

(7) Considering that the renovation process is a slow and uncomfortable process for the residents, the construction of an additional floor using prefabricated elements minimizes this discomfort of the building occupants and speeds up the process.

(8) To minimize disturbance to residents, it makes sense to build wooden additional floors from box-like module elements. However, module elements are difficult to design and construct because the upper ceiling tile cannot usually be loaded, so the loading must be aligned with the load-bearing walls below the floor, which may be few in number.

(9) To manage the renovation costs and use the insufficient autumn and winter capacities of the housing factories, wooden additional floor projects, especially during the winter months, face difficulties in the construction site conditions, especially in humidity control.

(10) Additional floor construction, which not only ensures energy efficiency but also brings the existing building to modern standards, is a complicated process consisting of several stages mentioned below that are technically more difficult.

(11) The socio-economic consequences of densification due to additional floors were brought to the agenda in several studies (e.g., [133–135]).

(12) Increased use of infrastructure, green space, and common appliances can be considered as the drawbacks of additional floor construction.

The construction of a wooden additional floor can be divided into four main stages [29,125]: (i) The feasibility phase, (ii) project planning, (iii) implementation planning, and (iv) construction. During the feasibility phase, construction professionals and real estate sectors are contacted, and construction conditions of additional floors are evaluated. At the project planning stage, the project conditions, course, and scope of the project are determined. During implementation planning, steps are taken to enhance the construction plans and expand the building rights, allowing the construction of additional floors. The terms and conditions of the city plan change, building rights, and the company managing the property are among critical issues for this phase. As the final stage, the construction of wooden additional floors begins. Effective flow of information between residents and stakeholders and the appointment of a representative of the housing or real estate firm to join site meeting organizations and discuss suitable schedules for the construction process are important considerations at this stage.

(**a**)

(**b**)

Figure 4. Wooden additional floor project examples: (**a**) An example where the shape of the roof has changed (Image courtesy of Samu Rantanen, Elissa Helminen, and Linnea Lindberg); (**b**) an example where the shape of the roof has not changed (Image courtesy of Johanna Partanen, Laura Lamberg, and Riina Hagren).

4. Conclusions and Prospects

This entry aimed to map out, organize, and collate scattered information on the current state of the art, as well as benefits and challenges of wooden additional floor projects with their different stages, focusing on the case of Finland, and presents it in an accessible and understandable discourse for non-technical readers.

Endorsing positive market development, informing people about efficient energy use and energy savings from building refurbishments including the construction of additional floors, providing different forms of financial support for these sustainable practices, conducting more research and creating more investment, and encouraging new modes of energy contracts are actions that play important roles for future applications of additional floor construction.

In terms of both building construction technology and tendering procedures, wooden additional floor projects have great development potential. Especially with regard to construction technology, the use of prefabricated wooden elements, e.g., volumetric modular elements, can make construction faster and reduce construction-related disturbances, which is a major problem for building occupants, contributing to the diffusion of additional floor construction, as in Finland. For example, the experience and expertise gained in these projects, which have become widespread in the last ten years in the Tampere region of Finland, paved the way for further improvement in contract and technical-based issues [29,125].

All stages of wooden additional floor projects need dedication and investment, as well as advanced communication and collaboration between all relevant stakeholders and housing or real estate professionals. This entry will contribute to the increase in awareness of the construction of wooden additional floors, which has environmental, financial, energy-efficient, and aesthetic benefits in Finland and other countries.

Author Contributions: Conceptualization, H.E.I., A.S. and M.K.; methodology, H.E.I., A.S. and M.K.; software, H.E.I. and A.S.; formal analysis, H.E.I., A.S. and M.K.; investigation, H.E.I. and A.S.; data curation, H.E.I., A.S. and M.K.; writing—original draft preparation, H.E.I.; writing—review and editing, H.E.I., A.S. and M.K.; visualization, H.E.I. and A.S.; supervision, A.S. and M.K.; project administration, A.S. and M.K.; All authors have read and agreed to the published version of the manuscript.

Funding: This research received no external funding.

Institutional Review Board Statement: Not applicable.

Informed Consent Statement: Not applicable.

Data Availability Statement: Not applicable.

Conflicts of Interest: The authors declare no conflict of interest.

Entry Link on the Encyclopedia Platform: https://encyclopedia.pub/21629.

References

1. European Commission. Energy Roadmap 2050 Impact Assessment and Scenario Analysis, Brussels. 2011. Available online: https://ec.europa.eu/energy/sites/ener/files/documents/roadmap2050_ia_20120430_en_0.pdf (accessed on 25 February 2022).
2. Donoghue, H. 2050 Energy Roadmap: Energy Policy & Innovation: Energy Roadmap 2050. *Eur. Energy Clim. J.* **2012**, *2*, 32–37.
3. Kuittinen, M.; Häkkinen, T. Reduced carbon footprints of buildings: New Finnish standards and assessments. *Build. Cities* **2020**, *1*, 182–197. [CrossRef]
4. Allard, I.; Nair, G.; Olofsson, T. Energy performance criteria for residential buildings: A comparison of Finnish, Norwegian, Swedish, and Russian building codes. *Energy Build.* **2021**, *250*, 111276. [CrossRef]
5. Energy Policies of The International Energy Agency (IEA) Countries: Finland 2018 Review. 2018. Available online: https://www.connaissancedesenergies.org/sites/default/files/pdf-actualites/situation_energetique_de_la_finlande.pdf (accessed on 25 February 2022).
6. European Commission. Long-Term Renovation Strategy 2020–2050 Finland, Report According to Article 2a of Directive (2010/31/EU) on the Energy Performance of Buildings, as Amended by Directive 2018/844/EU. 10 March 2020. Available online: https://ec.europa.eu/energy/sites/default/files/documents/fi_2020_ltrs_en.pdf (accessed on 25 February 2022).
7. Simson, R.; Fadejev, J.; Kurnitski, J.; Kesti, J.; Lautso, P. Assessment of Retrofit Measures for Industrial Halls: Energy Efficiency and Renovation Budget Estimation. *Energy Procedia* **2016**, *96*, 124–133. [CrossRef]
8. Kaasalainen, T.; Huuhka, S. Homogenous homes of Finland: 'standard' flats in non-standardized blocks. *Build. Res. Inf.* **2016**, *44*, 229–247. [CrossRef]
9. Hirvonena, J.; Jokisaloa, J.; Heljo, J.; Kosonena, R. Towards the EU emissions targets of 2050: Optimal energy renovation measures of Finnish apartment buildings. *Int. J. Sustain. Energy* **2019**, *38*, 649–672. [CrossRef]
10. The Housing Finance and Development Centre of Finland (ARA), The Suburban Innova Block of Flats Is Being Renovated into a Passive House. Available online: https://www.ara.fi/en-US/Housing_development/Development_projects/The_suburban_Innova_block_of_flats_is_be%2817681%29 (accessed on 25 February 2022).
11. Soikkeli, A.; Sorri, L. A New Suburb Renovation Concept. *Int. J. Archit. Environ. Eng.* **2014**, *8*, 647–655.
12. Soikkeli, A.; Hagan, H.; Karjalainen, M.; Koiso-Kanttila, J.; Kurnitski, J.; Viljakainen, M.; Hotakainen, T.; Jäntti, T.; Murtonen, N.; Sakki, R.; et al. Puun Mahdollisuudet Lähiöiden Korjauksessa. Oulu: Oulun Yliopisto, Arkkitehtuurin Osasto. Haettu Osoitteesta. 2011. Available online: https://www.oulu.fi/ark/tiedostot/puun_mahdollisuudet_lahioiden_korjauksissa_web.pdf (accessed on 25 February 2022). (In Finnish).
13. The Finnish Timber of Council (Puuinfo). Available online: https://puuinfo.fi/?lang=enç (accessed on 25 February 2022).
14. Huttunen, H.; Blomqvist, E.; Ellilä, E.; Hasu, E.; Perämäki, E.; Tervo, A.; Verma, I.; Ullrich, T.; Utriainen, J. The Finnish Townhouse as a Home. Starting Points and Interpretations. Habitat Components—Townhouse. Final Report. Aalto University Publication

Series CROSSOVER 8/2017. Helsinki, Finland. 2017. Available online: https://aaltodoc.aalto.fi/bitstream/handle/123456789/30185/isbn9789526071220.pdf?sequence=1&isAllowed=y (accessed on 25 February 2022).

15. United Nations Economic Commission for Europe United Nations Human Settlements Programme, Good Practices for Energy-Efficient Housing in the Unece Region, UNITED NATIONS, New York and Geneva. 2013. Available online: https://unece.org/fileadmin/DAM/hlm/documents/Publications/good.practices.ee.housing.pdf (accessed on 25 February 2022).
16. Streimikiene, D.; Balezentis, T. Innovative Policy Schemes to Promote Renovation of Multi-Flat Residential Buildings and Address the Problems of Energy Poverty of Aging Societies in Former Socialist Countries. *Sustainability* **2019**, *11*, 2015. [CrossRef]
17. Paiho, S.; Abdurafikov, R.; Hoang, H.; Castell-Rüdenhausen, M.Z.; Hedman, Å.; Kuusisto, J. Business Aspects of Energy Efficient Renovations of Sovietera Residential Districts A Case Study from Moscow, VTT Technology 154 ISSN-L 2242-1211 ISSN 2242-122X (Online), VTT Technical Research Centre of Finland, Espoo, Finland. 2014. Available online: https://www.vttresearch.com/sites/default/files/pdf/technology/2014/T154.pdf (accessed on 25 February 2022).
18. Official Statistics of Finland (OSF). Homeowners and Housing Companies Repaired by EUR 6.0 Billion in 2019. 2019. Available online: http://www.stat.fi/til/kora/2019/01/kora_2019_01_2020-06-11_tie_001_fi.html%20 (accessed on 25 February 2022).
19. Soikkeli, A.; Sorri, L. A New Suburb Renovation Concept. In Proceedings of the ICAE 2014: XII International Conference on Architectural Engineering, Copenhagen, Denmark, 12–13 June 2014; International Science Index 90; World Academy of Science, Engineering and Technology: Paris, France, 2014; pp. 636–644.
20. KTI Finland. The Finnish Property Market. 2019. Available online: https://kti.fi/wp-content/uploads/The-Finnish-Property-Market-2019.pdf (accessed on 25 February 2022).
21. Farahani, A.S. *Maintenance, Renovation and Energy Efficiency in the Swedish Multi-Family Housing Market*; The Division of Building Services Engineering, Chalmers University of Technology: Gothenburg, Sweden, 2017. Available online: https://core.ac.uk/download/pdf/198056482.pdf (accessed on 25 February 2022).
22. Ferrante, A.; Prati, D.; Fotopoulou, A. Triple A-Reno: Attractive, Acceptable and Affordable Deep Renovation by a Consumers Orientated and Performance Evidence Based Approach. In *WP4–Task 4.2 Analysis and Design of the Business Module*; Huygen Installatie Adviseurs: Maastricht, The Netherlands, 2018.
23. Soikkeli, A.; Sorri, L.; Koiso-Kanttila, J. New Concept for User-Orientated Suburb Renovation. In Proceedings of the World SB14, Barcelona, Spain, 28–30 October 2014; pp. 1–7.
24. Soikkeli, A. Additional floors in old apartment blocks. *Energy Procedia* **2016**, *96*, 815–823. [CrossRef]
25. Cronhjort, A.; Soikkeli, T.; Tulamo, T.; Junnonen, J. Urban Densification in Finland: Infill Development And Building Extensions With Timber Based Solutions. *WIT Trans. Ecol. Environ.* **2015**, *193*, 319–330.
26. Karjalainen, M.; Ilgın, H.E.; Metsäranta, L.; Norvasuo, M. Residents' Attitudes towards Wooden Facade Renovation and Additional Floor Construction in Finland. *Int. J. Environ. Res. Public Health* **2021**, *18*, 12316. [CrossRef] [PubMed]
27. Huuhka, S.; Vainio, T.; Moisio, M.; Lampinen, E.; Knuuttinen, M.; Bashmakov, S.; Köliö, A.; Lahdensivu, J.; Ala-Kotila, P.; Lahdenperä, P. To Demolish or to Repair? Carbon Footprint Impacts, Life Cycle Costs and Steering Instruments, Publications of the Ministry of the Environment 2021:9. Built Environment. 2021. Available online: https://julkaisut.valtioneuvosto.fi/bitstream/handle/10024/162862/YM_2021_9.pdf?sequence=4&isAllowed=y (accessed on 25 February 2022).
28. Bojić, M.; Miletić, M.; Malešević, J.; Djordjević, S.; Cvetković, D. Influence of additional storey construction to space heating of a residential building. *Energy Build.* **2012**, *54*, 511–518. [CrossRef]
29. Karjalainen, M.; Ilgın, H.E.; Somelar, D. Wooden Additional Floors in Old Apartment Buildings: Perspectives of Housing and Real Estate Companies from Finland. *Buildings* **2021**, *11*, 316. [CrossRef]
30. The National Building Code of Finland—Structural Fire Safety, Decree of the Ministry of the Environment. 2017. Available online: https://ym.fi/en/the-national-building-code-of-finland (accessed on 25 February 2022).
31. Green, M. *The Case for Tall Buildings: How Mass Timber Offers a Safe, Economical, and Environmentally Friendly Alternative for Tall Building Structures*; MGB Architecture and Design: Vancouver, BC, Canada; Toronto, ON, Canada, 2012.
32. Myers, F.; Fullera, R.; Crawford, R.H. The Potential to Reduce the Embodied Energy in Construction through the Use of Renewable Materials. In Proceedings of the ASA2012: The 46th Annual Conference of the Architectural Science Association (Formerly ANZAScA)—Building on Knowledge: Theory and Practice, Gold Coast, Australia, 14–16 November 2012.
33. Construction Industry Progress towards Sustainability with Renewable Materials. Recycling Magazine. 2020. Available online: https://www.recycling-magazine.com/2020/04/14/construction-industry-progress-towards-sustainability-with-renewablematerials/ (accessed on 25 February 2022).
34. Lammert, L. Circular Economy in Architecture—Sustainable Principles for Future Design. Master's Thesis, Oulu School of Architecture, Faculty of Technology, University of Oulu, Oulu, Finland, 2018. Available online: https://figbc.fi/wpcontent/uploads/sites/4/2020/05/nbnfioulu-201811233096.pdf (accessed on 25 February 2022).
35. Robati, M.; Oldfield, P.; Nezhad, A.A.; Carmichael, D.G.; Kuru, A. Carbon value engineering: A framework for integrating embodied carbon and cost reduction strategies in building design. *Build. Environ.* **2021**, *192*, 107620. [CrossRef]
36. Hart, J.; D'Amico, B.; Pomponi, F. Whole-life embodied carbon in multistory buildings: Steel, concrete and timber structures. *J. Ind. Ecol.* **2021**, *25*, 403–418. [CrossRef]
37. Soikkeli, A. Possibilities in the Renovation of Suburban Apartment Buildings. Case: Porvoonportti. In *Improving the Quality of Suburban Building Stock*; COST Action TU0701; Unifepress: Ferrara, Italy, 2012; pp. 127–140.

38. Churkina, G.; Organschi, A.; Reyer, C.P.O.; Ruff, A.; Vinke, K.; Liu, Z.; Reck, B.K.; Graedel, T.E.; Schellnhuber, H.J. Buildings as a global carbon sink. *Nat. Sustain.* **2020**, *3*, 269–276. [CrossRef]
39. Franzini, F.; Toivonen, R.; Toppinen, A. Why Not Wood? Benefits and Barriers of Wood as a Multistory Construction Material: Perceptions of Municipal Civil Servants from Finland. *Buildings* **2018**, *8*, 159. [CrossRef]
40. Geng, A.; Yang, H.; Chen, J.; Hong, Y. Review of carbon storage function of harvested wood products and the potential of wood substitution in greenhouse gas mitigation. *For. Policy Econ.* **2017**, *85*, 192–200. [CrossRef]
41. Hafner, A.; Schäfer, S. Comparative LCA study of different timber and mineral buildings and calculation method for substitution factors on building level. *J. Clean. Prod.* **2017**, *167*, 630–642. [CrossRef]
42. Dong, Y.; Qin, T.; Zhou, S.; Huang, L.; Bo, R.; Guo, H.; Yin, X. Comparative Whole Building Life Cycle Assessment of Energy Saving and Carbon Reduction Performance of Reinforced Concrete and Timber Stadiums—A Case Study in China. *Sustainability* **2020**, *12*, 1566. [CrossRef]
43. Bergman, R.; Puettmann, M.; Taylor, A.; Skog, K.E. The Carbon Impacts of Wood Products. *For. Prod. J.* **2014**, *64*, 220–231. [CrossRef]
44. Pierobon, F.; Huang, M.; Simonen, K.; Ganguly, I. Environmental benefits of using hybrid CLT structure in midrise nonresidential construction: An LCA based comparative case study in the U.S. Pacific Northwest. *J. Build. Eng.* **2019**, *26*, 100862. [CrossRef]
45. Ritter, M.; Skog, K.; Bergman, R. *Science Supporting the Economic and Environmental Benefits of Using Wood and Wood Products in Green Building Construction*; General Technical Report FPL-GTR-206; U.S. Department of Agriculture, Forest Service, Forest Products Laboratory: Madison, WI, USA, 2011; pp. 1–9.
46. Wang, L.; Toppinen, A.; Juslin, H. Use of wood in green building: A study of expert perspectives from the UK. *J. Clean. Prod.* **2014**, *65*, 350–361. [CrossRef]
47. Rinne, R.; Ilgın, H.E.; Karjalainen, M. Comparative Study on Life-Cycle Assessment and Carbon Footprint of Hybrid, Concrete and Timber Apartment Buildings in Finland. *Int. J. Environ. Res. Public Health* **2022**, *19*, 774. [CrossRef] [PubMed]
48. Skullestad, J.L.; Bohne, R.A.; Lohne, J. High-rise Timber Buildings as a Climate Change Mitigation Measure—A Comparative LCA of Structural System Alternatives. *Energy Procedia* **2016**, *96*, 112–123. [CrossRef]
49. CWC. *Energy and the Environment in Residential Construction*; Sustainable Building Series No.1; Canadian Wood Council: Ottawa, ON, Canada, 2007. Available online: https://cwc.ca/wp-content/uploads/publications-Energy-and-the-Environment.pdf (accessed on 25 February 2022).
50. European Commission. *A Sustainable Bioeconomy for Europe: Strengthening the Connection between Economy, Society and the Environment*; Publications Office of the European Union: Luxembourg, 2018.
51. Liang, S.; Gu, H.; Bergman, R.; Kelley, S. Comparative life-cycle assessment of a mass timber building and concrete alternative. *Wood Fiber Sci.* **2020**, *52*, 217–229. [CrossRef]
52. Sandberg, K.; Orskaug, T.; Andersson, A. Prefabricated Wood Elements for Sustainable Renovation of Residential Building Façades. *Energy Procedia* **2016**, *96*, 756–767. [CrossRef]
53. Gustavsson, L.; Joelsson, A.; Sathre, R. Life cycle primary energy use and carbon emission of an eight-storey wood-framed apartment building. *Energy Build.* **2010**, *42*, 230–242. [CrossRef]
54. Jussila, J.; Lähtinen, K. Effects of institutional practices on delays in construction—Views of Finnish homebuilder families. *Housing Stud.* **2020**, *35*, 1167–1193. [CrossRef]
55. Onyszkiewicz, J.; Sadowski, K. Proposals for the revitalization of prefabricated building facades in terms of the principles of sustainable development and social participation. *J. Build. Eng.* **2022**, *46*, 103713. [CrossRef]
56. Werner, F.; Taverna, R.; Hofer, P.; Richter, K. Carbon pool and substitution effects of an increased use of wood in buildings in Switzerland: First estimates. *Ann. For. Sci.* **2005**, *62*, 889–902. [CrossRef]
57. Kutnar, A.; Hill, C. Life Cycle Assessment—Opportunities for Forest Products Sector. *Bioprod. Bus.* **2017**, *2*, 52–64.
58. Bergman, R.D.; Falk, R.H.; Gu, H.; Napier, T.R.; Meil, J. *Life-Cycle Energy and GHG Emissions for New and Recovered Softwood Framing Lumber and Hardwood Flooring Considering End-of-Life Scenarios*; Research Paper FPL-RP-672; U.S. Department of Agriculture, Forest Service, Forest Products Laboratory: Madison, WI, USA, 2013; p. 35.
59. Karjalainen, M.; Ilgın, H.E. The Change over Time in Finnish Residents' Attitudes towards Multi-Story Timber Apartment Buildings. *Sustainability* **2021**, *13*, 5501. [CrossRef]
60. Jagarajan, R.; Abdullah Mohd Asmoni, M.N.; Mohammed, A.H.; Jaafar, M.N.; Lee Yim Mei, J.; Baba, M. Green retrofitting—A review of current status, implementations and challenges. *Renew. Sustain. Energy Rev.* **2017**, *67*, 1360–1368. [CrossRef]
61. Östman, B. National Fire Regulations for the Use of Wood in Buildings—Worldwide Review 2020. *Wood Mater. Sci. Eng.* **2021**, 1–4. [CrossRef]
62. Kincelova, K.; Boton, C.; Blanchet, P.; Dagenais, C. Fire Safety in Tall Timber Building: A BIM-Based Automated Code-Checking Approach. *Buildings* **2020**, *10*, 121. [CrossRef]
63. Xu, H.; Pope, I.; Gupta, V.; Cadena, J.; Carrascal, J.; Lange, D.; McLaggan, M.S.; Mendez, J.; Osorio, A.; Solarte, A.; et al. Large-scale compartment fires to develop a self-extinction design framework for mass timber—Part 1: Literature review and methodology. *Fire Saf. J.* **2022**, *128*, 103523. [CrossRef]
64. Östman, B. *Fire Safety in Timber Buildings—Technical Guideline for Europe*; SP Technical Research Institute of Sweden: Stockholm, Sweden, 2010; p. 19.
65. Östman, B.; Brandon, D.; Frantzich, H. Fire Safety Engineering in Timber Buildings. *Fire Saf. J.* **2017**, *91*, 11–20. [CrossRef]

66. Su, L.; Wu, X.; Zhang, X.; Huang, X. Smart performance-based design for building fire safety: Prediction of smoke motion via AI. *J. Build. Eng.* **2021**, *43*, 102529. [CrossRef]
67. Siddiqui, A.A.; Ewer, J.A.; Lawrence, P.J.; Galea, E.R.; Frost, I.R. Building Information Modelling for performance-based Fire Safety Engineering analysis—A strategy for data sharing. *J. Build. Eng.* **2021**, *42*, 102794. [CrossRef]
68. Qiu, J.; Anwar Orabi, M.; Usmani, A.; Li, G. A computational approach for modelling composite slabs in fire within OpenSees framework. *Eng. Struct.* **2022**, *255*, 113909. [CrossRef]
69. Liu, K.; Chen, W.; Ye, J.; Jiang, J. Full-scale fire and post earthquake fire experiments of CFS walls with new configurations. *Structures* **2022**, *35*, 706–721. [CrossRef]
70. Medved, S. Buildings Fires and Fire Safety. In *Building Physics*; Springer International Publishing: Berlin/Heidelberg, Germany, 2021; pp. 407–451. [CrossRef]
71. Liu, J.; Fischer, E.C. Review of large-scale CLT compartment fire tests. *Construct. Build. Mater.* **2022**, *318*, 126099. [CrossRef]
72. Gasparri, E. 9—Unitized Timber Envelopes: The future generation of sustainable, high-performance, industrialized facades for construction decarbonization. In *Woodhead Publishing Series in Civil and Structural Engineering, Rethinking Building Skins*; Gasparri, E., Brambilla, A., Lobaccaro, G., Goia, F., Andaloro, A., Sangiorgio, A., Eds.; Woodhead Publishing: Cambridge, UK, 2022; pp. 231–255. [CrossRef]
73. Tian, F.; Xu, D.; Xu, X. Synergistic Effect of APP and TBC Fire-Retardants on the Physico-Mechanical Properties of Strandboard. *Materials* **2022**, *15*, 435. [CrossRef]
74. Buchanan, A.H.; Östman, B.; Frangi, A. *Fire Resistance of Timber Structures*; NIST White Paper: Washington, DC, USA, 2014.
75. Caniato, M.; Bettarello, F.; Ferluga, A.; Marsich, L.; Schmid, C.; Fausti, P. Thermal and acoustic performance expectations on timber buildings. *Build. Acoust.* **2017**, *24*, 219–237. [CrossRef]
76. Asdrubali, F.; Ferracuti, B.; Lombardi, L.; Guattari, C.; Evangelisti, L.; Grazieschi, G. A review of structural, thermo-physical, acoustical, and environmental properties of wooden materials for building applications. *Build. Environ.* **2017**, *114*, 307–332. [CrossRef]
77. Olsson, L. Moisture safety in CLT construction without weather protection—Case studies, literature review and interviews. *E3S Web Conf.* **2020**, *172*, 10001. [CrossRef]
78. Mjörnell, K.; Olsson, L. Moisture Safety of Wooden Buildings—Design, Construction and Operation. *J. Sustain. Architect. Civil Eng.* **2019**, *24*, 29–35. [CrossRef]
79. Renner, J.S.; Mensah, R.A.; Jiang, L.; Xu, Q.; Das, O.; Berto, F. Fire Behavior of Wood-Based Composite Materials. *Polymers* **2021**, *13*, 4352. [CrossRef]
80. Robert, J.; Bush, P.A.A. Changes and Trends in the Pallet Industry: Alternative Materials and Industry Structure. *Memphis. Hardwood Mark. Rep.* **1998**, *LXXVI*, 11–14.
81. Trouy, M.C.; Triboulot, P. Materiau bois: Structure et caractéristiques. *Technol. Ingénieur Constr. Bois* **2019**, *253*, 11–18. [CrossRef]
82. Upton, B.; Miner, R.; Spinney, M.; Heath, L.S. The greenhouse gas and energy impacts of using wood instead of alternatives in residential construction in the United States. *Biomass Bioenergy* **2008**, *32*, 1–10. [CrossRef]
83. Fast, P.; Jackson, R. A Case Study for Tall Timber. *Struct. Mag.* **2017**, 50–52.
84. Gohlich, R.; Erochko, J.; Woods, J.E. Experimental testing and numerical modelling of a heavy timber moment-resisting frame with ductile steel links. *Earthq. Eng. Struct. Dynam.* **2018**, *47*, 1460–1477.
85. Andreolli, M.; Piazza, M.; Tomasi, R.; Zandonini, R. Ductile moment-resistant steel–timber connections. *Proc. Inst. Civil Eng. Struct. Build.* **2011**, *164*, 65–78. [CrossRef]
86. Li, Z.; Wang, X.; He, M. Experimental and analytical investigations into lateral performance of cross-laminated timber (CLT) shear walls with different construction methods. *J. Earthq. Eng.* **2020**, 1–23. [CrossRef]
87. Yang, R.; Li, H.; Lorenzo, R.; Ashraf, M.; Sun, Y.; Yuan, Q. Mechanical behaviour of steel timber composite shear connections, Construct. *Build. Mater.* **2020**, *258*, 119605. [CrossRef]
88. Iqbal, A. Developments in Tall Wood and Hybrid Buildings and Environmental Impacts. *Sustainability* **2021**, *13*, 11881. [CrossRef]
89. Yang, X.; Tang, X.; Ma, L.; Sun, Y. Plastic composite sound insulation performance of structural wood wall integrated with wood. *J. Bioresour. Bioprod.* **2019**, *4*, 115–122. [CrossRef]
90. Zhang, Y.; Xie, L. *Inspection and Evaluation of Wooden Frame Constructions*; Chinese Academy of Forestry: Beijing, China, 2011.
91. Ding, Y.; Zhang, Y.; Wang, Z.; Gao, Z.; Zhang, T.; Huang, X. Vibration test and comfort analysis of environmental and impact excitation for wooden floor structure. *Bioresources* **2020**, *15*, 8212–8234. [CrossRef]
92. Kankovsky, A.; Dedic, M. Wood flooring in combination with underfloor heating systems. *IOP Conf. Ser. Mater. Sci. Eng.* **2021**, *1203*, 22043. [CrossRef]
93. Sun, J.; Tu, K.; Büchele, S.; Koch, S.M.; Ding, Y.; Ramakrishna, S.N.; Stucki, S.; Guo, H.; Wu, C.; Keplinger, T.; et al. Functionalized wood with tunable tribopolarity for efficient triboelectric nanogenerators. *Matter* **2021**, *4*, 3049–3066. [CrossRef]
94. Song, Y.; Wang, H.B.; Cheng, X.L.; Li, G.K.; Chen, X.X.; Chen, H.T.; Miao, L.M.; Zhang, X.S.; Zhang, H.X. High-efficiency selfcharging smart bracelet for portable electronics. *Nano Energy* **2019**, *55*, 29–36. [CrossRef]
95. Parida, K.; Xiong, J.Q.; Zhou, X.R.; Lee, P.S. Progress on triboelectric nanogenerator with stretchability, selfhealability and bio-compatibility. *Nano Energy* **2019**, *59*, 237–257. [CrossRef]
96. Hao, S.F.; Jiao, J.Y.; Chen, Y.D.; Wang, Z.L.; Cao, X. Natural wood-based triboelectric nanogenerator as self-powered sensing for smart homes and floors. *Nano Energy* **2020**, *75*, 104957. [CrossRef]

97. Chandrasekhar, A.; Vivekananthan, V.; Khandelwal, G.; Kim, W.J.; Kim, S.J. Green energy from working surfaces: A contact electrification-enabled data theft protection and monitoring smart table. Mater. *Today Energy* **2020**, *18*, 100544. [CrossRef]

98. Tulonen, L.; Karjalainen, M.; Ilgın, H.E. *Tall Wooden Residential Buildings in Finland: What Are the Key Factors for Design and Implementation?* IntechOpen: London, UK, 2021.

99. Zwerger, K. Recognizing the Similar and Thus Accepting the Other: The European and Japanese Traditions of Building with Wood. *J. Tradit. Build. Archit. Urban.* **2021**, *2*, 305–317. [CrossRef]

100. Ilgın, H.E.; Karjalainen, M.; Koponen, O. Dovetailed Massive Wood Board Elements for Multi-Story Buildings. In Proceedings of the LIVENARCH VII Livable Environments & Architecture 7th International Congress OTHER ARCHITECT/URE(S), Trabzon, Turkey, 28–30 September 2021; Volume I, pp. 47–60.

101. Yusof, N.M.; Tahir, P.M.; Lee, S.H.; Khan, M.A.; James, R.M.S. Mechanical and physical properties of Cross-Laminated Timber made from Acacia mangium wood as function of adhesive types. *J. Wood Sci.* **2019**, *65*, 20. [CrossRef]

102. Karjalainen, M.; Ilgın, H.E.; Yli-Äyhö, M.; Soikkeli, A. *Complementary Building Concept: Wooden Apartment Building: The Noppa toward Zero Energy Building Approach*; IntechOpen: London, UK, 2021.

103. Li, M.; Zhang, S.; Gong, Y.; Tian, Z.; Ren, H. Gluing Techniques on Bond Performance and Mechanical Properties of Cross-Laminated Timber (CLT) Made from Larix kaempferi. *Polymers* **2021**, *13*, 733. [CrossRef]

104. Ilgın, H.E.; Karjalainen, M. Preliminary Design Proposals for Dovetail Wood Board Elements in Multi-Story Building Construction. *Architecture* **2021**, *1*, 56–68. [CrossRef]

105. Bahrami, A.; Nexén, O.; Jonsson, J. Comparing Performance of Cross-Laminated Timber and Reinforced Concrete Walls. *Int. J. Appl. Mech. Eng.* **2021**, *26*, 28–43. [CrossRef]

106. Ilgın, H.E.; Karjalainen, M.; Koponen, O. *Various Geometric Configuration Proposals for Dovetail Wooden Horizontal Structural Members in Multistory Building Construction*; IntechOpen: London, UK, 2022.

107. Sun, Z.; Chang, Z.; Bai, Y.; Gao, Z. Effects of working time on properties of a soybean meal-based adhesive for engineered wood flooring. *J. Adhes.* **2021**, 1–20. [CrossRef]

108. Karjalainen, M.; Ilgın, H.E. A Statistical Study on Multi-Story Timber Residential Buildings (1995–2020) in Finland. In Proceedings of the LIVENARCH VII Livable Environments & Architecture 7th International Congress OTHER ARCHITECT/URE(S), Trabzon, Turkey, 28–30 September 2021; Volume I, pp. 82–94.

109. Ilgın, H.E.; Karjalainen, M.; Koponen, O. *Review of the Current State-of-the-Art of Dovetail Massive Wood Elements*; IntechOpen: London, UK, 2021.

110. Aaltonen, A.; Hurmekoski, E.; Korhonen, J. What about Wood?—"Nonwood" Construction Experts' Perceptions of Environmental Regulation, Business Environment, and Future Trends in Residential Multistory Building in Finland. *For. Prod. J.* **2021**, *71*, 342–351. [CrossRef]

111. Ilgın, H.E.; Karjalainen, M. *Perceptions, Attitudes, and Interest of Architects in the Use of Engineered Wood Products for Construction: A Review*; IntechOpen: London, UK, 2021.

112. Roos, A.; Woxblom, L.; McCluskey, D. The influence of architects and structural engineers on timber in construction—Perceptions and roles. *Silva Fenn.* **2010**, *44*, 871–884. [CrossRef]

113. Karjalainen, M.; Ilgın, H.; Tulonen, L. Main Design Considerations and Prospects of Contemporary Tall Timber Apartment Buildings: Views of Key Professionals from Finland. *Sustainability* **2021**, *13*, 6593. [CrossRef]

114. Hemström, K.; Gustavsson, L.; Mahapatra, K. The sociotechnical regime and Swedish contractor perceptions of structural frames. Constr. *Manag. Econ.* **2017**, *35*, 184–195.

115. Karjalainen, M.; Ilgın, H.E.; Metsäranta, L.; Norvasuo, M. Suburban Residents' Preferences for Livable Residential Area in Finland. *Sustainability* **2021**, *13*, 11841. [CrossRef]

116. Markström, E.; Kuzman, M.K.; Bystedt, A.; Sandberg, D.; Fredriksson, M. Swedish architects view of engineered wood products in buildings. *J. Clean. Prod.* **2018**, *181*, 33–41. [CrossRef]

117. Karjalainen, M.; Ilgın, H.E.; Metsäranta, L.; Norvasuo, M. *Wooden Facade Renovation and Additional Floor Construction for Suburban Development in Finland*; IntechOpen: London, UK, 2022.

118. Ilgın, H.E.; Karjalainen, M.; Pelsmakers, S. Finnish architects' attitudes towards multi-storey timber-residential buildings. *Int. J. Build. Pathol. Adapt.* **2021**. ahead-of-print.

119. Häkkänen, L.; Ilgın, H.E.; Karjalainen, M. The Current State of the Finnish Cottage Phenomenon: Perspectives of Experts. *Buildings* **2022**, *12*, 260. [CrossRef]

120. Gold, S.; Rubik, F. Consumer attitudes towards timber as a construction material and towards timber frame houses—Selected findings of a representative survey among the German population. *J. Clean. Prod.* **2009**, *17*, 303–309. [CrossRef]

121. Lähtinen, K.; Harju, C.; Toppinen, A. Consumers' perceptions on the properties of wood affecting their willingness to live in and prejudices against houses made of timber. *Wood Mater. Sci. Eng.* **2019**, *14*, 325–331. [CrossRef]

122. Kylkilahti, E.; Berghäll, S.; Autio, M.; Nurminen, J.; Toivonen, R.; Lähtinen, K.; Vihemäki, H.; Franzini, F.; Toppinen, A. A consumer-driven bioeconomy in housing? Combining consumption style with students' perceptions of the use of wood in multi-story buildings. *Ambio* **2020**, *49*, 1943–1957. [CrossRef] [PubMed]

123. Wang, Z.; Yin, T. *Cross-Laminated Timber: A Review on Its Characteristics and an Introduction to Chinese Practices*; IntechOpen: London, UK, 2021.

124. Rahman, T.; Ashraf, M.; Ghabraie, K.; Subhani, M. Evaluating Timoshenko Method for Analyzing CLT under Out-of-Plane Loading. *Buildings* **2020**, *10*, 184. [CrossRef]

125. Karjalainen, M.; Ilgın, H.E.; Somelar, D. *Wooden Extra Stories in Concrete Block of Flats in Finland as an Ecologically Sensitive Engineering Solution, Ecological Engineering—Addressing Climate Challenges and Risks*; IntechOpen: London, UK, 2021.

126. Leskovar, V.Ž.; Nedelko, M.L.; Premrov, M. Building refurbishment by vertical extension with lightweight structural modules. In Proceedings of the 1st Latin American SDEWES Conference, Rio de Janeiro, Brazil, 28–31 January 2018.

127. Sundling, R.; Blomsterberg, Å.; Landin, A. Enabling energy-efficient renovation: The case of vertical extension to buildings. *Construct. Innov.* **2019**, *19*, 2–14. [CrossRef]

128. Artés, J.; Wadel, G.; Martí, N. Vertical Extension and Improving of Existing Buildings. *Open Construct. Build. Technol. J.* **2017**, *11*, 83–94. [CrossRef]

129. Dind, A.; Lufkin, S.; Rey, E. A Modular Timber Construction System for the Sustainable Vertical Extension of Office Buildings. *Designs* **2018**, *2*, 30. [CrossRef]

130. Janda, K.B. Building communities and social potential: Between and beyond organizations and individuals in commercial properties. *Energy Policy* **2014**, *67*, 48–55. [CrossRef]

131. Mills, B.; Schleich, J. Residential energy-efficient technology adoption, energy conservation, knowledge, and attitudes: An analysis of European countries. *Energy Policy* **2012**, *49*, 616–628. [CrossRef]

132. Nilsson, R. Vertical Extension of Buildings. Licentiate Thesis, Division of Construction Management, Lund University, Lund, Sweden, 2017.

133. Turok, I. Deconstructing density: Strategic dilemmas confronting the post-apartheid city. *Cities* **2011**, *28*, 470–477. [CrossRef]

134. Nabielek, K. Urban densification in The Netherlands: National spatial policy and empirical research of recent developments. In Proceedings of the 5th Conference of International Forum on Urbanism, Global Visions: Risks and Opportunities for the Urban Planet, National University of Singapore, Singapore, 24–26 February 2011.

135. Quastel, N.; Moos, M.; Lynch, N. Sustainability-as-density and the return of the social: The case of Vancouver, British Columbia. *Urban Geogr.* **2021**, *33*, 1055–1084. [CrossRef]

Entry

Cottage Culture in Finland: Development and Perspectives

Lotta Häkkänen, Hüseyin Emre Ilgın * and Markku Karjalainen

School of Architecture, Faculty of Built Environment, Tampere University, P.O. Box 600,
FI-33014 Tampere, Finland; lotta.k.h@gmail.com (L.H.); markku.karjalainen@tuni.fi (M.K.)
* Correspondence: emre.ilgin@tuni.fi

Definition: This entry provides an understanding of the past, present, and future of the Finnish cottage culture to create an overall picture of its development trajectory and its terminology, e.g., villa, in this context denoting a second home. Convenient, ready-made solutions, easy maintenance, a high level of equipment, year-round use, location, and modern and simple architectural styles are important selection criteria for (summer) cottages that belonged only to the wealthy bourgeois class in the 19th century and have taken their present form with a major transformation in Finland since then. Additionally, municipal regulations and increased attention to ecological concerns are other important issues regarding the cottage today. Cottage inheritance has changed over the generations, and the tightening of building regulations and increased environmental awareness are key drivers of the future transformation of cottage culture. Moreover, the increasing demand for single-family and outdoor spaces created by social changes such as remote working, which has become widespread with the COVID-19 pandemic, will make the summer cottage lifestyle even more popular in Finland. It is thought that this entry will contribute to the continuance of the Finnish cottage culture, which is essential for the vitality of countryside municipalities, local development, national culture, and the well-being of Finnish people.

Keywords: (summer) cottage; second home; (summer) villa; holiday home; Finland

Citation: Häkkänen, L.; Ilgın, H.E.; Karjalainen, M. Cottage Culture in Finland: Development and Perspectives. *Encyclopedia* **2022**, 2, 705–716. https://doi.org/10.3390/encyclopedia2020049

Academic Editors: Masa Noguchi, Antonio Frattari, Carlos Torres Formoso, Haşim Altan, John Odhiambo Onyango, Jun-Tae Kim, Kheira Anissa Tabet Aoul, Mehdi Amirkhani, Sara Jane Wilkinson, Shaila Bantanur and Raffaele Barretta

Received: 11 March 2022
Accepted: 11 April 2022
Published: 13 April 2022

Publisher's Note: MDPI stays neutral with regard to jurisdictional claims in published maps and institutional affiliations.

1. Introduction

Second homes, rooted in ancient civilizations, differ conceptually from region to region and culture to culture [1]. In many countries, people have a second home to access the geographical and cultural benefits that their primary property cannot give them. Among the major factors that make owning a second home attractive, is the desire to be in a natural environment, the search for an authentic atmosphere, or the inclination towards a countryside lifestyle [2]. In this sense, second homes are preferred not only by the privileged group but also by many people as weekend getaways in Nordic countries [3]. Partly as a consequence of the prosperity and abundance of the region, Scandinavian people generally buy a second home in the rural areas. Likewise, in countries like Turkey, second homes in mountain regions are used by many people as part of their culture to avoid the hot season in the summertime. For this reason, the importance of global borders are being lessened and second homes are gradually gaining an international dimension [4].

The Nordic lifestyle is closely linked to nature and the highly anticipated but short-lived summer season [5]. The most popular and ideal way to spend the summer months in Finland is in a lakeside cottage surrounded by nature [6] (see Figure 1). This tradition is grounded in traditional culture and lifestyle and forms an important part of the Finnish national landscape [7]. Many Finns consider the cottage to balance out city life [8].

Figure 1. The typical view from a lakeside cottage plot in Finland (Photo courtesy of Hüseyin Emre Ilgın).

'Cottage' is the best explanatory translation of the Finnish term 'mökki', a word with a robust cultural value for Finnish people, which is widely used in everyday life and formal contexts. The term has many meanings, from a humble cottage to a luxury villa like a 'second home'. A 'cottage' (see Figure 2) is constructed on a permanent basis and is defined as a residence used for recreational or leisure purposes [9]. The term 'summer cottage' is widespread and understood in Finnish, but it literally refers to dwellings used only in the summertime and is often linked to simple inhabitation. Additionally, terms that are functionally related to their meanings, e.g., holiday home, are mainly used for more official purposes and refer to leisure and vacation uses. Furthermore, second residence or second home is a common and modern notion that does not limit the use of the house as a term only for the holidays or a certain period of the year. In this entry, the term second home or place of residence is used to refer to a particularly well-outfitted year-round-use cottage.

(a)

(b)

Figure 2. *Cont.*

(c)

(d)

Figure 2. A cottage from the town of Ikaalinen, Finland: (**a**) entrance; (**b**) open kitchenette; (**c**) fireplace; (**d**) exterior view (Photos courtesy of Jenni Vilkman).

In the Finnish context, second homes play a vital role in the expression of the country's cultural landscape [10], illustrating the significance of outdoor leisure and conventional activities [11]. A large segment of the Finnish people defines second homes as very valuable [12]. Additionally, the second home in rural municipalities is critical to the regional economy and has considerable job potential in Finland [13]. Second homes and tourism potential have also been the focus of academic research for a long time and have become an established part of leisure time in various countries [14–20]. Both in Finland and internationally, second homes have become not only an essential part of people's leisure pursuits but also entire lifestyles [21].

The summer villa trend that started in the 19th century represented a model of luxury living that belonged only to the elite in the early years of Finland's independence [7]. Originally built near cities, the earliest ancestors of these cottages later spread to the coastlines and countryside. As industrialization increased in cities, holiday homes offered an escape to the natural environment. In the 1950s, spending time in the cottage became the pastime of almost all Finns, and in the 1980s cottage construction increased and reached its peak. The number of cottages in Finland at the end of 2018 exceeded half a million [9,12].

The size, comfort level, and prevalence of use in the winter months of the cottages, whose ownership is gradually transferred to new generations, is increasing [22]. In addition, global factors, e.g., the sharing economy, flexible working practices, and increased environmental awareness due to climate crises, also significantly affect the Finnish cottage culture [19]. Restrictions on coastline building and new wastewater guidelines are on the agenda among Finnish cottage owners [23]. On the other hand, although the interest in Finnish cottages still seems high, inherited traditional cottages located in Finnish forests are at risk of deterioration and are struggling to respond to changing trends and needs.

Wood, which has never lost its popularity among construction materials, has always been the most common construction material in cottage architecture [24]. Cottage buildings and newly built holiday homes are almost entirely wooden, 70% of which are log buildings (see Figure 3) [25]. Wood is considered a natural, warm, and ecological material that is compatible with the Finnish landscape [26–43]. Features such as breathability, moisture balance, allergy friendliness, and esthetics are other prominent benefits of log construction [44].

Figure 3. Log cottage example from northern Finland (Photo courtesy of Lotta Häkkänen).

Governance, development, mobility, and tourism are among the most discussed topics in the existing literature, terminologically based on the second home. To date, there are limited studies in the literature focusing on the Finnish cottage culture, which is of great social and economic significance for Finland [45].

The social aspect of the cottage phenomenon is accepted as one of the most critical dimensions and is associated with the themes of maintaining social facilities and infrastructure, increasing community welfare and social capital, and placing the commitments of second homeowners in the Finnish context [13]. Additionally, second home users are an important summer population group in various parts of Finland, creating demand for regional facilities and can be involved in community life and local politics. Considering the economic significance of the cottage culture in Finland, especially for countryside municipalities, multiple residences have a substantial effect on the regional economy and the second home provides significant business potential. In this context, second homes often make a large contribution to local economies, and this is especially significant as the population declines in countryside areas [13]. Other significant encouraging economic effects of second homes can be listed as follows: tax revenues to the local municipal economy, alleviating or rising estate prices, generating new jobs, creating new investments, and attracting new tourists. Thus, the cottage industry, which can be considered a historic initiative merging economic and cultural-political efforts in an institutional framework, is mainly defined as an economic activity that encourages an entrepreneurial and self-sufficient lifestyle in the countryside [46].

This entry focuses on the past, present, and future of the Finnish cottage, which together form an overall picture of the cottage's development trajectory. In doing so, it identifies the aspects of change that affect the architecture and usage patterns of Finnish cottages and assesses the possible prospects and their impact on cottage construction and cottage architecture at a generic level. The entry also clarifies the terminology used, e.g., second home, summer villa, in this regard. It is believed that this entry will contribute to the continuation of the cottage culture, which is critical for the vitality of rural municipalities, local development, national culture in Finland, and the well-being of Finnish people.

2. Literature Survey on the 'Second Home' Concept

As mentioned above, the second home is a common and modern notion that does not limit the use of the term to houses only occupied during holidays or a certain period of the year. 'Secondary dwelling' is progressively used as a corresponding phrase for the cottage. Second homes and secondary houses are referred to in leisure housing and media studies, but with frequently differing definitions [13,47]. The second home can be considered as a supplementary dwelling next to the primary house. On average, regarding the annual use of second homes, Scandinavian countries generally show an active pattern: they were reported to be used 75 days a year in Finland, 71 days in Sweden, and 47 days in Norway [48]. Second homes in Finland play a significant role in representing the national landscape [10], reflecting the significance of outdoor leisure and traditional activities [11]. Second homes and tourism are an established element of leisure in various countries and are the subject of scholarly studies (e.g., [15–17]).

There are a limited number of studies on the Finnish cottage culture, but there are numerous studies on the concept of a second home. Notable studies include Häkkänen et al. [49], who researched Finnish cottage culture from an experts' point of view through interviews, in which the main findings were based on key themes, including cottage buyers, characteristics of dream cottages, diversified cottages, the regulation of cottages in municipalities, and challenges in the regulation of cottages. Overvåg [50] explored the link between second homes and urban growth in the Oslo area, Norway. The study showed that second homes and urban growth are connected to some extent, but regulations by the government preclude a stronger link. In addition, Overvåg's work highlighted the impact of land scarcity and demand changes, noting that urban growth is one of the factors affecting the location of second homes owned by Oslo residents. Müller et al. [51] conducted a survey among Swedish holiday homeowners examining the relationships between vacationers and the Swedish countryside. The survey results showed that today's leisure accommodation business, do not considering settling in vacation home permanently. However, there were signs that residential use would increase, indicating that the boundaries between second homes and permanent residences might disappear after a while. Hiltunen and Rehunen [52] scrutinized the leisure-based mobile lifestyle between the urban home and rural second home in the Finnish context, mainly via GIS data and questionnaires. The results showed that social changes such as urbanization, modernization, rural restructuring, distribution, and mobility, as well as historical events and government decisions at different levels play an active role in second home tourism. Rinne et al. [53] examined second home management from the standpoint of public participation, using three group discussions focusing on three cottage-rich places in Finland. The results demonstrated that the traditional approaches of cottage owners are increasingly supported and reconfigured by heterogeneous and diverse second home users. The Finnish Environment Institute [13] analyzed how citizens and municipalities perceive the situation and development of second home tourism. While revealing that second homes play an important role in both leisure mobility and rural areas, the report underlined the views of Finnish municipalities that second homes contribute positively to regional economies.

During the COVID-19 pandemic, urbanization trends have changed, and the importance of housing conditions has increased drastically [54,55]. In the post-epidemic world, people began to prefer low-density rural areas over high-density urban centers [56]. In this context, the pandemic is an important social phenomenon influencing the meaning of 'second homes' and, in its initial stages, second homeowners thought of them as an escape from urban areas where the virus was spreading more wildly [57]. Shortly after the outbreak of the epidemic and the closure of workplaces, schools, and restaurants, people began to flee to their second homes as a better spot for isolation. Consequently, second home use has extended beyond the leisure industry and into 'protection from epidemics' to 'favored remote workplace' [58], as in the cases of Russia, Turkey, the United Kingdom, and France [59–61]. Furthermore, people focused on domestic tourism as many national borders were closed during the period of the COVID-19 pandemic. During the pandemic, second homes have become a more significant form of domestic tourism in many countries where a significant proportion of people have access to a second home, particularly in rural areas, as in the cases of the Nordic countries, Southern Europe, Russia, North America, and Australia [62]. In addition, outdoor areas, e.g., gardens and balconies, are encouraging leisure activities such as gardening, which were beneficial for well-being during the COVID-19 pandemic [45,52,63,64]. Thus, these crucial changes and new demands in daily life contribute to the importance of studies on cottage culture.

There are many studies on the COVID-19 pandemic and its impacts on second home ownership and tourism. Among the major studies, Zoğal et al. [57] examined the epistemic progression of the concept of 'second homes' and revealed possible consequences that could place second homes at the center of tourist activity. They concluded that in the initial phases of the COVID-19 pandemic, second homes were moving from high-density urban areas to low-density rural areas and a probable transformation in tourist preferences could place second homes at the center of tourism activity as soon as travel limitations were lifted, which could increase current housing commodification procedures by strengthening residential platforms. Pitkanen et al. [62] analyzed the safety of second homes during the COVID-19 pandemic in various locations, e.g., Sweden, Finland, Canada, and Russia, and found that even though the second home cultures of different locales may differ, in general, people were utilizing their second homes for escape and safety during the pandemic. Bieger et al. [65] discussed the relationship between the COVID-19 pandemic and second home prices in Switzerland, and the results indicated a considerable price increase in second homes—especially compared to flat prices—after the onset of the pandemic.

As can be understood from the literature review above, the concept of a second house, which has an important place in cottage terminology, is a research topic that has increased in importance and popularity in Finland and many other countries, especially with the pandemic. The next section will discuss the development, current status, and future of this important topic in the Finnish context.

3. Cottage Culture in Finland: Past, Present, and Future

Despite strong traditions, Finnish cottage culture is a relatively new phenomenon. Its development from past to present is summarized below [49,66]. The summer villa trend, which started in Finland in the 19th century, in big cities such as Helsinki and Turku, was only for the wealthy bourgeoisie. The number of villas, which was 1000 at the beginning of the 20th century, exceeded 3000 in the next two decades. These villas, which were the product of the urban culture of the past, were large and ornamental structures that only urban families bought, mostly used in their spare time during the summer months.

Towards the end of the 1910s, that is, after Finland gained its independence (1917), the effort towards modesty and practicality in the villas built in that period was striking. In the 1920s, cities' populations doubled and the search for summer residences expanded even more. With the widespread use of buses and private vehicle transportation, distances were no longer an obstacle to the purchase of villas, thanks to the rapidly developing transportation. By the late 1930s, the number of villas in Finland quadrupled to more than 20,000 by 1940. Extravagant decoration declined, and villas began to imitate detached houses in suburban areas, where gardening was an important part of life. Smaller weekend cottages were also being built in greater numbers than before.

By the 1950s, spending time in the summer cottage became a nationwide activity that could be experienced by the entire Finnish population. In those years, many people born in rural areas migrated to cities to work mostly in industry, which accelerated the increase in the number of summer cottages with urbanization. Simple summer cottages and weekend cottages began to become more and more common, and cottage ownership spread rapidly in rural municipalities. The size and equipment standard of the villas lost their importance and emphasis on access to the open air and nature allowed for more modest cottages.

However, in the 1960s, the opposite trend developed, namely summer cottages that were spacious, well-equipped, and could replace a second home for the family to use throughout the year. In those years, the time spent at the summer cottage increased at the weekends, as Saturday was a day off, and the number of villas exceeded 100,000, with one out of every 13 households owning a summer cottage.

In the following years, the number of summer houses continued to increase, and growth peaked in the 1980s, with more than 100,000 new cottages constructed over ten years. Cottage construction has slowed since then, only increasing 23 percent in the 1990s, and at the end of 2020, the number of cottages reached nearly 510,000 as shown in Figure 4 [67]. Additionally, cottage distribution on the basis of municipalities is indicated in Figure 5 [68].

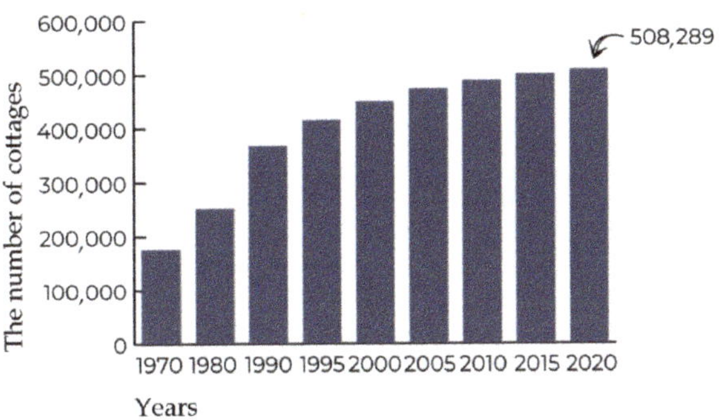

Figure 4. The number of cottages in Finland (1970–2020).

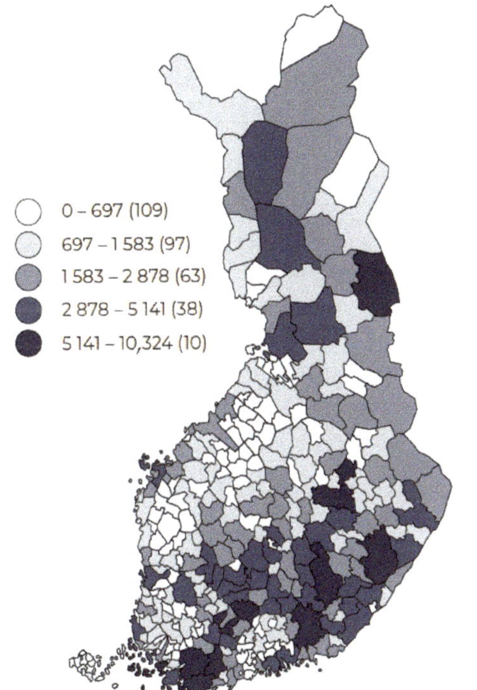

Figure 5. Distribution of cottages by municipalities in Finland.

Today, about fifty percent of the Finnish population regularly visit a cottage [3], and summer cottage owners are only one part of this group. Grandchildren and children often use their parents' cottage, and about three million Finns travel to these homes at least once a year [68]. Even though cottage preferences vary according to socioeconomic status and income level [47], Finnish cottage culture remains widespread throughout the country. Given the demands of key stakeholder groups, particularly contemporary consumers, who play a crucial role in understanding the current state of Finnish cottage culture, the following stand out, as stated in Häkkänen et al. [49]: (i) convenience, ready-made solutions, easy maintenance, the high level of equipment, year-round use, location, and modern and simple architectural style were among the main selection criteria in cottages, which were generally preferred by a wealthy buyer profile over 50 years old; (ii) dryland cottages were also evaluated as a possible alternative to waterfront cottages, due to their advantages such as ease of access to services; (iii) environmental issues were among the topics that buyers were interested in; (iv) when certain requirements were met, municipalities with different approaches and procedures to the cottages showed positive attitudes towards coastal development, regarding recreation areas where building rights and wastewater treatment come to the fore; (v) there were some regulatory barriers to meeting user needs, e.g., watered toilets in cottages that could be converted into permanent residences under certain conditions.

Considering the above-mentioned current demands of cottage buyers, modern, comprehensively equipped holiday homes designed for year-round-use are likely to become an increasingly important part of the future. Due to the limited selection of affordable and suitable waterfront properties, the many inland countryside houses may become an alternative for future cottages. It is thought that the younger generations especially will be more prone to such second homes, which are different from the traditional lakeside cottages. Additionally, rental cottages and other less binding forms of holiday homes will likely continue to gain popularity. The cottage will likely be further modernized in the future. However, there are also some ways that cottages which are too traditional could be changed. Identifying these and combining them with the demands and requirements of modern cottage owners is likely to increase the popularity of cottages.

On the other hand, it can be predicted that the increasing demand for detached houses and outdoor use, which has developed as a reflection of the COVID-19 pandemic, will further expand the summer cottage lifestyle. The serious price increase in second homes in Finland after the epidemic also supports this expectation. Remote work, spending more time in a second home, and therefore the increasing demand for houses with larger gardens to be more intertwined with nature, will significantly affect the Finnish cottage market, which will push the demand for year-round-use cottages to a much higher level than before the pandemic.

Table 1 summarizes the cottage culture in Finland from past, present, and future perspectives.

Table 1. The evolution of Finnish cottage culture.

The Evolution of the Finnish Cottage Culture		
Period	**Terminology (Mostly) Used**	**Featured Notes**
The 1800s	Summer villa	Villas, large and decorative buildings designed for the wealthy bourgeoisie, for leisure and summer use, began to grow around big cities.
1900–1920	Summer villa	The number of villas, which was 1000 in the early 1900s, increased to over 3000 in the next two decades. After the independence of Finland, modesty and practicality came to the fore in villa designs.
The 1920s	Summer villa/residence	Thanks to rapidly developing transportation conditions, distances were no longer an obstacle to the purchase of villas.
1930s & 1940s	Summer villa/residence & (weekend) cottage	By the late 1930s, the number of villas in Finland quadrupled to more than 20,000 by 1940. Extravagant decoration became even less popular, and villas began to imitate suburban detached houses. Demand for smaller weekend cottages increased.

Table 1. *Cont.*

The Evolution of the Finnish Cottage Culture		
Period	**Terminology (Mostly) Used**	**Featured Notes**
The 1950s	Summer/ weekend cottage	Spending time in summer cottages became a nationwide activity, and with increased urbanization, the number of cottages increased rapidly. The size and equipment level of the villas lost their importance.
The 1960s	(Summer) cottage	The cottages, whose number exceeded 100,000, were starting to turn into year-round-use second homes that were spacious and well equipped.
1970s & 1980s	(Summer) cottage	The number of cottages continued to increase, and growth peaked in the 1980s, with more than 100,000 new ones built within a decade.
The 1990s	(Summer) cottage	Cottage construction slowed, increasing only 23% in the 1990s.
The 2000s	(Summer) cottage	At the end of 2020, the number of cottages exceeded half a million.
Today	(Summer) cottage	Equipment level, year-round use, and location are the main selection criteria for cottages, which are often bought by people over 50 with a wealthy profile. Municipal regulations and environmental concerns are other important issues.
Future	(Summer) cottage	Demand for the well-equipped year-round-use cottage will increase. Inland cottages will be an alternative to lakeside properties. Less binding types of cottages will continue to gain popularity. The epidemic will further expand the summer cottage lifestyle.

4. Conclusions

This entry concentrated on the past, present, and future of the Finnish cottage, which together form an overall picture of the cottage's development trajectory. In doing so, the entry identified the aspects of change that affect the architecture and usage patterns of Finnish cottages and assessed their possible prospects and impact on cottage construction and cottage architecture at a generic level. It also clarified the terminology used, e.g., second home, summer villa, in this regard. The main purpose of this entry was to contribute to the continuation of the Finnish cottage culture, which has undergone social, cultural, and architectural changes throughout its transformation from summer villas used only by the elite in the 1800s to summer cottages used by all Finns today.

Although cottage life is becoming less and less like the traditional concept of spending long summer weeks in a modest cottage, it still holds a special place in the hearts of the Finnish people. Finns' interest in cottages does not seem to be waning, but the ways of implementing it are changing. The ease of solutions and maintenance of the cottage, as well as the flexibility of use, are important for buyers. In addition, the number of contemporary, well-equipped summer cottages suitable for year-round-use will continue to increase in Finland. The contrast between permanent urban residences and holiday homes remains the main reason Finns flee to cottages.

The generational change of cottage owners, tightening of building regulations, and global megatrends, such as growing environmental awareness, are key drivers of the future transformation of Finnish cottage culture. Additionally, as one of the consequences of the COVID-19 pandemic, increasing demand for cottages for single-family and outdoor use, created by changing priorities, such as remote working and spending more time in a second home, will make the summer cottage lifestyle even more popular in Finland. To preserve the cottage phenomenon, which is culturally and economically important to Finland, sustainable solutions must be developed to meet the wishes and needs of modern cottage owners and regulatory requirements.

Author Contributions: Conceptualization, L.H., H.E.I. and M.K.; methodology, L.H., H.E.I. and M.K.; software, L.H. and H.E.I.; formal analysis, L.H., H.E.I. and M.K.; investigation, L.H., H.E.I. and M.K.; data curation, L.H., H.E.I. and M.K.; writing—original draft preparation, H.E.I.; writing—review and editing, L.H., H.E.I. and M.K.; visualization, L.H.; supervision, M.K. and H.E.I.; project administration, M.K. All authors have read and agreed to the published version of the manuscript.

Funding: This research received no external funding.

Institutional Review Board Statement: Not applicable.

Informed Consent Statement: Not applicable.

Data Availability Statement: Not applicable.

Conflicts of Interest: The authors declare no conflict of interest.

References

1. Coppock, J.T. (Ed.) *Second Homes: Curse or Blessing?* Pergamon: Oxford, UK, 1977.
2. Eriksen, T.H. Second Homes in the Nordics. Available online: https://nordics.info/show/artikel/second-homes-in-the-nordics#:~:text=Second%20homes%20are%20used%20as,and%20an%20abundance%20of%20space (accessed on 10 March 2022).
3. Müller, D.K. Second Homes in the Nordic Countries: Between Common Heritage and Exclusive Commodity. *Scand. J. Hosp. Tour.* **2007**, *7*, 193–201. [CrossRef]
4. Müller, D.K. Reinventing the countryside: German second-home owners in Southern Sweden. *Curr. Issues Tour.* **2002**, *5*, 426–446. [CrossRef]
5. Hall, C.M.; Müller, D.K.; Saarinen, J. *Nordic Tourism*; Issues and Cases; Channel View Publications: Bristol, UK, 2009.
6. Reijo, M. Suomalaisilla käytössä kakkosasuntoja useimmin EU: N alueella. *Hyvinvointikatsaus* **2002**, *2*, 20–23. (In Finnish)
7. Jetsonen, J.; Jetsonen, S. *Finnish Summer Houses*; Princeton Architectural Press: New York, NY, USA, 2008.
8. Pitkänen, K.; Kokki, R. Publications of the Savonlinna Institute for Regional Development and Research No. 11, Non-Faculty Institutes, Savonlinna Centre for Continuing Education and Regional Development, University of Joensuu Publications. 2005. Available online: https://erepo.uef.fi/bitstream/handle/123456789/8449/urn_isbn_952-458-701-7.pdf?sequence=1&isAllowed=y (accessed on 10 March 2022). (In Finnish)
9. Concepts and Definitions, Statistics Finland. Available online: https://www.stat.fi/til/rakke/kas_en.html (accessed on 10 March 2022).
10. Vepsäläinen, M.; Pitkänen, K. Second home countryside: Representations of the rural in Finnish popular discourses. *J. Rural. Stud.* **2010**, *26*, 194–204. [CrossRef]
11. Hiltunen, M.J.; Pitkänen, K.; Vepsäläinen, M.; Hall, C.M. Second home tourism in Finland: Current trends and eco-social impacts. In *Second Home Tourism in Europe: Lifestyle Issues and Policy Responses*; Roca, Z., Ed.; Ashgate: Farnham, UK, 2013; pp. 165–198.
12. Pitkänen, K.; Puhakka, R.; Semi, J.; Hall, C.M. Generation Y and second homes: Continuity and change in Finnish outdoor recreation. *Tour. Rev. Int.* **2014**, *18*, 207–221. [CrossRef]
13. Adamiak, C.; Vepsäläinen, M.; Strandell, A.; Hiltunen, M.J.; Pitkänen, K.; Hall, C.H.; Rinne, J.; Hannonen, O.; Paloniemi, R.; Åkerlund, U. *Second Home Tourism in Finland Perceptions of Citizens and Municipalities on the State and Development of Second Home Tourism*; Finnish Environment Institute Publications: Helsinki, Finland, 2015.
14. Hiltunen, M.J. Environmental impacts of rural second home tourism–Case Lake district in Finland. *Scand. J. Hosp. Tour.* **2007**, *7*, 243–265. [CrossRef]
15. Casado-Diaz, M.A. Socio-demographic impacts of residential tourism: A case study of Torrevieja, Spain. *Int. J. Tour. Res.* **1999**, *1*, 223–237. [CrossRef]
16. Mottiar, Z.; Quinn, B. Shaping leisure/tourism places–The role of holiday homeowners: A case study of Courtown, Co. Wexford, Ireland. *Leis. Stud.* **2003**, *22*, 109–127. [CrossRef]
17. Hall, C.M.; Müller, D.K. (Eds.) *Tourism, Mobility and Second Homes: Between Elite Landscape and Common Ground*; Channel View Publications: Clevedon, UK, 2004.
18. McIntyre, N.; Williams, D.R.; McHugh, K.E. (Eds.) *Multiple Dwelling and Tourism: Negotiating Place, Home and Identity*; Cabi: Wallingford, UK, 2006.
19. Marjavaara, R. *Second Home Tourism: The Root to Displacement in Sweden. Gerum 2008: 1*; Department of Social and Economic Geography, Umeå University: Umeå, Sweden, 2008.
20. Karisto, A. Kesämökki ja arjen ympäristöpolitiikka. In *Arkielämän Ympäristöpolitiikka*; Massa, I., Ahonen, S., Eds.; Gaudeamus: Helsinki, Finland, 2006; pp. 122–137.
21. OSF. Official Statistics of Finland. Kotitalouksien Käytettävissä Olevat Rahatulot Vuonna 2011. 2013. Available online: http://www.stat.fi/til/tjt/2011/03/tjt_2011_03_2013-04-10_kat_001_fi.html>28.6.2013 (accessed on 10 March 2022).
22. Free-Time Residences 2018, Statistics Finland. Available online: https://www.stat.fi/til/rakke/2018/rakke_2018_2019-05-21_kat_001_en.html (accessed on 10 March 2022).

23. *Finnish Industrial Wastewater Guide, Conveying Non-Domestic Wastewater to Sewers Publication Series No. 69*; The Finnish Water Utilities Association: Helsinki, Finland, 2018; Available online: https://www.vvy.fi/site/assets/files/1110/finnish_industrial_wastewater_guide.pdf (accessed on 10 March 2022).

24. Laapotti, S. Wood Is a Sustainable Building Material–But There Is a Money Problem to Solve, Unit Magazine, Tampere University. Available online: https://www.tuni.fi/unit-magazine/en/articles/wood-sustainable-building-material-there-money-problem-solve (accessed on 10 March 2022).

25. The Finnish Timber Council (Puuinfo). Available online: http://www.puuinfo.fi/puutieto (accessed on 10 March 2022).

26. Karjalainen, M.; Ilgın, H.E. The Change over Time in Finnish Residents' Attitudes towards Multi–Story Timber Apartment Buildings. *Sustainability* **2021**, *13*, 5501. [CrossRef]

27. Karjalainen, M.; Ilgın, H.E.; Metsäranta, L.; Norvasuo, M. Suburban Residents' Preferences for Livable Residential Area in Finland. *Sustainability* **2021**, *13*, 11841. [CrossRef]

28. Ilgın, H.E.; Karjalainen, M. *Perceptions, Attitudes, and Interest of Architects in the Use of Engineered Wood Products for Construction: A Review*; IntechOpen: London, UK, 2021.

29. Karjalainen, M.; Ilgın, H.E.; Metsäranta, L.; Norvasuo, M. *Wooden Facade Renovation and Additional Floor Construction for Suburban Development in Finland*; IntechOpen: London, UK, 2022.

30. Karjalainen, M.; Ilgın, H.E.; Metsäranta, L.; Norvasuo, M. Residents' Attitudes towards Wooden Facade Renovation and Additional Floor Construction in Finland. *Int. J. Environ. Res. Public Health* **2021**, *18*, 12316. [CrossRef]

31. Rinne, R.; Ilgın, H.E.; Karjalainen, M. Comparative Study on Life-Cycle Assessment and Carbon Footprint of Hybrid, Concrete and Timber Apartment Buildings in Finland. *Int. J. Environ. Res. Public Health* **2022**, *19*, 774. [CrossRef] [PubMed]

32. Ilgın, H.E.; Karjalainen, M. Preliminary Design Proposals for Dovetail Wood Board Elements in Multi-Story Building Construction. *Architecture* **2021**, *1*, 56–68. [CrossRef]

33. Karjalainen, M.; Ilgın, H.E.; Somelar, D. Wooden Additional Floors in old Apartment Buildings: Perspectives of Housing and Real Estate Companies from Finland. *Buildings* **2021**, *11*, 316. [CrossRef]

34. Ilgın, H.E.; Karjalainen, M.; Koponen, O. *Review of the Current State-of-the-Art of Dovetail Massive Wood Elements*; IntechOpen: London, UK, 2021.

35. Karjalainen, M.; Ilgın, H.E.; Tulonen, L. Main Design Considerations and Prospects of Contemporary Tall Timber Apartment Buildings: Views of Key Professionals from Finland. *Sustainability* **2021**, *13*, 6593. [CrossRef]

36. Tulonen, L.; Karjalainen, M.; Ilgın, H.E. *Tall Wooden Residential Buildings in Finland: What Are the Key Factors for Design and Implementation?* IntechOpen: London, UK, 2021.

37. Karjalainen, M.; Ilgın, H.E.; Yli-Äyhö, M.; Soikkeli, A. *Complementary Building Concept: Wooden Apartment Building: The Noppa toward Zero Energy Building Approach*; IntechOpen: London, UK, 2021.

38. Ilgın, H.E.; Karjalainen, M.; Koponen, O. *Various Geometric Configuration Proposals for Dovetail Wooden Horizontal Structural Members in Multistory Building Construction*; IntechOpen: London, UK, 2022.

39. Karjalainen, M.; Ilgın, H.E.; Somelar, D. *Wooden Extra Stories in Concrete Block of Flats in Finland as an Ecologically Sensitive Engineering Solution, Ecological Engineering–Addressing Climate Challenges and Risks*; IntechOpen: London, UK, 2021.

40. Ilgın, H.E.; Karjalainen, M.; Koponen, O. Dovetail Massive Wood Board Elements for Multi-Story Buildings. In Proceedings of the LIVENARCH VII Livable Environments & Architecture 7th International Congress OTHER ARCHITECT/URE(S), Trabzon, Turkey, 28–30 September 2021; Volume I, pp. 47–60.

41. Karjalainen, M.; Ilgın, H.E. A Statistical Study on Multi-story Timber Residential Buildings (1995–2020) in Finland. In Proceedings of the LIVENARCH VII Livable Environments & Architecture 7th International Congress OTHER ARCHITECT/URE(S), Trabzon, Turkey, 28–30 September 2021; Volume I, pp. 82–94.

42. Ilgın, H.E.; Karjalainen, M.; Koponen, O.; Soikkeli, A. *A Study on Contractors' Perception of Using Wood for Construction*; IntechOpen: London, UK, 2022.

43. Soikkeli, A.; Ilgın, H.E.; Karjalainen, M. Wooden Additional Floor in Finland. *Encyclopedia* **2022**, *2*, 578–592. [CrossRef]

44. Polina, I. The Designing, Construction, and Maintenance of Honka Log Houses. Guidelines for Russian Customers. Master's Thesis, Lappeenranta Degree Program in Civil and Construction Engineering, Saimaa University of Applied Sciences Technology, South Karelia, Finland, 2010.

45. Poortinga, W.; Bird, N.; Hallingberg, B.; Phillips, R.; Williams, D. The role of perceived public and private green space in subjective health and wellbeing during and after the first peak of the COVID-19 outbreak. *Landsc. Urban Plan.* **2021**, *211*, 104092. [CrossRef]

46. Kraatari, E. Finnish Cottage Industry and Cultural Policy: A Historical View. *Nord. Kult. Tidskr.* **2013**, *16*, 137–152. [CrossRef]

47. Aho, S.; Ilola, H. *Another Home in the Country? Second Housing and Rural Vitality*; Faculty of Economics and Tourism, University of Lapland Publications: Rovaniemi, Finland, 2006. (In Finnish)

48. Lithander, J.; Tynelius, U.; Malmsten, P.; Råboc, I.; Fransson, E. Rural Housing–Landsbygdsboende i Norge, Sverige och Finland. Tillväxtanalys, Rapport 05. Östersund. Nordic Council of Ministers. 2012. Available online: http://www.tillvaxtanalys.se/download/18.56ef093c139bf3ef89029cb/1349864058705/Rapport_2012_05.pdf (accessed on 10 March 2022).

49. Häkkänen, L.; Ilgın, H.E.; Karjalainen, M. The Current State of the Finnish Cottage Phenomenon: Perspectives of Experts. *Buildings* **2022**, *12*, 260. [CrossRef]

50. Overvåg, K. Second homes and urban growth in the Oslo area, Norway. *Nor. Geogr. Tidskr. Nor. J. Geogr.* **2009**, *63*, 154–165. [CrossRef]

51. Müller, D.K.; Nordin, U.; Marjavaara, R. *Fritidsboendes Relationer Till den Svenska Landsbygden. Resultat av en Enkät Bland Svenska Fritidshusägare 2009*; GERUM Kulturgeografisk Arbetsrapport 2010-04-15; Kulturgeografiska Institutionen, Umeå Universitet: Umeå, Sweden, 2010.
52. Hiltunen, M.J.; Rehunen, A. Second home mobility in Finland: Patterns, practices and relations of leisure oriented mobile lifestyle. *Fennia* **2014**, *192*, 1–22. [CrossRef]
53. Rinne, J.; Paloniemi, R.; Tuulentie, S.; Kietäväinen, A. Participation of second-home users in local planning and decision-making—A study of three cottage-rich locations in Finland. *J. Policy Res. Tour. Leis. Events* **2015**, *7*, 98–114. [CrossRef]
54. Mouratidis, K. How COVID-19 reshaped quality of life in cities: A synthesis and implications for urban planning. *Land Use Policy* **2021**, *111*, 105772. [CrossRef] [PubMed]
55. Ilgın, H.E.; Karjalainen, M.; Pelsmakers, S. Finnish architects' attitudes towards multi–storey timber–Residential buildings. *Int. J. Build. Pathol. Adapt.* **2021**, *in press*. [CrossRef]
56. Batty, M. The Coronavirus crisis: What will the post-pandemic city look like? *Environ. Plan. B Urban Anal. City Sci.* **2020**, *47*, 547–552. [CrossRef]
57. Zoğal, V.; Domènech, A.; Emekli, G. Stay at (which) home: Second homes during and after the COVID-19 pandemic. *J. Tour. Futures* **2020**, in press. [CrossRef]
58. Gallent, N. Covid-19 and the flight to second homes. *Town Ctry. Plan.* **2020**, *89*, 141–144.
59. Grigoryev, L.M. Global social drama of pandemic and recession. *Popul. Econ.* **2020**, *4*, 18–25. [CrossRef]
60. Nikolaeva, U.G.; Rusanov, A.V. Self-isolation at the dacha: Can't? Can? Have to? *Popul. Econ.* **2020**, *4*, 182–198. [CrossRef]
61. Seraphin, H.; Dosquet, F. Mountain tourism and second home tourism as post COVID- 19 lockdown placebo? *Worldw. Hosp. Tour. Themes* **2020**, *12*, 485–500. [CrossRef]
62. Pitkänen, K.; Hannonen, O.; Toso, S.; Gallent, N.; Hamiduddin, I.; Halseth, G.; Hall, C.M.; Müller, D.K.; Treivish, A.; Nevedova, T. Second homes during Corona—Safe or unsafe haven and for whom? Reflections from researchers around the world. *Finn. J. Tour. Res.* **2020**, *16*, 20–39. [CrossRef]
63. Corley, J.; Okely, J.A.; Taylor, A.M.; Page, D.; Welstead, M.; Skarabela, B.; Russ, T.C. Home garden use during COVID-19: Associations with physical and mental wellbeing in older adults. *J. Environ. Psychol.* **2021**, *73*, 101545. [CrossRef]
64. Lehberger, M.; Kleih, A.-K.; Sparke, K. Self-reported well-being and the importance of green spaces—A comparison of garden owners and non-garden owners in times of COVID-19. *Landsc. Urban Plan.* **2021**, *212*, 104108. [CrossRef]
65. Bieger, T.; Weinert, R.; Klumbies, A. COVID-19 and Second Home Prices in Switzerland: An Empirical Insight. *Z. Für Tour.* **2021**, *13*, 375–386. [CrossRef]
66. From Villa Ownership to National Leisure-Time Activity, Statistics Finland. Available online: https://www.stat.fi/tup/suomi90/kesakuu_en.html (accessed on 10 March 2022).
67. Free-Time Residences 2020, Statistics Finland. Available online: https://www.stat.fi/til/rakke/2020/rakke_2020_2021-05-27_kat_001_en.html (accessed on 10 March 2022).
68. FCG Finnish Consulting Group Oy. Advisory Board for Finnish Archipelago Affairs Ministry of Agriculture and Forestry, Cottage Barometer. 2016. Available online: https://mmm.fi/documents/1410837/1880296/Mokkibarometri+2016/7b69ab48-5859-4b55-8dc2-5514cdfa6000 (accessed on 10 March 2022). (In Finnish)

Entry

Bagnoli Urban Regeneration through Phytoremediation

Clelia Cirillo, Barbara Bertoli, Giovanna Acampora and Loredana Marcolongo *

Research Institute on Terrestrial Ecosystem, National Research Council of Italy, 80131 Naples, Italy; cirilloclelia@gmail.com (C.C.); barbara.bertoli@cnr.it (B.B.); giovanna.acampora@cnr.it (G.A.)
* Correspondence: loredana.marcolongo@cnr.it

Definition: The Bagnolidistrict in Naples has needed urban redevelopmentfor many years. The area is not only affected by pollution caused by many industries but also by environmental pollutants, according togeognostic surveys that have found numerous contaminantsin the subsoil and water.Currently, the combination of an urban rehabilitation processwith the phytodepuration technique may represent a successful idea for obtaining bothurban regenerationand environmental remediation. Phytoremediation, a biologically based technology, has attracted the attention of both thepublic and scientists as a low-cost alternative for soil requalification. The use of plants as well as the microorganisms present in their root systems plays an important role in the ecological engineering field in controlling and reducing pollutants present in theair, water and soil.The result is efficient, sustainable and cost-effective environmental recovery compared to conventional chemical–physical techniques. In this way, not only the environmental recovery of SIN Bagnoli-Corogliocan be obtained, but also the regeneration of its landscape.

Keywords: urban regeneration; phytoremediation; Bagnoli former area

Citation: Cirillo, C.; Bertoli, B.; Acampora, G.; Marcolongo, L. Bagnoli Urban Regeneration through Phytoremediation. *Encyclopedia* **2022**, *2*, 882–892. https://doi.org/10.3390/ encyclopedia2020058

Academic Editors: Masa Noguchi, Antonio Frattari, Carlos Torres Formoso, Haşim Altan, John Odhiambo Onyango, Jun-Tae Kim, Kheira Anissa Tabet Aoul, Mehdi Amirkhani, Sara Jane Wilkinson, Shaila Bantanur and Raffaele Barretta

Received: 22 February 2022
Accepted: 17 April 2022
Published: 24 April 2022

Publisher's Note: MDPI stays neutral with regard to jurisdictional claims in published maps and institutional affiliations.

1. Introduction

The alteration of the natural balance of thecoastal landscape of theBagnoli urban district in Naples, caused by urbanization but even more so by industrialization processes (Ilva, Eternit, Cementir and Federconsorzi), will be mitigated by an urban regeneration plan. The Bagnoli district, located along the highly urbanized coast of the Gulf of Pozzuoli, is included in the volcanic system of CampiFlegrei. In 1905,the construction of the Ilva plant (flat rolled and similar industrial products) marked the conversion of Bagnoli-Coroglio from a residential to an industrial center, which lasted until the end of 1990 when the first phase of closure of the industry occurred. In 2018, the Environmental and Urban Regeneration Plan was developed [1], which recognizes Bagnoli/Coroglio as aNational Interest Site (SIN) and was a variant of the urban planning instruments already in force. In order to plan the environmental rehabilitation and urban regeneration of the site, an international competition of ideas was organized by INVITALIA (National Agency for Inward Investment and Economic Development), the implementing body of the program, which is owned by the Italian Ministry of Economy. The projectproposes the realization of integrative environmental characterization, the restoration of bathing facilities, waterfront requalification and reconnection with the city. The decommissioning and reclamation of the steel plants began in 1994, following the decision of the Interministerial Committee for Economic Planning and Sustainable Development.The Italian Governmentfunded the remediation plans viaspecific laws (N. 582—18 November 1996 and N.388—23 December 2000). The project proposed the utilization of theIlvaand Eternit sites for nonindustrialactivities. At theFederconsorzi site, the "Cityof Science" was built, an institute for the dissemination and enhancement of scientific culture(IDIS foundation). INVITALIAhas announced the completion of the asbestos remediation at the former Eternit site. The plan provided forthe carrying out of the environmental recovery of disused industrial sites by dismantling them and creating a park characterized by botanical species suitable for aiding the reduction in environmental pollution [1]. Afforestation will form part of theenvironmental mitigation and phytoremediation measures intended to respond to the need to reduce the pollution caused by anthropogenic pressures [2]. In this regard, it is crucial to design green areasfor the Bagnoli districtwith theintroductionof botanical species selected especiallyforsoil depollution. In fact, in the sector in front of the industrial plant, which is now decommissioned, high concentrations of polycyclic aromatic hydrocarbons (PAHs), heavy metals, such as arsenic and beryllium, and trace elements, such as lead, zinc, cadmium, copper and mercury, have been found, even at deep levels [3].

2. The Bagnoli Plain in Historical Cartography:between Reality and Utopia

The evolution of the coastal landscape in the western area of Naples can be analyzed using historical cartography to understand howanthropization has modified the coast over two centuries. The historical cartography of thePhlegraean area is scarce. To obtain cartographic information, we must refer to the most complete maps of the city of Naples, whose landscape outlines [4] continue towardsCampiFlegrei and its most famous areas, such as Nisida, Agnano and Astroni. These representations show an uninhabited and marshy agricultural area, with few architectural additions. The circular tower of Nisida, represented in the cartographies of the 17th century, stands out in the well-known representation of De Fer, "Les Merveilles de Pozzuoli", from 1701 [5]. In the map of Giovanni Carafa Duke of Noja (Figure 1) [6], Bagnoli is characterized by its seashore and by thecultivated areas of Coroglio. The road system that would influence the evolution of the 19th century road network is also clearly discernible, based on two lines that connect Pozzuoli with Naples:"the road that leads to the city of Pozzuoli to the marina" and "the road from Pozzuoli to the mountain".

Figure 1. Topographical map of the city of Naples and its outlines (Giovanni Carafa Duke of Noja, Naples, 1775) [6].

To observe the evolution during the 19th century, an important instrument is offered by the Kingdom of Naples cartography (from 1817 to 1823), where the topography of the plain shows a green countryside separated from the sea by a coastline characterized by hydrothermal springs. In the pre-unification map, a long straight line represents a road thatbegan the incorporation of the Phlegraean coast into thecity area, extending the line of urban development towards the west.

The Bagnoli area assumed the role of an entry point to the CampiFlegrei [7]. In the 19th century, the cartography showed a shooting range and a sandy coast.In this period, factories for chemical products and the well-known Lefevre glassworks were built, whose buildings now house the "City of Science" scientific center [8]. InTopographical Engineerscartography, there is no trace ofa residential settlement whose construction [9] began at the end of 1800 at the hands of the Marquis Candido Giusso, making Bagnoli a residential area suitable for vacation. Between 1883 and 1888, a valid urban planning idea for the Bagnoli area was proposed by Lamont Young [10], which exhibited the drawings of his futuristic project

(Figure 2). The utopia of the Anglo-Neapolitan architect will be overturned by a reality that has decreed the environmental destruction of one of the most evocative places in Naples.

Figure 2. Lamont Young Project (1883–1888). GIS cartography realized by at the Laboratory on the Landscape of the CNR/ IRET implementing orto-imagine 2007 of the Province of Naples, Extract n° 8 of the Cartography of 1817–1823. Graphical elaboration by Marina Russo.

3. Brief History of Bagnoli, a District with Great Disregarded Potential

When in 1853 the Bournique and Damiani glassworks and the sulfuric acid, alum and iron sulphate factory of Lefevre were established near the beach of Coroglio, at the foot of the Posillipo hill, the whole area still preserved its fascinatingpotential. Today what remains of the coast tells a reality very far from one of peace and healthiness. The Italsider plant in Bagnoli represents the emblem of one of the largest urban voids in Europe, a symbol of environmental degradation caused by development policies implemented without taking into account the environmental and landscape impacts. The 19th century photographs by Alinari and Sommers [11] (Figure 3a) portrayed an agricultural territory that reached almost to the coastline. The rediscovery of the first thermal spring in 1827 (TermeMasullo) led to the progressive rebirth of the thermal culture, promoting the recovery of several spas with tourist vocations. Manganella (1831), Cotroneo (1831), Rocco (1850) and Tricarico (1882) beaches, born as an extension of the thermal bath, played a significant role in the development of the local economy. As documented by the postcards of the time, since the 1920s, Bagnoli has been a holiday destination full of tourist facilities, hotels and restaurants. Subsequently, the urbanization growth and the opening of new industrial sites sanctioned the definitive closure of the thermal complexes. The first significant urbanization in Bagnoli occurred at the end of 19th century (1880–1885) with the Marquis Giusso's plan, which foreshadowed the design of an area characterized by cottages and spas [12,13].

(a)   (b)

Figure 3. (a) Nisida and lazzaretto—Bagnoli (G. Sommers) "Isle of Nisida"(Catalogue #: 1191) [11]; (b) Ilva 1905: Tanks and pier [14].

The Marquis made a first parceling, inspired by coeval European experiments of settlements with tourist and residential purposes. In 1914, Giusso organized a town planestablishingthe building parameters, and the plan maintained its effectiveness until the last post-war reconstruction.The city plan allowed the construction of the municipal roads and decreed compliance with the legislation to preserve "a dignified appearance" for the Bagnoli district [13].

Launched under Saredo inquiry (1902), Francesco Saverio Nittiapplied the first special law (1904) for the industrialization of Naples, glimpsing the potential ofthe steel industry for the development of the local economy. The Ilva company [15], established in Genoa in 1905, had acquired the area behind the beach of Coroglio at a low cost, building in 1906 the first Italian continuous cycle plant [14] (Figure 3b), directed by a German technician teams led by the engineer FrizLührmann [16].

After the Second World War, Ilva began to expand the facilities, advancingalong the coastline. The landscape and environmental damage was being inexorably consumed [13]. In 1961, the industry, with state participation, assumed the name of Italsider, implementing the production of steel that contributed to the Italian "economic miracle" [17].

However, from the early 1970s onwards, the economic situation rapidly worsened.The economic crisis was caused by problems with the supply of the raw materials and by the increase in energy costs.After numerous restructuring hypotheses, in the second half of the 1970s, a world steel crisis led toa progressive closure of the steel mills. Theplants' closures met with opposition from the workers, who began a long period of clashes and demands against the management [13].

On 20 October 1990, with the last casting, the hot area ofthe Bagnoli steel center was finally switched off [18]. The dismantling of the Italsider plant was placed at the center ofcultural and political debate surrounding the problems related to the recovery of abandoned areas, seeking to find abalance between conservation and demolition. In 1993, the symbol of the waterfront rebirth of Coroglio was the "City of Science". The transformation andredevelopment of the Naples industrial areas constituted the guidelines of the extensive process of revision of the General Town Plan, begun by the urban manager De Lucia with the administration of Mayor Bassolino. Naples disused industrial areas were subject to the guidelines ofthe Territorial Pact of CampiFlegrei. The variant for the western area of Naples, which includes Bagnoli, covers a large area of about 1300 hectares [19].

Within this complex territorial system, the decommissionedareas represent about 330 hectares, in a position of great value for the surrounding landscape: Coroglio overlooking the Gulf of Pozzuoli between the hilly edge of Posillipo and the Bagnoli urban fabric of the 20th century. The strategic plan provided for the creation of a "green city", replacing chimneys, rolling mills and sheds, with a large city park stretching out to the sea, where the construction ofa beach reopened to tourismwas planned. The strong signal that the Municipality of Naples intended to send to its citizens was to restore the Bagnoli area to its condition prior to indiscriminate industrialization, breaking any kind of compromise with the past [1].

With the approval of the variants, this purpose was gradually blunted and the will to preserve a part of the numerous industrial buildings, as evidence of industrial archeology, prevailed (Figure 4). The most significant projects are aimed at enhancing the industrial components in order to preserve their identity, and at the creation of a technology park with services for leisure, sporting and cultural activities [1]. In recent years, the theme of the reconversion and redevelopment of abandoned industrial areas has played an important role inprocesses of transformation and urban regeneration that stimulates a new season of reflection.

Figure 4. View of the urban void left by the decommissioning of steel plants.

4. The Environmental Recovery of Bagnoli and Its Coast Regeneration

In the Bagnoli area, there has been an overlap of pollution of chemical elements (e.g., arsenic) transported by hydrothermal fluids characteristic of the Gulf of Pozzuoli, linked to industrial activities that have released metals and fossil fuels into the environment.

The volcanic activity is still present today, as evidenced behind the former Cementir plant, by the thermal water that flows from the ground.

As previously highlighted, the Bagnoliurban district is located close to the active volcanic caldera of the Campi Flegrei. Despite its tourist vocation, during the twentieth century numerous and different industrial complexes were established in the area, such as the well-known Italsider steel mill, cement factories (Cementir) and companies producing fertilizers (Federconsorzi) and asbestos (Eternit).

In this context, about 40% of the samples, taken from surface areas of Bagnoli up to about 20 years after the dismantling of the industrial plants, showed a moderate to high degree of contamination. The sediments recently deposited are still influenced by contributions of contaminants released by the Italsider plant [3].

Several studies on the pollution factors affecting the Bagnoli area have been carried out over the years [20–24]. In particular, the source of heavy metal pollution in seawater and soil of the brownfield [25] site derivedfrom hydrothermal fluids related to the volcanic activity (natural source) and from the use of fossil fuels and industrial emissions for the steel production (anthropogenic source). De Vivo and Lima [3] have shown that the hydrothermal fluids that rise in the underground waters inthe Bagnoli-Coroglio area are the main cause of heavy metal pollution. The groundwatercontent analysis highlighted the presence of metals such as As, Cd, Cu, Hg and Pb normally represented in areas of high volcanic activity.In addition, the sea sediment samples collectedalong the Bagnoli coastline, near the former discharges of the Ilva plant, contained metals (Cd, Cu, Hg, Pb, Zn and Fe) and PAHs of high molecular weight at concentrations well above the tolerable levels for health safety [26]. The studies carried on sediment cores have traced the presence of metals such as As, Cr, Ni and V to a mainly natural contribution, being associated with the presence ofclay [27].

However, what is evident from the analysis of metals is the overlapping of elements of natural and industrial pollution. Emblematic is the arsenic case, which some studies attribute to a prevalentgeogenic nature [26]. Arsenic pollution has also been related to brownfield sites (steelworks and glass factories used arsenic in their production cycle), as high concentrations of the metal have been found in thesewage drains along the coast [28]. This evidence showsthe difficulty of attributing the origin of pollution to a single phenomenon.

In contrast, PAHs and PCBs (polychlorinated biphenyls) percolating through soils and landfills, are certainly ascribable to industrial activities [22].

The prevention of contaminant migration becomes crucial and requires the recovery of the brownfield sites. Conventional technologies for reclamation and redevelopment of polluted areas are consideredto be underperforming. Today, it is necessary to define strategies based on urban ecology capable of preserving the urban system over time without producing degradation but triggering regenerative processes. The use of suitable technologies, materials and systems can become the tool to transform dismissed areas, preserving the urban and natural heritage with a conscious management of the resources.

The current demand is to achieveeconomic, nondestructive and eco-friendly redevelopments [7]. In this context, phytoremediation, a well-known plant-based technique, is considered very promising and widely used. The main utility lies in the removal of pollutants from contaminated water and soils by restoring part of the self-purifying capacity typical of the ecosystems themselves. Numerous advantages are derived from the use of this biotechnology [29–31]. The costs of realization and management are certainly reduced. The landscape insertion of species, especially the native ones, could be optimal and applicablefor vast polluted areas. Furthermore, the aspect strictly linked to an economic circularity must be considered.The waste biomass can be exploited as a renewable energy supply or, when adequately treated, as a source of phytoproducts. Considering the use of the phytoremediation, the evaluation of sustainability takes on fundamental importance, as it is the tool that enhances the phyto approach. In this regard, compared to other technological approaches, phytoremediation already has an inherent idea of sustainability at an environmental, economic and social level.

Plants' behavior in contaminated soils depends on several factors. The plant species have different abilities to tolerate pollutants, and therefore depend on their resistance and the possibility of absorbing and moving them. The feasibility can be established even before the success of the phytoremediation intervention. An environmental analysis of the site examining the characteristics of the soil, pH, texture, organic matter and nutrients, hydrology and climatic conditions to identify the most suitable plant species for this purposeis crucial. The information about the real mobility and bioavailability of contaminants is fundamental in the phytoremediation technique, which uses living organisms and can act on the concentration of bioavailable contaminants.

There are numerous reports [30–32] that highlight the ability of selected plants to sequester contaminants present in amended soils.Interestingly, some plants capable of growing naturally in contaminated areas will be selected for their characteristics as possible candidates for phytostabilization, phytoremediation and revegetation processes. The physical and chemical properties of the soil are decisive in influencing the distribution of contaminants in the different chemical forms in which they are present.

The plants play a fundamental role in the environmental decontamination of sites heavily polluted by organic substances or heavy metals, through their absorption, degradation and stabilization. In areas contaminated by heavy metals, shrubs and herbs, such as *Armeria marittima* and *Minuatariavernato* (detects Cu), *Alyssum bertoloniia* (Ni accumulator) and *Viola calaminare* (accumulates high levels of minerals in the leaves) [31], can act as indicators of pollution.In Mediterranean areas, *Arundodonax*, a perennial cane, showed strong tolerance to heavy metals [33]. The presence of high concentrations of cadmium and nickel, indeed, has not had depressive effects on the photosynthetic rate and growth of the cane. In addition, *A. donax* does not enter the food chain, thus reducing the spread of substances harmful to health [34].

The reclamation of the Bagnoli soil is made problematic because of its fragility, instability and the contaminated seabed. Despite the pollution, insmall artificial lakes used to collect the wastewater of steel processing, several botanical species have been observed [35], such as *Phragmites australis* and perennial and rhizomatous marsh plant, together with *Typha latifolia* and *Arundo donax*, characterized by phyto vegetating action [36] (Table 1).

Table 1. Plants useful for the remediation of polluted sites (Adapted from [36]).

Filtration	Rhizofiltration	Phytostabilization	Phytodegradation	Phytovolatilization
Cupressus sempervirens	*Panicum virgatum*	*Picea abies*	*Sorghas trumnutans*	*Brassica oleracea*
Platanus spp.	*Festuca arundinacea*	*Agrostis capillaris*	*Panicum virgatum*	*Beta vulgaris*
Taxus baccata	*Carexelata*		*Agropyron desertorum*	*Oryza sativa*
Thuya occidentalis	*Phragmites australis*		*Medicago sativa*	*Brassica juncea*
Acer campestre	*Typha latifolia*		*Bromus inermis*	
Chamae cyperispisifera	*Salix* spp.		*Festucaarundinacea*	
Quercus robur	*Populus* spp.		*Morus* spp.	
Sambucus racemosa	*Eichhornia crassipes*		*Malus* spp.	
Sorbus aucuparia	*Hydrocotyle umbellata*		*Kochia scoparia*	
Acer campestre	*Lemna minor*		*Nepeta cataria*	
Populus deltoidestrichocarpa	*Azolla pinnata*		*Carduus nutans*	
Pinusnigra				
Cupresso cyparisleylandii				

An ongoing in situ bioremediation aims to cleanup the soils in the contaminated zone of Bagnoli. In the brownfield site area, the presence of metals in the roots and tissues of *Bituminaria bituminosa* and *Daucus carota* plants, which growth spontaneously, was tested [37]. Theconcentration ofmetalsin the root was higher than in leaves and shoots, indicating the immobilization ofmetals in the roots of plants.

Guarino et al. [38] identified plant families more represented in the polluted area of Bagnoli. Poaceae, Fabaceae, Asteraceae and Apiaceae areadapted to grow and survive also under contaminated soil condition. The analysis of the composition of pollutantsin the site's soil showed that the mainsources of PAHs derived from the combustion of oil and coal. Several native plants present in the studied site accumulated PAHs in the roots, limiting theirtranslocationinto the soil [38].

The field test on Bagnoli soil was carried out in two different places: (i) inside the SIN; and (ii) in a cold greenhouse with polyethylene cover [39]. The species used, reported in Table 2,were selected considering the positive results obtained in situ.

Table 2. Plants used in mesocosms experiment (Adapted from [39]).

Family	Species
Fabaceae	*Lotus corniculatus*
	Bituminaria bituminosa
	Medicago sativa
Poaceae	*Festuca arundinacea*
	Dactylis glomerata
	Piptatherum miliaceum
	Arundo donax.
Scrophulariaceae	*Verbascum sinantum*
Asteraceae	*Ditrichia viscosa*
	Helianthus annuus
Salicaceae	*Salix purpurea*
	Populus alba

Interestingly, the phytoremediation processes take place mostly in the rhizosphere, where the soil microenvironments are among the most dynamic and biologically diverse on the earth, due to the presence of numerous nematodes, bacteria, protozoa, algae, actinomycetes and mushrooms [40]. The multi-contaminated soil, containing As, Pb, Zn and Cd, revealed an enrichment of bacteria tolerant to these metals, which include *Paenibaciluus* spp., *Mycobacterium* spp., *Pseudomonas aeruginosa*, *Pseudomonas fluoresens* and *Rhodococcus* spp., which are also involved in the degradation and adsorption of polycyclic aromatic hydrocarbons [41]. In addition to bacteria, in the rhizosphere a particular type of fungi is represented, mycorrhiza, living in symbiosis with plant roots and producing hyphae that branch and penetrate the roots, reaching a length of between 5 km and 20 km. In the presence of heavy metals, mycorrhiza maintain high availability of essential nutrients for the plant, and at the same time dilute heavy metals, reducing their toxic effects.

The result ofthe SIN Bagnoli analysis highlighted that the remediation process also involved rhizosphere microbes [38]. *P. aeruginosa*, *P. fluoresens*, *Mycobacterium* spp., *Rhodococcus* spp. and *Paenibaciluus* spp. are involved in PAHs' degradation. 85% of the microorganisms detected belong to the phylum of Proteobacteria. At lower levels, Alphaproteobacteria and Gammaproteobacteria were also represented [38].

A method to increasethe degradation capacity andtolerance to the contamination stress was the integration of a beneficial consortium of microorganisms and bacteria of the rhizosphere. In this perspective, in addition to ligninolytic fungi, mycorrhizal fungi and rhizosphere bacteria that promote plant growth have also been inoculated into the soil [38].The widespread contamination by persistent organic pollutants in the SIN Bagnoli soils makes remediation particularly complicated. The degradation of these xenobiotics occurs by enzymes mainly produced by ligninolytic fungi able to degrade lignin. The microorganism–plant associations improve the transformation and degradation of pollutingcompounds, ultimately resulting in their removal and therebyin lower levels of abiotic stress for plants. It can be concluded that hydrocarbons' degradationin SIN Bagnoli may be based on an integrated rhizodegradation. In fact, microbiotaenzymes and those produced by the roots contribute to PHAs' degradation [39].

Interestingly, the benthic foraminifera sea population was also characterized by microorganisms resistant topollution [42]. Some species, such as *Miliolinellasubrotunda* and *Elphidium advena*, among the 113 recognized atBagnoli, showed percentages of abnormal specimens correlated with pollutants such as Mn, Pb, ZnandPAHs. Hence, in SIN Bagnoli, the environmental stress on foraminiferalcommunities may actually be attributable to the pollution of brownfield sites.

Therefore, the use of phytotechnologyinvolves allbiological, chemical and physical processes that allow the absorption, the seizure, the biodegradation and the metabolization of contaminants, both byplants and bymicroorganisms of the rhizosphere [43].

In the Bagnoli case, the soils polluted by nonbiodegradable heavy metals, not metabolized by plants by direct absorption, can be decontaminated by phytoextraction with removal of pollutants, followed by phytostabilization to reduce their mobility.The reported species can be used to allow an effective and durable removal of pollutants, producing biomass useful for bioenergy production [44]. The biomass used to extract pollutants takes on an important economic value when used to produce energy, for combustion or through gasification followed by cogeneration. The ash produced, rich in metals, can be applied in extractive processes to recover the metals themselves. In addition, some plants, which do not move metals to the reproductive organs, will be used in nonfood productions (such as glues, plastics, biodiesel or industrial oils).In conclusion, phytoremediation, as well as being a technique capable of regenerating polluted areas, is ableto provide the spaces reserved for ecosystem services in a newly conceived city, promoting the redevelopment of the concept of green in an urban context.

5. Conclusions

The SIN Bagnoli-Coroglio reclamation plan has a long and troubled history, which still has yet to be fullyimplemented. After the industrial activities'closure, over the years, the requirements for the environmental recovery of the site have been monitored and evaluated. The plan for recovery must consider thatthe pollution originated from industrial activities over a century ago. The major contaminants are represented by heavy metals and PHAs derived from fossil fuels' combustion, industrial waste, dumps, slag and scum. However, Bagnoli is affected by the geothermal activity of the CampiFlegrei, in whose caldera is located. Therefore, the brownfield site represents an overlap between anthropogenic and natural contaminationcomponents.

Among the different typologiesof applicable remediationtreatments, phytoremediation is a sustainable option. Phytotechnology involves agricultural practices and integrated biological systems, allowing the reduction in pollutant concentrations and restoring the functions of the soil over time. In the SIN Bagnoli-Coroglio, numerous analysis and studies conducted over the years have verified the feasibility of using phytoremediation for the reclamation of polluted soils and the creation ofa green space in anurban parkcontext. For thisreason, the utilization of native plants in conjunction with microbe-assisted phytoremediation was found to be an advantageousapproach, obtaining high rates of degradation also suitable for the environmentalecological balance. The plant species selected were characterized by high tolerance to specific contaminants, extensive root systems and the presence of a rich rhizosphere. In particular, Fabacee and Poaceae have a fibrous rootsystem that improvesthe contact between contaminants and degrading microbes.

The phytoremediation-integrated systems used forcontaminants' reduction in soils are an interesting form of green technology with great potential. At the same time, the improvement of environmentalquality, the restoration of soil functions and the protection of human health can be achieved. Crucial to obtaining positive results through bioremediation is the right combination of different biological elements and a well-designed agronomic practice.

While we wait for the completion of the urban redevelopment of the western suburbs of Naples, the reclamation of the decommissioned industrial area of Bagnoli would represent a dynamic reality of landscape reconversion, restoring historical identity to the territory of the Campi Flegrei.

Author Contributions: Conceptualization, C.C.; writing—original draft preparation, C.C., B.B. and L.M.; writing—review and editing, C.C., B.B. and L.M.; visualization, G.A.; supervision, C.C., B.B. and L.M.; project administration, G.A.; funding acquisition, C.C. and G.A. All authors have read and agreed to the published version of the manuscript.

Funding: This research received no external funding.

Acknowledgments: Special thanks to Marina Russo (CNR—IRET) for the graphic support in drafting the article.

Conflicts of Interest: The authors declare no conflict of interest.

References

1. Environmental Rehabilitation and Urban Regeneration Program on the Site of National Interest of Bagnoli—Coroglio. Available online: http://presidenza.governo.it/AmministrazioneTrasparente/Organizzazione/CommissariStraordinari/nastasi/RelazioneConclusiva_Sett2018.pdf (accessed on 3 September 2018).
2. Zalesny, R.S.; Casler, M.D.; Hallett, R.A.; Lin, C.; Pilipović, A. Chapter 9—Bioremediation and soils. In *Soils and Landscape Restoration*; Stanturf, J.A., Callaham, M.A., Eds.; Academic Press: Cambridge, MA, USA, 2021; pp. 237–273.
3. De Vivo, B.; Lima, A. Characterization and remediation of a brownfield site: The Bagnoli case in Italy. In *Environmental Geochemistry Site Characterization, Data Analysis and Case Histories*; De Vivo, B., Belkin, H.E., Lima, A., Eds.; Elsevier: Amsterdam, The Netherlands, 2008; pp. 355–385.
4. Cardone, V. *Bagnoli nei Campi Flegrei. La Periferia Anomala di Napoli*; CUEN: Napoli, Italy, 1989; 263p.
5. De Fer, N. *Les merveilles de Pozzoli ou Pouzzol Cume et Baia ou Bayes Dan le Voisinage de Naples*; A Paris: Chez l'Auteur dans l'Isle du Palais sur le Quay de l'Orloge a la Sphere Royale: Pozzuoli, Italy, 1705.
6. Biblioteca Nazionale di Napoli—Biblioteca Digitale—(3756carafa_22.tif). Available online: http://digitale.bnnonline.it/index.php?it/149/ricerca-contenuti-digitali/show/85/ (accessed on 17 December 2010).
7. Cirillo, C.; Acampora, G.; Scarpa, L.; Russo, M.; Bertoli, B. Napoli e il paesaggio costiero: Il recupero ambientale di Bagnoli e la rigenerazione del litorale flegreo. In Proceedings of the atti VI Simposio Il Monitoraggio Costiero Mediterraneo: Problematiche e Tecniche di Misure, Livorno, Italy, 6 November 2017; pp. 112–117.
8. Imbesi, P.N. *Il Riqualificar Facendo e le Aree Dismesse: Il Senso di Un'esperienza di Progettazione Partecipata*; Gangemi: Roma, Italy, 2012.
9. Luongo, G.; Cubellis, E.; Obrizzo, F.; Petrazzuoli, S.M. The mechanics of the Campi Flegrei resurgent caldera—a model. *J. Volcanol. Geotherm. Res.* **1991**, *45*, 161–172. [CrossRef]
10. Alisio, G.; Young, L.M. *Utopia e Realtà Nell'Urbanistica Napoletana dell'Ottocento*; Officina Edizioni: Roma, Italy, 1978.
11. Available online: https://commons.wikimedia.org/wiki/File:Sommer,_Giorgio_(1834-1914)_-_n._1191_-_Nisida.jpg (accessed on 11 April 2018).

12. Bertoli, B. Le utopie smarrite della 'Bagnoli jungle' nella rappresenta-zione delle arti visive. In *La Città Altra/The Other City Storia e Immagine della Diversità Urbana: Luoghi e Paesaggi dei Privilegi e Del Benessere, Dell'isolamento, del Disagio, della Multiculturalità*; di Capano, F., Pascariello, M.I., Visone, M., Eds.; Cirice: Napoli, Italy, 2018.
13. Brancaccio, S. *Bagnoli una Proposta Operativa*; Arte Tipografica: Napoli, Italy, 1995.
14. Available online: https://it.wikipedia.org/wiki/File:Cartolina_postale_dello_Stabilimento_Ilva_di_Bagnoli._Vasche_e_pontile,_Napoli_1905.jpg (accessed on 27 February 2018).
15. Gravagnuolo, B.; Miriello, R. *Bagnoli una Fabbrica*; Liberetà S.p.A.: Napoli, Italy, 1991.
16. Dall'Occhio, G. *Napoli, La Città del Sole*; Editori: Napoli, Italy, 2009.
17. Strazzullo, M.R. *L'archivio Ilva di Bagnoli: Una Fabbrica tra Passato e Presente*; CLS: Napoli, Italy, 1992.
18. Soverina, F. *La Memoria D'acciaio. Una Fabbrica, un Quartiere, una Città*; ICSR: Napoli, Italy, 2010.
19. Corona, G.; De Lucia, V. Bagnoli oggi: Quale futuro? Conversazione con Vezio De Lucia Aree deindustrializzate. In *Meridiana*; Viella: Roma, Italy, 2016; pp. 269–275.
20. Armiento, G.; Caprioli, R.; Cerbone, A.; Chiavarini, S.; Crovato, C.; De Cassan, M.; De Ros, L.; Montereali, M.R.; Nardi, E.; Nardi, L.; et al. Current status of coastalsedimentscontamination in the former industrial area of Bagnoli-Coroglio (Naples, Italy). *Chem. Ecol.* **2020**, *36*, 579–597. [CrossRef]
21. Passaro, S.; Gherardi, S.; Romano, E.; Ausili, A.; Sesta, G.; Pierfranceschi, G.; Tamburrino, S.; Sprovieri, M. Coupled geophysics and geochemistry to record recent coastal changes of contaminated sites of the Bagnoli industrial area, Southern Italy. *Estuar. Coast. Shelf Sci.* **2020**, *246*, 107036. [CrossRef]
22. Trifuoggi, M.; Donadio, C.; Mangoni, O.; Ferrara, L.; Bolinesi, F.; Nastro, R.A.; Stanislao, C.; Toscanesi, M.; Di Natale, G.; Arienzo, M. Distribution and enrichment of trace metals in surface marine sediments in the Gulf of Pozzuoli and off the coast of the brownfield metallurgical site of Ilva of Bagnoli (Campania, Italy). *Mar. Pollut. Bull.* **2017**, *124*, 502–511. [CrossRef] [PubMed]
23. Tamburrino, S.; Passaro, S.; Salvagio Manta, D.; Quinci, E.; Ausili, A.; Romano, E.; Sprovieri, M. Re-shaping the "original SIN": A need to re-think sediment management and policy by 635 introducing the "buffer zone". *J. Soils Sediments* **2020**, *20*, 2563–2572. [CrossRef]
24. Buccino, M.; Daliri, M.; Calabrese, M.; Somma, R. A numerical study of arsenic contamination at the Bagnoli bay seabed by a semi-anthropogenic source. Analysis of current regime. *Sci. Total Environ.* **2021**, *782*, 146811. [CrossRef] [PubMed]
25. Tarzia, M.; De Vivo, B.; Somma, R.; Ayuso, R.A.; McGill, R.A.R.; Parrish, R.R. Anthropogenic vs. natural pollution: An environmental study of an industrial site under remediation (Naples, Italy). *Geochem. Explor. Environ. Anal.* **2002**, *2*, 45–56. [CrossRef]
26. Giglioli, S.; Colombo, L.; Contestabile, P.; Musco, L.; Armiento, G.; Somma, R.; Vicinanza, D.; Azzellino, A. Source apportionmentassessment of Marine sediment contamination in a post-industrial area (Bagnoli, Naples). *Water* **2020**, *12*, 2181. [CrossRef]
27. Romano, E.; Bergamin, L.; Celia Magno, M.; Pierfranceschi, G.; Ausili, A. Temporal changes of metal and trace element contamination in marine sediments due to a steel plant: The case study of Bagnoli (Naples, Italy). *Appl. Geochem.* **2018**, *88*, 85–94. [CrossRef]
28. Sbarbati, C.; Barbieri, M.; Barron, A.; Bostick, B.; Colombani, N.; Mastrocicco, M.; Prommer, H.; Passaretti, S.; Zheng, Y.; Petitta, M. Redox dependent arsenic occurrence and partitioning in an industrial coastal aquifer: Evidence from high spatial resolution characterization of groundwater and sediments. *Water* **2020**, *12*, 2932. [CrossRef]
29. Cervelli, E.; Pindozzi, S.; Capolupo, A.; Okello, C.; Rigillo, M.; Boccia, L. Ecosystem services and bioremediation of polluted areas. *Ecol. Eng.* **2016**, *87*, 139–149. [CrossRef]
30. Pandey, V.C.; Bajpai, O. Chapter 1—Phytoremediation: From Theory Toward Practice. In *Phytomanagement of Polluted Sites*; Pandey, V.C., Bauddh, K., Eds.; Elsevier: Amsterdam, The Netherlands, 2019; pp. 1–49.
31. Solomou, A.D.; Germani, R.; Proutsos, N.; Petropoulou, M.; Koutroumpilas, P.; Galanis, C.; Maroulis, G.; Kolimenakis, A. Utilizing Mediterranean Plants to Remove Contaminants from the Soil Environment: A Short Review. *Agriculture* **2022**, *12*, 238. [CrossRef]
32. Meuser, H. *Soil Remediation and Rehabilitation: Treatment of Contaminated and Disturbed Land*; Springer Science & Business Media: Dordrecht, The Netherlands, 2013; Volume 23.
33. Gou, Z.H.; Miao, X.F. Growth changes and tissues anatomical characteristics of giant reed (*Arundodonax*, L.) in soil contaminated with arsenic, cadmium and lead. *J. Cent. South Univ. Technol.* **2010**, *17*, 770–777.
34. Fiorentino, N.; Impagliazzo, A.; Ventorino, V.; Pepe, O.; Piccolo, A.; Fagnano, M. Biomass accumulation and heavy metal uptake of giant reed on polluted soil in southern Italy. *J. Biotechnol.* **2010**, *150*, 261. [CrossRef]
35. Di Nunno, F. Bagnoli: La Natura Prende il Sopravvento nei Laghi Dell'italsider. 2016. Available online: altrimondinews.it/2016/03/02/bagnoli-laghi-italsider (accessed on 2 March 2016).
36. Marchiol, L.; Fellet, G.; Zerbi, G. Phytoremediation of soils polluted by heavy metals and metalloids using crops: The state of the art. *Ital. J. Agron.* **2008**, *3*, 3–14. [CrossRef]
37. Adamo, P.; Mingo, A.; Coppola, I.; Motti, R.; Stinca, A.; Agrelli, D. Plant colonization of brownfield soil and post-washing sludge: Effect of organic amendment and environmental conditions. *Int. J. Environ. Sci. Technol.* **2015**, *12*, 1811–1824. [CrossRef]
38. Guarino, C.; Zuzolo, D.; Marziano, M.; Conte, B.; Baiamonte, G.; Morra, L.; Benotti, D.; Gresia, D.; Stacul, E.R.; Cicchella, D.; et al. Investigation and Assessment for an effectiveapproach to the reclamation of Polycyclic Aromatic Hydrocarbon (PAHs) contaminated site: SIN Bagnoli, Italy. *Sci Rep.* **2019**, *9*, 11522. [CrossRef]

39. Zuzolo, D.; Guarino, C.; Postiglione, A.; Tartaglia, M.; Scarano, P.; Prigioniero, A.; Terzano, R.; Porfido, C.; Morra, L.; Benotti, D.; et al. Overcome the limits of multi-contaminated industrial soils bioremediation: Insights from a multi-disciplinary study. *J. Hazard.* **2022**, *421*, 126762. [CrossRef]

40. Wenzel, W.W. Rhizosphere processes and management in plant-assisted bioremediation (phytoremediation) of soils. *Plant Soil* **2009**, *321*, 385–408. [CrossRef]

41. Buee, M.; De Boer, W.; Martin, F.; Overbeek, L.V.; Jurkevitch, E. The rhizosphere zoo: An overview of plant-associated communities of microorganisms, including phages, bacteria, archaea, and fungi, and of some of their structuring factors. *Plant Soil* **2009**, *321*, 189–212. [CrossRef]

42. Romano, E.; Bergamin, L.; Ausili, A.; Pierfranceschi, G.; Maggi, C.; Sesta, G.; Gabellini, M. The impact of the Bagnoli industrial site (Naples, Italy) on sea-bottom environment. Chemical and textural features of sediments and the related response of benthic foraminifera. *Mar. Pollut. Bull.* **2009**, *59*, 245–256. [CrossRef] [PubMed]

43. De Paolis, M.R.; Pietrosanti, L.; Capotorti, G.; Massacci, A.; Lippi, D. Eco-physiological characterization of the culturable bacterial fraction of a heavy-metal contaminated soil subjected to phytoremediation. *Water Air Soil Pollut.* **2011**, *216*, 505–512. [CrossRef]

44. Zacchini, M.; Iori, V.; Mugnozza, G.S.; Pietrini, F.; Massacci, A. Cadmiumaccumulation and tolerance in Populusnigra and Salix alba. *Biol. Plant.* **2011**, *55*, 383–386. [CrossRef]

Entry

Conductive Heat Transfer in Thermal Bridges

Mathias Fuchs

Mathias Fuchs Services, Hallgartenstr. 6, 81375 Munich, Germany; mathias@mathiasfuchs.com

Definition: A thermal bridge is a component of a building that is characterized by a higher thermal loss compared with its surroundings. Their accurate modeling is a key step in energy performance analysis due to the increased awareness of the importance of sustainable design. Thermal modeling in architecture and engineering is often not carried out volumetrically, thereby sacrificing accuracy for complex geometries, whereas numerical textbooks often give the finite element method in much higher generality than required, or only treat the case of uniform materials. Despite thermal modeling traditionally belonging exclusively to the engineer's toolbox, computational and parametric design can often benefit from understanding the key steps of finite element thermal modeling, in order to inform a real-time design feedback loop. In this entry, these gaps are filled and the reader is introduced to all relevant physical and computational notions and methods necessary to understand and compute the stationary energy dissipation and thermal conductance of thermal bridges composed of materials in complex geometries. The overview is a self-contained and coherent expository, and both physically and mathematically as correct as possible, but intuitive and accessible to all audiences. Details for a typical example of an insulated I-beam thermal bridge are provided.

Keywords: thermal bridge; sustainable architecture; finite element modeling; energy performance; heat transfer; heat equation; simulation; computational design; thermal modeling; computational methods

Citation: Fuchs, M. Conductive Heat Transfer in Thermal Bridges. *Encyclopedia* **2022**, *2*, 1019–1035. https://doi.org/10.3390/encyclopedia2020067

Academic Editors: Masa Noguchi, Antonio Frattari, Carlos Torres Formoso, Haşim Altan, John Odhiambo Onyango, Jun-Tae Kim, Kheira Anissa Tabet Aoul, Mehdi Amirkhani, Sara Jane Wilkinson, Shaila Bantanur and Raffaele Barretta

Received: 4 April 2022
Accepted: 20 May 2022
Published: 24 May 2022

Publisher's Note: MDPI stays neutral with regard to jurisdictional claims in published maps and institutional affiliations.

1. Introduction

Building energy performance simulation, modeling and optimization has become an ever more important discipline at the multidisciplinary junction between building physics, design, architecture, engineering, numerical mathematics and optimization. Soaring energy costs and an increased awareness of the urgency of sustainability amid a global energy crisis have led to growing interest far beyond the original scope of an engineering sub-discipline. Physically, heat transfer is described by the heat equation—a century old and well studied partial differential equation (henceforth abbreviated PDE). In principle, it determines the heat distribution in a given geometry, for given material constants and boundary conditions, either in a time-dependent or time-independent manner. In practice, however, an accurate solution requires a finite element discretization of the heat equation. A finite element solution solves an approximation to the heat equation on a tetrahedralization of the input domain; see Zienkiewicz et al. [1] for a general introduction to finite element (FE) methods; Ciarlet [2] for an introduction to FE on the particular class of elliptic PDEs relevant here, and Wilson and Nickell [3], Lewis et al. [4] for the application of FE to thermal modeling. Effectively, this places simulation and modeling at the heart of every modern building performance simulation.

There is no treatment in the literature that coherently covers the engineering aspects together with the full three-dimensional FE method in a way that is accessible to an audience from, say, a design background. Most expositions focus either on basic one-dimensional quantities such as the U-value [5], on numerical methods for the full 3D heat transfer, or on computational design in view of architectural and artistic aspects [6]. A thermal bridge is defined as a building component with particularly high thermal loss; as such, they are the most important objects of modeling [7,8]. This entry aims to outline a complete introduction to put the reader in a position to understand all steps necessary to compute the energy loss incurred by a particular thermal bridge situation in design.

As a rule of thumb, heating costs constitute one of the largest shares of the per capita energy profile of the average Western inhabitant of moderate latitudes. For example, it is estimated that the entire average annual primary energy consumption in developed countries is roughly 100–200 GJ per capita [9]. On the other hand, for example in Germany, the annual per capita heating energy profile is estimated to be more than 3000 kWh $\approx$ 10 GJ per year [10]. Thus, a considerable fraction of about 5–10 percent of the energy is spent on heating alone. This places the heating energy profile among the top private household budget items.

Consequently, there are plenty of legal incentives, policies and regulations aimed at instigating energy efficiency certifications in most countries in the middle latitudes. Energy certification, consulting, contracting, and modernization have all become their own industry segments. This increased pressure on home owners, real estate agents and developers has recently led to the previously unthinkable situation where the energy certificate has the highest or one of the highest impacts on a real estate's (rental) value, among all value predictors [11].

Apart from these environmental and monetary incentives for having good insulation, thermal performance and insulation are, of course, highly relevant in regards to building safety due to their importance for vapor condensation and mold prevention. Consequently, estimating and optimizing a building's energy performance has become one of the cornerstones of sustainable architecture, modernization, renovation, maintenance and building safety.

There is an abundance of commercial and non-commercial FE software packages available on the market. It would seem difficult to even attempt to cite a comprehensive list of the most important ones, due to the far-reaching impact they have, the multitude of application scenarios, and the long amount of time these packages have already been under development for. For example, one of the most popular commercial software packages capable of modeling thermal conductance is ANSYS [12], and one of the most popular open source software packages if Freefem++ [13]. However, it is often desirable to understand and implement a basic FE algorithm from scratch, rather than treating the FE method as a black box. This entry aims to introduce students of all the aforementioned disciplines, in particular those without prior exposure to FE methods and computational physics from scratch to a state where the reader could, at least in principle, implement their own FE solution. In this text, no prior exposure to computational design, programming, or mathematics beyond an early undergraduate level are assumed. In particular, it will be explained how the seemingly arcane physical description of heat transfer translates into an amenable pseudo-code which can readily be employed and adapted to more specialized optimization questions.

Most architecturally inspired texts, or even those for engineering students, tend not to describe the full volumetric heat equation, and focus on one or two-dimensional case studies [14], or even just a description of insulation indicators such as U-values, avoiding differential equations altogether. On the other hand, mathematically oriented textbooks tend to ignore materials altogether, and just treat the "heat equation" in the form of $\partial_t u - \Delta u = \rho$ where u is the temperature distribution and ρ is a heat source term, ∂_t is the time derivative, and Δ is the Laplacian [15]. However, the relevant equation for use in our case is when $\partial_t u = 0$ is the steady state or stationary property and $\rho = 0$. In fact, there is usually no heating inside the thermal bridge. In contrast, the term Δu needs to be replaced with the slightly more general expression $\nabla(\kappa \nabla u)$. In this entry, the following convention is used: The operator ∇ applied to a scalar field such as u is the gradient of the scalar field, such as the temperature gradient, and ∇ applied to a vector field, such as the heat flux, is the divergence of the vector field. The reader is invited to recall that the divergence of a vector field is a scalar field that describes how much the vector field converges or diverges.

These variations of the heat equation will be explained in Section 3.2. For the purely mathematical and theoretical treatment, omitting κ from the picture makes almost no difference, but here, all the material properties are expressed through, and have an impact on the modeling, in terms of their conductivity κ. All of the information about the materials is in the quantity κ. It is isotropic but also non-uniform and hence can not be "pulled in front" as a simple constant number. Therefore, one can not simplify the term $\nabla(\kappa \nabla u)$ to Δu as almost all mathematically oriented textbooks tend to do.

In this sense, the PDE literature equation is both an oversimplification of material continuity and a complication of time dependence. Due to these small differences, the classical PDE literature is almost useless for the architect and engineer who focuses on modeling thermal bridges. The passage from the PDE $\partial_t u - \Delta u = \rho$ to the PDE $\nabla(\kappa \nabla u) = 0$—Equation (2) below—makes both the analytical understanding and the practical numerical treatment much easier to comprehend, but it is quite different and not entirely trivial.

Similarly, there do not seem to be many treatments of FE methods for this particular case. One of the reasons being that in an engineering context, FE methods are most often associated with structural engineering solving the linear elasticity equations—or simplifications derived from it. Likewise, the second most frequent field of PDEs is computational fluid dynamics. Consequently, the term "FE methods" is sometimes mistakenly understood to refer only to structural analysis or computational fluid dynamics.

It should be noted that in the last few years a large amount of work was dedicated to "mesh-free methods", with the heat equation being a prominent example in the context of geometry processing [16]. However, in the context of actual engineering, FE methods remain the gold standard and method of choice due to their proven accuracy and versatility. The same holds true for the finite difference method [17]—as opposed to the FE method. The former places a grid over the entire domain and solves the heat equation with a simple differencing scheme. However, the drawback of the finite difference method is the scale dependency, and the fact that complex geometries can hardly be resolved accurately with an equi-spaced grid. The finite difference method will thus not be treated.

Likewise, instationary (time-dependent) computations will not be treated, nor will those that require taking radiative effects into account. For instance, the treatment of overheating in summer requires taking day-night temperature curves, as well as radiative transmission and absorption characteristics into account. Even though these considerations tend to have a large impact on the design of glass facades as well as shadow casting interior components, etc., these topics go well beyond the accurate treatment of thermal bridges and their insulation. Thus, since descriptions of the FE method just for the relevant Equation (2) are rare, this gap will be filled in this entry.

This entry does not aim to enable the reader to draft DIN-conformant engineering reports. For instance, the specifications of effective thermal surface resistances, as required, for example, in DIN 4180-2 [18] are technically intricate and very country-specific. Instead, this entry will assume the extreme case where the room facing surface is kept at $20\,°C$ and the outward facing surface is kept at $-13.1\,°C$ as is often done in Germany and Austria. Compared with the computation that takes surface resistances into account, this gives a conservative bias. Such a bias is often acceptable for a designer or architect who wishes to estimate the thermal feasibility during the design process. It is possible to implement the class of algorithms described in this entry in such a way that it runs in real-time. The most important ISO norms relevant for thermal modeling will be referenced in Section 4.5. Further details on norms, reporting, regulations, and certification would go beyond the scope of the entry.

One of the most important applications of thermal modeling is the simulation of hygrothermal properties: water vapor diffusion and condensation; see Taylor et al. [19], Gasparin et al. [20] and the references therein.

The entry is structured as follows. Section 2 gives all of the relevant definitions. Then, the physical and mathematical concepts will be introduced as thoroughly as necessary and as concisely as possible in Section 3 and some analytical examples of the heat equation will also be provided. The actual FE formulation is described in Section 4. Section 5 presents a typical case study of an I-beam penetrating a wall. Then, the entry will be finish with a conclusion in Section 6.

2. Definition of Important Terms And Quantities

2.1. Thermal Bridge

A thermal bridge is defined to be a part of the structure that is separated from the surrounding structure by a material boundary or air pocket, and that has an exceedingly high thermal energy loss. A particular example is the case of an I-beam penetrating a wall as seen in Section 5, or any openings such as doors and windows.

2.2. Types of Heat Transfer and Their Impact on Thermal Bridge Modeling

There are three types of heat transfer: Convective, radiative, and conductive. This survey will focus on the latter since it is the most relevant for understanding solid thermal bridges with a low volume portion of gaseous or fluid media, and a small portion of the surface exposed to sunlight. Usually, the main structural features of a building are not affected by fluid or gaseous dynamics inside void volumes within the structure. Likewise, the outside is exposed to sunlight and a large amount of radiative loss of thermal energy.

In the extreme—worst-case—scenario with temperatures of $-13.1\,°C$ outside and $20\,°C$ inside, one can posit that neither convective nor radiative effects lead to an even higher temperature spread. Direct sunlight exposure can easily heat up interior structures to more than $20\,°C$, but modeling radiative overheating is beyond the scope of this entry. In general, the approach is to work with assumptions that generally lead to overestimate the conductive energy loss as this scenario is most likely to be relevant for regulatory bodies and certifications.

2.3. The Main Problem

Using that definition, the problem being addressed in this entry is the following: Given the geometry of a thermal bridge and perfect knowledge of all occurring materials and their thermal conductivities, compute the dissipation (energy loss per time) in kW, assuming extreme interior and exterior air temperatures keeping the exposed parts of the thermal bridge at $T_1 = 20\,°C$ and $T_2 = -13.1\,°C$. By definition, those surfaces of the thermal bridge that are not exposed to exterior or interior air can be assumed to face a much better insulator than the thermal bridge. It will thus be permissible to model the thermal flux through these surfaces as zero. Thus, the relevant approximation is that in which the thermal flux from interior to exterior is the predominant one, as opposed to the flux in the direction along the wall. Since no country-specific regulatory requirements will be discussed here, one can generally assume that the main point in answering this problem, beyond optimizing the energy performance, is to decide which parts of the surface are below the local dew point and would therefore be susceptible to vapor condensation and mold growth, not taking vapor diffusion into account.

2.4. Essential Physical Quantities

2.4.1. Temperature Gradient

The temperature gradient indicates a rate of spatial change of temperature and therefore has unit Kelvin per meter. As a direction and a magnitude, it is a vectorial quantity—a vector field $\nabla u = (\partial_x u, \partial_y u, \partial_z u)^T$ —the T denotes transposition, i.e., a column vector. Each point in space, even the material boundaries anddiscontinuities, have such a vector associated with it. As a vector field, it has integral curves leading from the exterior surface of the thermal bridge to the interior and leading to the interior. As a geometrical picture of the temperature field, they serve as visual representations of the "direction of the energy loss". To give a simplistic example, a solid uniform timber door of 0.05 m thickness has a temperature gradient of about $(20 - (-13.1))\,K/0.05\,m = 662\,K/m$ of magnitude under the conditions of Section 2.3. Its direction points inwards everywhere.

2.4.2. Thermal Conductivity

The most important material property is the thermal conductivity, which helps to compute the thermal power dissipation from the temperature gradient. It is reasonable to assume that the thermal energy transmitted orthogonally through a cross section of area B is directly proportional both to B and to the temperature gradient. Thus, the ratio of transmitted power in Watt, divided by the product of B and the temperature gradient, should not depend on anything other than the material. It turns out this assumption is realistic, and this ratio is called the thermal conductivity. By consequence, it is measured in $W/((K/m) \cdot m^2) = W/(mK)$, i.e., Watt per meters and Kelvin. As a quantity, it assigns a value to each point (x, y, z) for a given geometry with given materials. Typical values are on the order of magnitude of about 0.02 W/mK for a very good insulator (hemp), and about 2 W/mK for a bad one such as steel. It is important not to conflate thermal conductivity with thermal diffusivity. The latter is measured in square meters per second and only plays a role in time-dependent modeling. It will therefore not play a role for this entry.

2.4.3. Thermal Flux

The next most important quantity necessary to understand conductive heat transfer is "thermal flux", defined as the negative product of the preceding two quantities, temperature gradient with the thermal conductivity. It has unit $W/(mK) \cdot K/m = W/m^2$ and describes the dissipation power per area cross section. Since temperature gradient is a vector field, and thermal conductivity is a scalar field, the product is also a vector field. Using the simplistic example of the door, one would compute a magnitude of 0.1 W/(mK) $\cdot$ 662 K/m $= 66.2$ W/m^2. The "negative" in the definition makes the direction point outward instead of inward. Thus, every square meter of the timber door accounts for 66 Watt power dissipation. Like the temperature gradient, thermal flux is a vector field. It assigns to each point (x, y, z) a direction and a quantity. The direction of the thermal flux is inverse to that of the temperature gradient but the magnitude is not; instead, it depends on the material.

2.4.4. Thermal Dissipation

Thermal dissipation is measured in Watt, corresponding to a power. Since its unit is that of energy per time, the term "building energy performance" should instead be called "building's thermal dissipation performance". It can be computed or measured for a given temperature difference and thermal bridge, given complete material knowledge. The consumed energy is obtained from the thermal dissipation by multiplying with the duration; it can then be given in Joule or converted to kWh. From the definition of the thermal flux, it should be clear that the thermal dissipation is the product (or, more precisely and more generally, surface integral) of thermal flux with area; and the area is understood be small and orthogonal to the temperature gradient.

For instance, a badly insulated building could have a typical thermal dissipation value of 10 kW in extreme winter conditions, which corresponds to an energy consumption of 240 kWh in a time span of 24 h. The unit of kWh is convenient since it can immediately be converted to money at a typical cost of about 0.2 EUR/kWh in Europe. The SI unit of energy being Joule, the conversion of kWh to Joule is 1 kWh $= 3.6$ MJ. This allows for putting the energy values given in GJ in relation to those given in kWh. For instance, on an area of 100 m^2, a structure has an energy loss of 100 m$^2 \cdot 66.2$ W/m$^2 = 6.62$ kW which can be compared with the typical energy consumption of a household electric radiator of about 1 kW. In the example of the timber door that suffers a thermal flux of 66.2 W/m^2, a sectional area of two square meters implies a thermal dissipation of 132.4 Watt.

2.4.5. Thermal Transmittance or U-Value and Thermal Resistance

There is a large number of practical quantities that may be reported for a particular constructive element. They are mostly motivated by the one-dimensional heat equation. In it, the thermal flux is constant wherever the constructive element is not heated or cooled.

On a material boundary where the thermal conductivity increases by a certain factor, the temperature gradient decreases by the same factor, for instance, at a boundary between steel and timber, this factor is about ten. This is consistent with the intuition that within steel which is a better "thermal conductor", the temperatures are more "spread out" than within timber.

Therefore, it makes sense to define a new quantity, the local thermal resistance of a material, as the reciprocal of the thermal conductivity. A highly conductive material has little resistance, and vice versa. Like conductivity, it is a scalar field, and associates to each point (x, y, z) a value without direction, having the unit mK/W. A closely related notion is the important quantity "U-value" whose unit is the same as that of thermal flux divided by temperature. It simply indicates the entire thermal flux through a one-dimensional construction, in relation to a given applied external temperature spread. In the one-dimensional point of view, an insulator compound of several materials achieves a U-value which can be computed as the reciprocal sum of resistances, where the resistance of a construction component of one material is its thickness times its resistance, i.e., thickness divided by conductivity; and the resistance of a component consisting of several materials is the sum of the individual resistances.

For practical purposes, it is sufficient to keep in mind that the U-value of a one-dimensional component indicates the thermal dissipation per area cross section and per degree Celsius applied temperature spread. Its unit is $W/(m^2\,K)$.

2.4.6. Thermal Conductance

Proceeding in a straightforward way, one can arrive at the definition of a thermal bridge's thermal conductance. The preceding sub-subsection has established that for a particular given geometry, Watt per Kelvin is the correct unit. It turns out that this is the correct unit to generalize to an arbitrary three-dimensional thermal bridge: One can accurately associate a value in Watt per Kelvin to a given geometry with materials, the thermal conductance. It indicates the amount of thermal dissipation per degree Kelvin temperature difference between inside and outside. For instance, one easily calculates a thermal conductance of 4 W/K for the timber door, implying that an increase of one degree Celsius in the temperature spread leads to 4 more Watt required to keep the interior air temperature the same. All other thermal bridges of the house add up to the total thermal dissipation.

Other quantities are also thinkable. A long and thin component might have a conductance per length, resulting in W/mK. However, these only play a role in very special situations or geometries, such as insulations along a rod, and so on. Another value that is often reported is the temperature factor f_{Rsi} [21]; however, its exact definition depends on the regulation-specific surface resistances.

3. Physical Background

3.1. The Physics of Heat Transfer

Consider a geometrical area A that transmits the thermal energy ΔW in a short amount of time Δt. Then, $P = \Delta W / \Delta t = W'$, the time derivative of W, is the energy dissipation through the surface S. One can express this dissipation in terms of the heat flux $\kappa \nabla u$. For small surface patches of area A perpendicular to ∇u, it is simply $P = \kappa ||\nabla u|| A$. For instance, consider a brick wall of area A and thickness l along which there is a temperature gradient between T_1 and T_2. Then, one ends up at $P = A\kappa(T_1 - T_2)/l$. For larger or curved surface patches, the dissipation is the surface integral $\oint \kappa \nabla u \, dS$ but that expression will only be needed once at the end of the entry. Note that the unit is always Watt: $W/(mK) \cdot K/m \cdot m^2 = W$.

3.2. Relationship with the "Heat Equation" from PDE Literature

Here, the focus is on the time-dependent heat equation $\alpha \, \partial_t u - \Delta u = \rho$ where α is a material constant, $u = u(t, x, y, z)$ is the temperature at time t at the location (x, y, z), $\partial_t u$ is its time-derivative, Δ is the Laplace operator $\partial_x^2 + \partial_y^2 + \partial_z^2$ and $\rho = \rho(t, x, y, z)$ is the amount of heating in (measured in the unit power per volume). However, this is just a special case of the full heat equation for isotropic non-uniform materials

$$\alpha\,\partial_t u - \nabla\left(\kappa \nabla u\right) = \rho \tag{1}$$

Ref. ([22], Equation (5.50) inserted into the equation preceding (5.51)), taking into account that $\nabla(\kappa\nabla u) = \nabla^2 u = \Delta u$ if κ is just ignored by setting $\kappa = 1$ everywhere. Note that when κ does depend on the location, as is the case here, it can not be pulled out of the first ∇. However, the most important application for engineering is establishing insulation properties for the worst-case scenario of a constant steep temperature difference between inside and outside, so in that case the steady state equation where $\partial_t u = 0$ can be considered. Moreover, since a wall is usually not heated from within the material but only from its surfaces—described mathematically by the boundary conditions, see Section 4.1—one can set $\rho = 0$. With these two simplifications, Equation (1) simplifies to the steady state (stationary) heat equation

$$\nabla\left(\kappa\nabla u\right) = 0 \tag{2}$$

which is the equation whose numerical practical solution is the topic of this entry.

3.3. The Steady State Heat Equation

In this subsection, the physical laws will be delineated. Instead of following the historical developments of the subject, as in Joseph Fourier's theory of heat, this section will directly state the facts and focus on making the terms amenable to non-physicists as quickly as possible. The meaning of the term $\kappa\nabla u$ inside Equation (2) is the negative of the heat flux. Therefore, Equation (2) simply states that the divergence of the heat flux is zero. Note that the meaning of the term "divergence" of a vector field is both the intuitive one—where it expresses the amount of incoming vs outgoing heat in every small infinitesimal volume. Therefore, the heat equation simply expresses the fact that there is no heating inside the thermal bridge—only at its boundaries, the inner and outer surfaces. Actually, the physical derivation of Equation (2) is just a formalization of the energy conversation in the thermal context. As noted, the right hand side of Equation (2) being zero amounts to the absence of heating or exothermic reactions, or phase transitions within the structure. Heat is only applied from outside, i.e., by means of boundary conditions. An example of phase transition that is relevant to building physics is condensing air humidity or its inverse, evaporation. However, these will not be treated in this entry.

The unit of Equation (2) is that of "local heating power", i.e., Watt per cubic meter and expresses the scalar field expressing the exterior heating. The inner operator ∇ is the gradient operator and simply transforms the temperature into the temper its gradient by taking the three directional derivatives along the axes. Equation (2) assumes that all occurring materials are isotropic: The heat is conducted equally in all directions. This is not to be confused with the notion of homogeneity or uniformity. As explained in Section 3.1 above, the energetic loss (dissipation) incurred by thermal conductivity along a surface S is the surface integral of the heat flux $\oint \kappa\nabla u\,dS$.

3.4. Simple Analytical and Computational Examples

Before passing to the description of the discretized heat equation, it is instructive to give examples of analytical solutions of the heat equation. These are instructive because they can be understood without appealing to meshes or to any computational methods. In addition, the rate of decay of the heat away from the hot boundary condition is in some cases analytically amenable but hard to understand from a purely numerical solution.

3.4.1. In One Dimension

In one dimension, the heat equation can easily be solved analytically. This covers the case of all rods, but also walls, etc., where the heat flux does not change perpendicular to its direction. Indeed, let $[A, B]$ be an interval with $A < B$, let $\kappa = \kappa(x)$ be a function on that interval, indicating the thermal conductivity, and let u_1 be the prescribed temperature at the lower end, and u_2 the prescribed temperature at the upper end. In one dimension, there is no "room" for Neumann boundary conditions. Let x be the variable along the interval. Then, Equation (2) becomes

$$\partial_x\big(\kappa(x)\partial_x(u)\big) = 0, \tag{3}$$

In addition, with the substitution $y := \partial_x u$ for the temperature gradient and the product rule, one arrives at $\kappa' y + \kappa y' = 0$, where one can now use the prime symbol for the x-derivative. Thus, $y' = -\kappa'/\kappa \cdot y$ which is an ordinary linear differential equation of separable type with non-constant coefficients ([23], Chapter 1, VII) and one can easily write down the analytical solution in terms of a definite integral given there, but it would not be very instructive for architectural applications. Some care has to be taken for material boundaries where κ is not differentiable. Indeed, in our case, κ is constant except at material boundaries where it is not differentiable. In this case, the solution to Equation (3) consists in first finding a solution to the temperature gradient while taking into account that at material boundaries where the conductivity changes by a factor of say α, the temperature gradient changes by a factor of $1/\alpha$. This provides a solution to y that is unique except for the choice of a scalar factor. The boundary conditions on u determine this scalar factor, taking into account that $u(B) = T_2 = T_1 + \int_A^B u'(x)dx$. A graphical representation showing this way of solving the one-dimensional heat equation is well-known in the engineering community and sometimes called the "Glaser method" [24]. It is not difficult to set the Glaser method in relation with the fact that the rod's total thermal resistance is the sum of its material components' resistances. Due to its intuitiveness and graphical representation, the Glaser method is very popular. The reader can easily find simple analytical examples. However, the Glaser method falls short of providing reliable results even for simple three-dimensional geometries, as the following two examples show.

3.4.2. In Two Dimensions

In two dimensions, finding analytical solutions takes a very different approach. To find the simplest case of an analytical solution to the heat equation that is not simply an extrusion of a one-dimensional solution, it suffices to consider just one material, leading to κ being constant everywhere. As already noted, Equation (2) simplifies to $\nabla(\kappa\nabla u) = \kappa\nabla^2 u = \kappa\Delta u$ and this being zero is equivalent to

$$\Delta u = 0, \tag{4}$$

where u is a function on a planar domain—since $\kappa \neq 0$ everywhere. Consider a ring-shaped region between two concentric circles of radii $r_1 \leq r \leq r2$, and assume there are the temperature constraints $u = T_1$ on the inner ring, $u = T_2$ on the outer ring. Then, the solution u is obviously also rotationally symmetric around the circles' center, so it is just a function of r, and this function turns out to be $u(r) = C - D\log(r)$ where C and D are constants that only depend on r_1, r_2, T_1 and T_2 but not on r and not on κ (the expression for C and D can be worked out but is not instructive). It is easy to verify that this is a solution to Equation (4) using the expression for the two-dimensional Laplacian in polar coordinates. This example also works for segments of the ring cut along rays through the circles' centers, imposing Neumann boundary conditions on the cuts (see Section 4.1). The important fact is that the temperature decays very slowly—namely, with the negative growth rate of the logarithm.

To illustrate a computational solution on a geometry that is somewhat similar but relevant in architecture, Figure 1 shows the solution of the heat equation on an interior edge. The effects of the boundary conditions are clearly visible: The hot and the cold boundary surfaces were given Dirichlet boundary conditions corresponding to 20 °C and −13.1 °C, respectively. The remaining two boundaries are given zero Neumann boundary conditions, resulting in the thermal isolines meeting those boundaries perpendicularly.

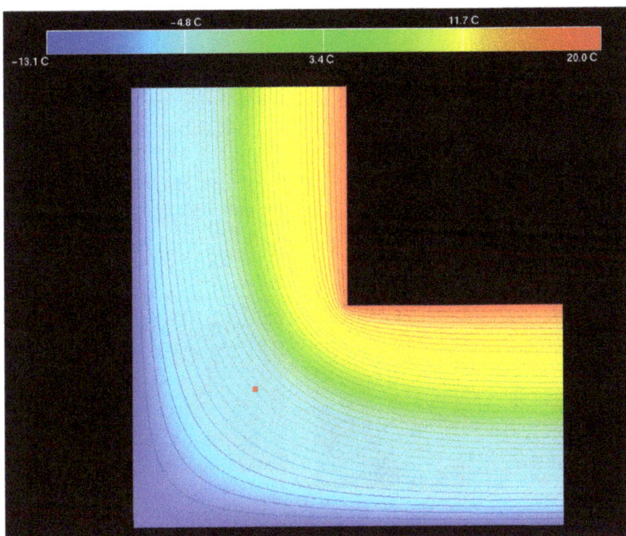

Figure 1. A two-dimensional solution on an interior edge, the point $(0.25, 0.25)$ marked with a red cross.

3.4.3. In Three Dimensions

Analogously to the preceding example, it is instructive to give the simplest case of a truly three-dimensional constant-time scalar field in a solid thermal bridge, such that it can not be simply reduced to the case of extrusion of a one- or two-dimensional scalar field along a perpendicular axis. Again, the value of an analytical solution is that it can be instructively obtained without reference to discretization, meshes and any numerical or computational aspects. It turns out that the simplest truly three-dimensional case of conductive heat transfer is that of an "igloo"—put simply, it is a semi-sphere between interior radius r_1 and exterior radius r_2, kept at temperatures T_1 and T_2. To find the simplest example, it suffices again to assume the same material everywhere, and the temperature field must obey Equation (4) where now Δ is the three-dimensional Laplacian. Similarly to the preceding example, one finds that by the symmetry of the situation the temperature field can only depend on the distance r from the igloo's center of symmetry, so that $u = u(r)$ where $r = \sqrt{x^2 + y^2 + z^2}$. When using the expression for the Laplacian in spherical coordinates, it is easy to verify that the solution is $u(r) = C + D/r$ where C and D are constants that depend only on r_1, r_2, T_1, T_2 but not on r or κ.

Along the ground level $x = 0$ of the "igloo", Neumann boundary conditions (see Section 4.1) are then obeyed as if the ground was a perfect insulator and also kept at temperatures T_1 and T_2 inside and outside, respectively, making this example not entirely unrealistic. Again, perhaps surprisingly, the temperature field does not depend on the material of the "igloo". However, the total thermal dissipation, i.e., the required total thermal output of the igloo's heating needed to keep the inside at temperature T_1, does depend on the material's conductivity and would be an easy and interesting exercise to calculate, using just the notions introduced in Section 2.

The relevance of this simplistic example is that the quantitative behavior of the heat flux—decaying with $1/r$—is very different from the preceding example—where the decay is $-\log(r)$, highlighting the strong impact of passing from two to three dimensions. The same contrast can be observed when passing from an interior edge to an interior corner. This semi-analytical reasoning is even more important due to interior edges and corners being very fundamental geometries.

A similar and practically very relevant geometry is that of an interior corner. Figure 2 shows three views of the full three-dimensional temperature solution in a corner. Dirichlet boundary conditions were applied on the hottest (interior) and coldest (exterior) surfaces, and Neumann boundary conditions on all other surfaces. Note that the temperature isosurfaces meet the Neumann boundary surfaces perpendicularly. One can now compare the impact of passing from two to three dimensions: In Figure 1, the point $(0.25, 0.25)$, marked red, had temperature $-6.5\,^{\circ}\text{C}$, whereas in Figure 2B, the corresponding point—also marked red—only has temperature $-10.7\,^{\circ}\text{C}$. This difference between 2D and 3D modeling highlights the necessity to treat thermal bridges volumetrically to achieve a realistic result. Under the assumption that the entire wall is of the same material the temperature does not depend on the material.

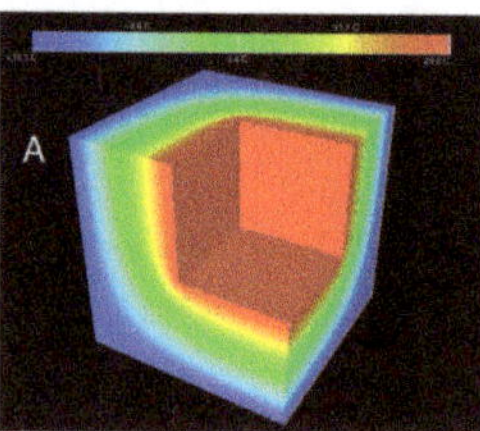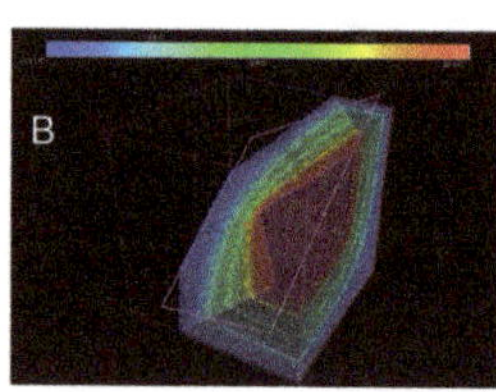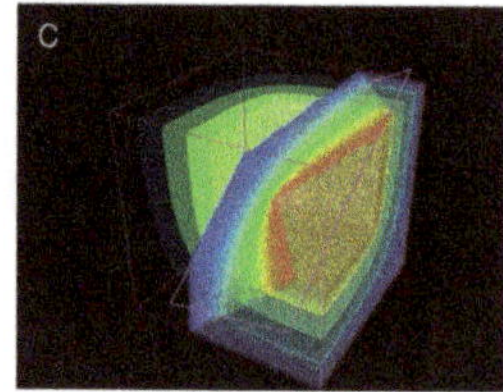

Figure 2. (**A**) A three-dimensional solution on an interior corner. (**B**) A section of the solution containing the point $(0.25, 0.25, 0.25)$, marked with a red cross. (**C**) A generic section.

3.5. Existence and Uniqueness Of Solutions

For completeness, one should indicate why Equation (2) does have any solutions at all—possibly with boundary conditions. To explain the steps of the proof one would require explaining not only the concept of weak solutions but also that of Sobolev spaces which would go beyond the scope of the entry. Suffice to say that the PDE Equation (2) is written as $Du = 0$ where D is the operator $\nabla \kappa \nabla$ in the sense of functional analysis. One then verifies that this operator is strongly elliptic in the sense of Nirenberg [25] since its "symbol" is the expression $\kappa ||\zeta||^2 + < \nabla \kappa, \zeta >$ and its "principal symbol", its leading term is $\kappa ||\zeta||^2$ which is non-zero whenever ζ is non-zero. Having a non-zero principal symbol whenever $\zeta \neq 0$ is what makes an operator "elliptic"; from there one can apply the existence and uniqueness theory of solutions to strongly elliptic differential operators as in (Evans [15], Chapter 6.2). This shows that there is a solution to our elliptic boundary value problem Equation (2) with Dirichlet boundary conditions on the interior and exterior surfaces and Neumann boundary conditions on the remaining surfaces. Moreover, this solution is unique.

4. Finite Elements—The Numerical and Computational Fundamentals

4.1. Boundary Conditions

4.1.1. Exterior and Interior Facing Surfaces—Dirichlet Boundary Conditions

When solving the temperature field, one applies the prior knowledge on exterior conditions. As mentioned earlier, one simplifies the simulation by erring on the conservative side and assuming that the surfaces that face the exterior and interior air masses are cooled and heated to the extreme temperatures $T_2 = -13.1\,^{\circ}\text{C}$ and $T_1 = 20\,^{\circ}\text{C}$. The entire geometry of the thermal bridge shall be denoted by Ω—the "problem domain". Those surfaces of Ω that are exposed to exterior or interior shall be denoted with S. The temperature u on S is known a priori. These kinds of boundary conditions are called Dirichlet boundary conditions in mathematical literature [15].

4.1.2. Structure Facing Surfaces—Neumann Boundary Conditions

The boundary of the domain Ω shall be denoted by $\partial\Omega$. Theses are all surfaces of Ω of S but there are more surfaces that are part of $\partial\Omega$, namely those that face the surrounding walls. Thus, the boundary of Ω is a "Boolean union"—in the design and CAD sense—of S and its complement $\partial\Omega - S$. The prior knowledge is that by definition a thermal bridge's energy loss towards its surroundings in the structure are small compared to those from inside to outside. It therefore makes sense to have the integral lines of the heat flux to go from inside to outside but never into the neighboring or surrounding structure. This is equivalent to demanding that the isocurves of the temperature only go into surrounding structures since they are always orthogonal to the heat flux; see Figure 1. These kinds of boundary conditions are expressed by $< \nabla u, n >= 0$—the exterior normal derivative of u vanishes. This kind of boundary condition is called Neumann boundary condition.

4.2. Tetrahedral Meshes

The FE algorithm described above relies essentially on the existence of an accurate tetrahedral mesh decomposing the problem domain, i.e., the geometry of the thermal bridge. Accurate meshing of polyhedral domains into tetrahedra is an entire business on its own. A good starting point is the work of Shewchuk on constrained Delaunay triangulations; see Shewchuk and Brown [26], Hang [27], Geuzaine and Remacle [28] and the references therein. A good tetrahedral mesh of a given domain is:

- a collection of vertices—points in three-dimensional space, together with
- a collection of tetrahedra—quadruples of indices into the vertex list, together with
- a way of associating each tetrahedron with one and only one material.

Such that the following requirements are satisfied as best as possible:

1. The union of all tetrahedra covers the entire thermal bridge but no other volumes.
2. Each material corresponds uniquely to a subset of all tetrahedra.
3. The surfaces constituting the material boundaries are surface sub-meshes in the sense that faces that lie on the boundary are a surface mesh without topological defects. These surfaces appear as the "constraints" of the Delaunay triangulation.
4. Each tetrahedron should be "as equilateral as possible" in the sense that no interior edge angle among all vertices is less than a typical threshold of about $20°$.

4.3. Derivation of Algorithm for the Temperature Coefficients

This subsection will only introduce as many mathematical notions and notations as necessary for understanding just the particular case in question. The reader who is applying but not deriving the algorithm can gloss over the following details. While an introduction to FE methods can easily lead into deep waters of numerical and computational mathematics, it turns out that the case at hand allows for significant simplification with regards to the usual treatment in the literature. In numerical linear algebra, the simplest case of a problem is described by a collection of linear constraints. In this case, the columns of a matrix describing a linear problem correspond to the unknowns, and the rows correspond to the constraints. One can immediately translate this setup to the steady heat transfer problem, obtaining the following problem. Find the scalar field u on Ω such that:

$$\begin{cases} \nabla(\kappa \nabla u) = 0 \text{ in } \Omega \\ u = u_0 \text{ in the interior and exterior facing surface } S \\ \partial_n u = 0 \text{ in the remaining boundary } \partial\Omega - S \end{cases} \tag{5}$$

where $\partial_n u =< \nabla u, \mathbf{n} >$ is the exterior normal derivative—the derivative of the temperature in the outward facing direction at those parts of the boundary of Ω that do not face exterior or interior air, for instance, the neighboring walls, or air pockets inside the wall. One needs Green's first identity in the form

$$\int_{\Omega} \left(\Psi \nabla \Gamma + < \Gamma, \nabla \Psi > \right) dV = \oint_{\partial\Omega} \Psi < \Gamma, \mathbf{n} > dS \tag{6}$$

where Γ is a vector field and Ψ is a scalar function. This variation of Green's first identity is derived by applying the classical divergence theorem from vector calculus to the vector field $\Psi\Gamma$ and expanding the divergence with elementary rules. Applied to the heat flux $\Gamma = \kappa\nabla u$, this identity yields:

$$\int_\Omega \left(\Psi\nabla(\kappa\nabla u) + <\kappa\nabla u, \nabla\Psi> \right) dV = \oint_{\partial\Omega} \Psi <\kappa\nabla u, \mathbf{n}> dS \tag{7}$$

Splitting this boundary integral over the two different boundary components $\partial\Omega = S \cup (\partial\Omega - S)$, one further obtains:

$$\oint_{\partial\Omega} \Psi <\kappa\nabla u, \mathbf{n}> dS = \oint_{\partial\Omega} \Psi\kappa\,\partial_n u\,dS = \oint_S \Psi\kappa\,\partial_n u\,dS + \oint_{\partial\Omega-S} \Psi\kappa\,\partial_n u\,dS \tag{8}$$

where the function Ψ is not yet specified further—it holds for all such Ψ. After inserting our knowledge on the boundary conditions from Equation (5), one obtains

$$\int_\Omega \kappa <\nabla u, \nabla\Psi> dV = \oint_S \Psi\kappa\,\partial_n u\,dS \tag{9}$$

So far, this holds for all (reasonable) functions Ψ. One can now restrict the treatment to all functions Ψ where $\Psi = 0$ on S.

For all such functions Ψ that vanish on S, one indeed arrives at

$$\int_\Omega \kappa <\nabla u, \nabla\Psi> dV = 0 \tag{10}$$

Therefore, one has already arrived at what is commonly called the "weak" formulation of the problem Equation (5): Find u such that

$$\begin{cases} u = u_0 \text{ on } S \\ \int_\Omega \kappa <\nabla u, \nabla\Psi> dV = 0 \text{ for all } \Psi \text{ such that } \Psi = 0 \text{ on } S \end{cases} \tag{11}$$

Note that the boundary condition on the extreme temperatures at the interior and exterior facing walls needs to be imposed "externally"—it is called an essential boundary condition. In contrast, the Neumann boundary condition is taken care of automatically—which is known as a natural boundary condition. The role of Ψ is a test function—in analogy with the FE method in structural mechanics with the concept of "virtual displacement", it could be called a "virtual temperature". Let n be the number of vertices of the tetrahedral mesh. It makes sense to seek the values of the solution u at those vertices. Write u_i for the unknown value of u at vertex i where $i = 1, \ldots, n$. Once those values are found, the value of u at an arbitrary point $x \in \Omega \subset \mathbb{R}^3$ will be approximated by simple linear interpolation.

$$u(x) = \sum_{i=1}^n u_i t_i(x) \tag{12}$$

where t_i is a function on Ω that is obtained by linearly interpolating a hypothetical solution that takes the value one at vertex i and the value zero at all other vertices. If the mesh was just a two-dimensional mesh, one could imagine plotting the "graph" of an analogously constructed function and that graph would look like a tent held up by a "pole" at vertex i, whence the choice of the letter t. In the mathematical literature, the functions t_i would be called a "partition of unity" because $\sum_i t_i = 1$ everywhere on the mesh. To be explicit, the function t_i can be constructed with the help of barycentric coordinates. Denote the barycentric coordinate of the point x inside a tetrahedron that is adjacent to vertex i with x_i. Then

$$t_i(x) = \begin{cases} x_i & \text{if } x \text{ is in a tetrahedron adjacent to vertex } i \\ 0 & \text{otherwise.} \end{cases} \tag{13}$$

The numbers u_i are the weights of the functions t_i towards the solution u. In order to find an algorithm that approximates the numbers u_i, one simply inserts Equation (12) into

Equation (11). At the same time, it makes sense to only consider all of the tent functions that do not belong to vertices on S, as test functions. Henceforth, indices of vertices that are not on S will be denoted by j. Then, the weak formulation becomes: Find the coefficients u_i on all vertices not on S such that

$$\sum_i u_i \int_\Omega \kappa < \nabla t_i, \nabla t_j > dV = 0 \text{ for all } j \tag{14}$$

Note that those t_j indeed vanish on S; hence, they satisfy the condition. Note that the sum over $_i$ involves all vertices, including those corresponding to the vertices in S. The corresponding coefficients u_i are non-zero. Therefore, Equation (14) is not just solved by the zero solution. To properly bring it into the form of a well-posed in-homogeneous linear problem, one applies the usual numerical procedure to "bring all known values to the right hand side". Thus, one can subtract those summands, indexed by i, that correspond to vertices in S. Each index j determines a joint linear constraint on all the numbers u_i. Conveniently, there are exactly as many constraints as unknowns, namely n minus the number of vertices in S. Now, it lends itself to abbreviate $A_{ij} = \int_\Omega \kappa < \nabla t_i, \nabla t_j > dV$ for a matrix A with n columns and n rows, one has arrived at the simple linear condition $Au = 0$ where, by a slight abuse of notion, the letter u refers to the vector $(u_1, \ldots, u_n)^T$. The matrix A is called the "stiffness" matrix, corresponding to the intuitive idea that if one of the coefficients is large and if for the sake of explanation the right hand side was not exactly zero but small and non-zero, then the corresponding term in the expansion was forced to be small; thus, a large stiffness permits only a small variation in the unknowns. One can assume that the vertices of the mesh are ordered in such a way that the ones in S come first. Then, the matrix A has a block structure: the upper left block of size $p \times p$, the lower right block of size $q \times q$, where $n = p + q$ are the numbers of vertices in S and in $\partial\Omega - S$, respectively. From these formulas, one elementarily derives the following numerical solution scheme:

- Compute the lower right block A_{lr} of the stiffness matrix A as well as an appropriate sparse matrix factorization.
- Compute the lower left block A_{ll} - a rectangular sub-matrix with q rows and p columns.
- Assemble the vector of Dirichlet boundary components b; this is a vector of size p; solve the linear system of equations $A_{lr}u = -A_{ll}b$, resulting in the desired coefficients u_i.

This completes the description of a basic FE scheme.

4.4. Extraction of Dissipation and U-Value from the Discretized Temperature

Given a temperature field u, it is easy to derive numerically all desired quantities. For instance, the discrete gradient operator [29] gives ∇u. Multiplication with κ interpolated on tetrahedra gives the thermal flux $\kappa \nabla u$. Summation of all scalar products between outward facing normals at all interior surfaces with the thermal flux gives an approximation to $\oint \kappa \nabla U dS$ which is the thermal dissipation. Division by $T_1 - T_2$ gives the thermal conductance in W/K as defined in Section 2.4.6.

4.5. Norms and Standards

This subsection will gather the most important ISO norms pertinent to the physical quantities discussed above. It will focus on the ISO norms as these are the most widespread ones, with the largest international recognition.

The definitions of the physical quantities are given in ISO 7345:2018 [30]. ISO norm ISO 6946:2017 [31] treats the topic of surface resistance definitions in Section 2.4.5 above. There are two ISO norms just for particular thermal bridges: windows, doors and shutters: ISO 10077-1:2017 [32] and ISO 10077-2:2017 [33]. The definitions of assumed surface temperatures can be found in ISO 10211:2017 [21]. Calculation methods for non-stationary (dynamic) thermal properties are given in ISO 13786:2017 [34].

The industrial norm for measuring thermal conductivity as described in Section 2.4.2 is ISO 13787:2003 [35]. The minimal required interior surface temperatures necessary for vapor condensation and mold growth prevention is given in ISO 13788:2012 [36]. Finally, ISO 13789:2017 [37] describes methods used to determine effective thermal transmission coefficients for estimating convective heat transfer due to ventilation.

To give examples of the large differences between national regulations, the Canadian norm CSA Z5010 [38] describes the thermal bridging calculation methodology, whereas the German standard for thermal bridge insulation is DIN 4180-2 [18]. Further discussion of national and local norms and standards would go beyond the scope of this entry.

5. I-Beam

One of the most frequently occurring examples of thermal bridges are structurally required steel components that penetrate walls and are exposed both to exterior and interior air. Due to steel's high conductivity, they are particularly prone to leaking heat. Figure 3 gives an overview of the geometry of an I-beam of type IPE400. A typical situation is where an I-beam penetrates a wall perpendicularly. The I-beam has an end on the outside—supporting a balcony, for instance, and the other end is on the inside, supporting a continuing wall slab. The ends are exposed to interior and exterior air masses at a distance of one meter each from the thermal break.

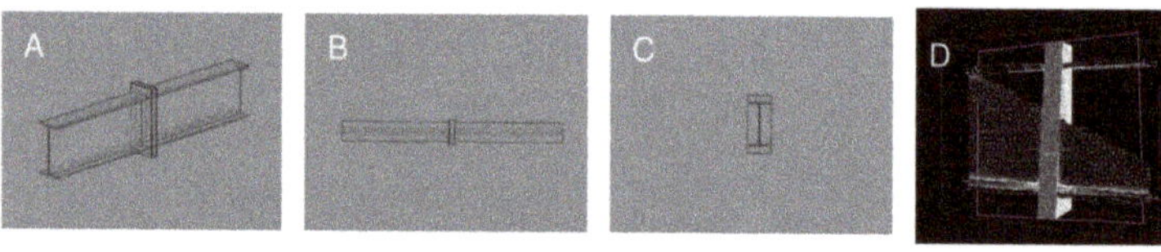

Figure 3. (**A**) The I-beam IPE400 with a thermal break in parallel perspective projection. (**B**) Top view. (**C**) Right view. (**D**) Section of the tetrahedral mesh.

There are commercially available thermal breaks interrupting the I-beam orthogonally in such a way that the load bearing is guaranteed by screws joining two plates. For instance, a commercial solution is the popular "Schöck Isokorb" [39]. A typical question is whether this commercially available thermal break could be replaced with a potentially cheaper plastic insulator component without aggravating the vapor condensation risk. This thermal break consists of two large steel plates. The geometry was modeled in Rhino3D and meshed with the Gmsh meshing software [28], see Figure 3D. FE analysis, as described in the preceding section, reveals that this is not the case. The factor f_{Rsi} does not reach the critical threshold. The temperature at the interior facing side of the steel plate is only 11.9 °C. See Figure 4 for a visualization of the resulting temperature field. A 3D visualization is available at Fuchs [40].

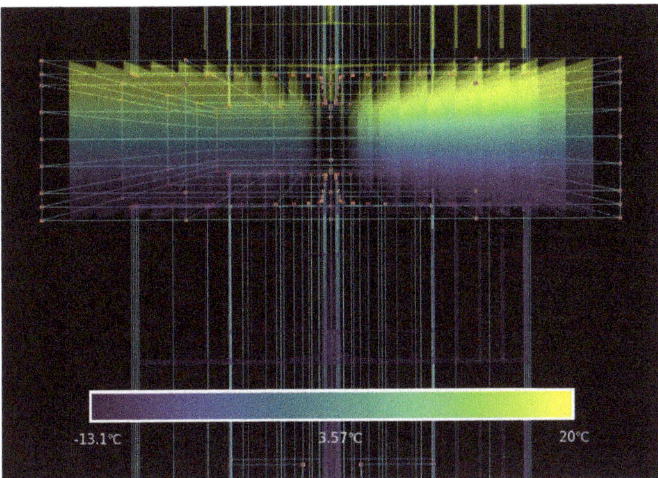

Figure 4. The solution of the I-beam temperature field.

6. Conclusions

In this entry, the paramount importance of accurate thermal modeling for energy-aware architecture has been pointed out. Filling this gap in the existing literature motivates this entry: Engineering texts often content themselves with approximations—that are often just one-dimensional such as the Glaser method or two-dimensional—without giving a full FE formulation. Texts on numerical methods describe the FE method in much more generality than is needed here, and mathematical literature on PDEs tend to describe only a variation of the equation that differs from the relevant one in two aspects: They often treat the time-dependent variant, and they consider a uniform material—this has the effect of abbreviating the divergence of the heat flux to a constant multiple of the Laplacian operator of the temperature. This obscures the equation's physical meaning, and makes it pointless for engineering applications. In this entry, this gap has been filled, introducing the reader to the FE solution of the relevant steady state and isotropic but non-uniform heat equation. This entry first defined all relevant terms: Temperature gradient, heat flux, energy dissipation, U-values, thermal conductance. It then explained why the relevant equation is Equation (2) and that its physical meaning is that the heat flux does not "diverge", in accordance with the absence of heating or cooling within the material, and that the intuitive notion of divergence actually corresponds to the divergence operator.

The notion of tetrahedral mesh was introduced, and how to obtain a computer-amenable discretization of the relevant steady state heat equation was also explained. This should have put the reader in a position to easily implement their own FE solver of the steady state heat equation, given a mesh that describes the given geometry in sufficient detail.

In a case study, a typical thermal bridge is an I-beam penetrating a wall. The question as to whether the insulation with plastic in the middle of the thermal break is sufficient was answered negatively—it turned out the interior facing surface temperature was not high enough to reliably prevent vapor condensation and mold growth.

The author hopes that the perusal of this entry has helped the reader gain an active understanding of the basic building blocks of the FE method for energy performance modeling, thereby putting them in a position to increase their maturity regarding sustainable architecture.

Funding: This research received no external funding.

Institutional Review Board Statement: Not applicable.

Informed Consent Statement: Not applicable.

Data Availability Statement: The software is available from the author. A software implementation of the core of the method—the solver—is contained in the materials to review.

Acknowledgments: English proofreading was done by Jamie Stangroom.

Conflicts of Interest: The author declares no conflict of interest.

Abbreviations

The following nomenclature, symbols and abbreviations are used in this manuscript:

FEM	finite element method
PDE	Partial differential equation
thermal dissipation	thermal energy loss per unit duration
Laplacian	the operator $\partial_x^2 + \partial_y^2 + \partial_z^2$
scalar field	the assignment of a real number to each point (x, y, z)
vector field	the assignment of a vector (three real numbers) to each point (x, y, z)
∂_t	time derivative of a scalar field
∇	on a scalar field, its gradient; on a vector field, its divergence
gradient	the vector field given by the three partial derivatives of a scalar field
divergence	the scalar field given by the sum of the three separate partial derivatives of a vector field
stationary heat equation	the heat equation with $\partial_t = 0$; synonym with steady state

References

1. Zienkiewicz, O.C.; Taylor, R.L.; Zhu, J.Z. *The Finite Element Method: Its Basis and Fundamentals*, 7th ed.; Elsevier/Butterworth Heinemann: Amsterdam, The Netherlands, 2013.
2. Ciarlet, P.G. *The Finite Element Method for Elliptic Problems*; SIAM: Philadelphia, PA, USA, 2002.
3. Wilson, E.L.; Nickell, R.E. Application of the finite element method to heat conduction analysis. *Nucl. Eng. Des.* **1966**, *4*, 276–286. [CrossRef]
4. Lewis, R.W.; Morgan, K.; Thomas, H.; Seetharamu, K.N. *The Finite Element Method in Heat Transfer Analysis*; John Wiley & Sons: Hoboken, NJ, USA, 1996.
5. Häupl, P.; Homann, M.; Kölzow, C.; Riese, O.; Maas, A.; Höfker, G.; Christian, N. *Lehrbuch der Bauphysik: Schall-Wärme-Feuchte-Licht-Brand-Klima*; Springer: Berlin/Heidelberg, Germany, 2017.
6. Cody, B. *Form Follows Energy: Using Natural Forces to Maximize Performance*; Birkhäuser: Basel, Switzerland, 2017.
7. Asdrubali, F.; Baldinelli, G.; Bianchi, F. A quantitative methodology to evaluate thermal bridges in buildings. *Appl. Energy* **2012**, *97*, 365–373. [CrossRef]
8. Zalewski, L.; Lassue, S.; Rousse, D.; Boukhalfa, K. Experimental and numerical characterization of thermal bridges in prefabricated building walls. *Energy Convers. Manag.* **2010**, *51*, 2869–2877. [CrossRef]
9. Arto, I.; Capellán-Pérez, I.; Lago, R.; Bueno, G.; Bermejo, R. The energy requirements of a developed world. *Energy Sustain. Dev.* **2016**, *33*, 1–13. [CrossRef]
10. Kleinhückelkotten, S.; Neitzke, H.; Moser, S. *Repräsentative Erhebung von Pro-Kopf-Verbräuchen natürlicher Ressourcen in Deutschland (nach Bevölkerungsgruppen)*; Umweltbundesamt: Dessau-Roßlau, Germany, 2016.
11. Khazal, A.; Sønstebø, O.J. Valuation of energy performance certificates in the rental market—Professionals vs. nonprofessionals. *Energy Policy* **2020**, *147*, 111830. [CrossRef]
12. *Terms and Conditions*, Ansys fluent; Ansys: Canonsburg, PA, USA, 2015. Available online: https://www.ansys.com/academic/terms-and-conditions (accessed on 1 April 2022).
13. Hecht, F. New development in FreeFem++. *J. Numer. Math.* **2012**, *20*, 251–265. [CrossRef]
14. Déqué, F.; Ollivier, F.; Roux, J. Effect of 2D modelling of thermal bridges on the energy performance of buildings: Numerical application on the Matisse apartment. *Energy Build.* **2001**, *33*, 583–587. [CrossRef]
15. Evans, L.C. *Partial Differential Equations*; American Mathematical Society: Providence, RI, USA, 2010.
16. Sawhney, R.; Crane, K. Monte Carlo Geometry Processing: A Grid-Free Approach to PDE-Based Methods on Volumetric Domains. *ACM Trans. Graph.* **2020**, *39*, 123. [CrossRef]
17. Grossmann, C.; Roos, H.G.; Stynes, M. *Numerical Treatment of Partial Differential Equations*; Springer: Berlin/Heidelberg, Germany, 2007; Volume 154.
18. Din Norm. *DIN Norm 4180-2: Thermal Protection and Energy Economy in Buildings—Part 2: Minimum Requirements to Thermal Insulation*; Din Norm: Berlin, Germany, 2012.
19. Taylor, B.J.; Cawthorne, D.; Imbabi, M.S. Analytical investigation of the steady-state behaviour of dynamic and diffusive building envelopes. *Build. Environ.* **1996**, *31*, 519–525. [CrossRef]

20. Gasparin, S.; Berger, J.; Dutykh, D.; Mendes, N. Solving nonlinear diffusive problems in buildings by means of a Spectral reduced-order model. *J. Build. Perform. Simul.* **2019**, *12*, 17–36. [CrossRef]
21. ISO. *Norm 10211: Thermal Bridges in Building Construction–Heat Flows and Surface Temperatures–Detailed Calculations*; ISO Norm, Vernier: Geneva, Switzerland, 2017.
22. Gerthsen, C.; Vogel, H. *Physik, 17. Auflage*; Springer: Berlin/Heidelberg, Germany, 1993.
23. Walter, W. *Gewöhnliche Differentialgleichungen. Eine Einführung*; Springer: Berlin/Heidelberg, Germany, 2000.
24. Glaser, H. Vereinfachte Berechnung der Dampfdiffusion durch geschichtete Wände bei Ausscheidung von Wasser und Eis. *Kältetechnik* **1958**, *10*, 358–364.
25. Nirenberg, L. Remarks on strongly elliptic partial differential equations. *Commun. Pure Appl. Math.* **1955**, *8*, 649–675. [CrossRef]
26. Shewchuk, J.R.; Brown, B.C. Fast segment insertion and incremental construction of constrained Delaunay triangulations. *Comput. Geom.* **2015**, *48*, 554–574. [CrossRef]
27. Hang, S. TetGen, a Delaunay-based quality tetrahedral mesh generator. *ACM Trans. Math. Softw.* **2015**, *41*, 11.
28. Geuzaine, C.; Remacle, J.F. Gmsh: A 3-D Finite Element Mesh Generator with built-in Pre- and Post-Processing Facilities. *Int. J. Numer. Methods Eng.* **2009**, *79*, 1309–1331. [CrossRef]
29. Mancinelli, C.; Livesu, M.; Puppo, E. A comparison of methods for gradient field estimation on simplicial meshes. *Comput. Graph.* **2019**, *80*, 37–50. [CrossRef]
30. ISO. *Norm 7345: Thermal Performance of Buildings and Building Components—Physical Quantities and Definitions*; ISO Norm, Vernier: Geneva, Switzerland, 2018.
31. ISO. *Norm 6946: Building Components and Building Elements—Thermal Resistance and Thermal Transmittance—Calculation Methods*; ISO Norm, Vernier: Geneva, Switzerland, 2017.
32. ISO. *Norm 10077-1: Thermal Performance of Windows, Doors and Shutters—Calculation of Thermal Transmittance—Part 1: General*; ISO Norm, Vernier: Geneva, Switzerland, 2017.
33. ISO. *Norm 10077-2: Thermal Performance of Windows, Doors and Shutters—Calculation of Thermal Transmittance—Part 2: Numerical Method for Frames*; ISO Norm, Vernier: Geneva, Switzerland, 2017.
34. ISO. *Norm 13786:Thermal Performance of Building Components—Dynamic Thermal Characteristics—Calculation Methods*; ISO Norm, Vernier: Geneva, Switzerland, 2017.
35. ISO. *Norm 13787: Thermal Insulation Products for Building Equipment and Industrial Installations—Determination of Declared Thermal Conductivity*; ISO Norm, Vernier: Geneva, Switzerland, 2003.
36. ISO. *Norm 13788: Hygrothermal Performance of Building Components and Building Elements—Internal Surface Temperature to Avoid Critical Surface Humidity and Interstitial Condensation—Calculation Methods*; ISO Norm, Vernier: Geneva, Switzerland, 2012.
37. ISO. *Norm 13789: Thermal Performance of Buildings — Transmission and Ventilation Heat Transfer Coefficients — Calculation Method*; ISO Norm, Vernier: Geneva, Switzerland, 2017.
38. ISO. *Norm CSA Z5010: Thermal Bridging Calculation Methodology*; ISO Norm: Toronto, Canada, 2021.
39. Müller, M.; Li, Z.; Standeker, J. Schöck Isokorb®–Gebrauchstauglichkeit unter Berücksichtigung der Langzeiteigenschaften. *Beton-Und Stahlbetonbau* **2020**, *115*, 35–43. [CrossRef]
40. Fuchs, M. I Beam Thermal Break Visualization. Available online: https://mathiasfuchs.com/rawtetgen.html (accessed on 31 March 2022).

 *encyclopedia*

Entry

Bioplastic as a Substitute for Plastic in Construction Industry

Ilaria Oberti * and Alessia Paciello

Dipartimento di Architettura, Ingegneria delle Costruzioni e Ambiente Costruito, Politecnico di Milano, 20133 Milan, Italy; alessia.paciello@mail.polimi.it
* Correspondence: ilaria.oberti@polimi.it; Tel.: +39-0223995147

Definition: Bioplastics have proven to be a viable substitute for plastics in some sectors, although their use in construction is still limited. The construction sector currently uses 23% of the world's plastic production, both for the materials themselves and for their packaging and protection. A considerable part is not recycled and is dispersed into the environment or ends up in landfills. In response to the environmental problems caused by oil-based plastic pollution, the development of biocomposite materials such as bioplastics represents a paradigm shift. This entry aims to explain what bioplastics are, providing a classification and the description of the different properties and applications. It also lays out the most interesting uses of these materials in the construction field.

Keywords: bio-based material; construction; environment; circular economy; life cycle

Citation: Oberti, I.; Paciello, A. Bioplastic as a Substitute for Plastic in Construction Industry. *Encyclopedia* **2022**, *2*, 1408–1420. https://doi.org/10.3390/encyclopedia2030095

Academic Editors: Masa Noguchi, Antonio Frattari, Carlos Torres Formoso, Haşim Altan, John Odhiambo Onyango, Jun-Tae Kim, Kheira Anissa Tabet Aoul, Mehdi Amirkhani, Sara Jane Wilkinson, Shaila Bantanur and Raffaele Barretta

Received: 30 June 2022
Accepted: 25 July 2022
Published: 28 July 2022

Publisher's Note: MDPI stays neutral with regard to jurisdictional claims in published maps and institutional affiliations.

1. Introduction

The problem of plastic pollution has become one of the most pressing global environmental issues. It is estimated that there are more than 269 thousand tons of plastic in the oceans, and if urgent prevention measures are not taken, there will be more plastic than fish by 2050 in terms of weight [1]. The immoderate production of plastics is growing year by year and is expected to grow by another 70% in the next 30 years [2]. The accumulation of plastics and microplastics in the oceans and on land over the decades is due to the reckless production and use of this material, which is particularly valued for its cheapness and variety of use. The causes of the plastic pollution problem can be summarized as follows: inordinate production, short use, polymer deterioration, inadequate waste management and disposal, and insufficient production chain for alternative materials. This is a great threat to the environment because already, millions of tons are dispersed in nature and end up amassing along shorelines or creating real garbage islands. This accumulation of products is also aggravated by the non-management or inadequate management of waste; in fact, it is often scattered in nature or abandoned in illegal dumps, polluting soil, fresh water, and oceans. Accumulation is also caused by the excessive use of single-use plastics: these are mainly polyethylene, found mostly in packaging and disposable products such as cutlery, glasses, or food containers. All these items have a very short life span, but a very long decomposition time that allows them to remain in the environment for years. For instance, a plastic bag has an average use of 12 min but takes 500 years to fully decompose [3]. Plastic pollution is aggravated by the deterioration of polymers, which produce a high amount of plastics/microplastics in the environment. Contributing to the waste management problem is the fact that only a small percentage of plastic is recycled. This is due to both the high economic cost of recycling processes and the low financial incentives for recycling. It is much cheaper to produce most types of plastic from scratch than to recycle old plastic. In addition, the complexity of recycling processes is an obstacle to achieving circular economy goals. Finally, the scarcity of production chains of more sustainable materials must be considered. Ecological alternatives are still limited and mainly related to products such as packaging, containers, and single-use plastic items, for which the raw materials used are plant-based polymers or polymers made from animal proteins. In most cases, the extraordinary properties of plastics mean that other, more environmentally friendly options are in short supply. However, a shift to more sustainable, non-petroleum-derived materials is the key step that production chains should take to reverse the trend of increasing plastic use and to transform the current linear economic model into a circular one. In fact, the circular economy is based on the different cycles of the production chain of discarded products to limit waste, minimize the use of totally new raw materials, maximize the potential of a material, and reduce the amount of waste that is dispersed into the environment. In this economic and production model, bioplastics assume high importance, particularly all those that are compostable since, after being used in different production cycles, they can be biodegraded, avoiding remaining in nature for years as opposed to fossil plastics. With this in mind, biopolymers, and biomaterials more generally, should be considered as a means for the ecological transformation of the economy and the construction sector. However, as Figure 1 shows, only 8% of the bioplastics produced are used in the buildings and construction sector, compared to 23% of the use, in the same sector, of plastic produced worldwide [4].

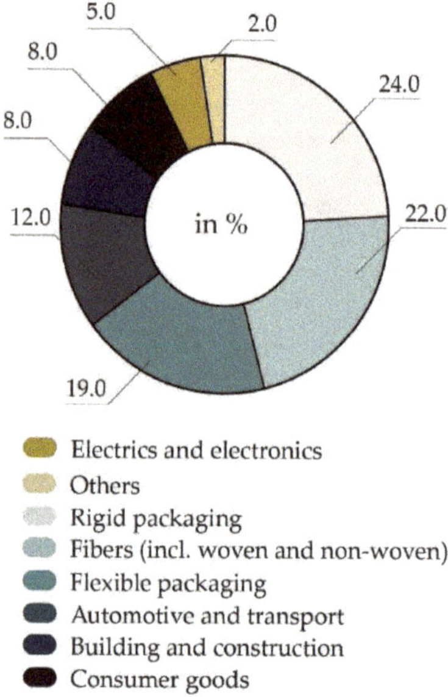

Figure 1. Use of bioplastics by market segment (2021). Reworked by the authors from [5].

The inordinate consumption of non-renewable raw materials and the production of waste that is difficult to degrade are the main problems associated with the use of plastics. Adherence to a production model using bioplastics would counteract both of these problems by switching from the exploitation of non-renewable raw materials derived from oil to renewable raw materials often derived from the agricultural sector. However, reference to the two aspects, raw materials and waste, is not sufficient to determine whether bioplastics are really a solution to the problem of plastic pollution.

2. Bioplastics

2.1. Bioplastic Definition

To understand the concept of bioplastics, it must first be made clear that this, like "traditional" plastics, includes materials composed of polymers. Polymers can be divided into two categories according to the origin: They are either of natural origin when they generate naturally and are, therefore, derived from biological processes, such as cellulose or chitin; or of synthetic origin, when they are produced artificially by humans through chemical processes. From this, it can be seen that not all plastics composed of synthetic polymers are part of the category of materials commonly referred to as "plastics", but rather fall under the category of bioplastics, since, being biodegradable, they meet the second characteristic of the official definition given by European Bioplastics. The terms "polymer" and "biopolymer" are commonly misused, using the first as a synonym for "plastic" and the second (with the intention of indicating a natural polymer) as a synonym for bioplastic. In general, bioplastics offer additional benefits over plastics obtained from oil, such as reduced carbon footprint, better functionality, and provide additional options in waste management, such as organic recycling. The use of bio-based plastics can reduce the consumption of non-renewable raw materials, the price of which may increase significantly in the coming decades due to increasingly limited availability. In addition, these plastics have the advantage of reducing greenhouse gas emissions because the plants that are grown for their production contribute to the absorption of CO_2 from the atmosphere during their growth period. Certainly, all of the aspects mentioned above are positive elements, but in order to be able to fully assess the situation, other aspects must also be taken into consideration. In fact, if the sector of bio-based plastics were to become widespread, one would have to fight from the outset against the exploitation of land used for the intensive cultivation of cereals [6] such as corn, to the detriment of agricultural food production and with the strong risk of compromising natural habitats. The Drivers of Deforestation and Degradation study [7] shows that among the main causes of global deforestation is agriculture, especially intensive agriculture for industrial purposes, which causes the alteration of local ecosystems, with possible consequences for fauna and flora. In addition to environmental concerns, the use of bio-based plastics raises some doubts regarding the durability, rigidity, and strength of these materials. It is possible, in fact, that these tend to lose their functional properties under certain environmental conditions such as high humidity. Therefore, there is still a need to improve the performance of bio-based materials to make them suitable for applications in the construction field.

2.2. Classification

The term bioplastic refers to a set of materials with different properties and applications, which can be grouped into three macro categories. The official classification follows the definition given by European Bioplastics, which distinguishes between bio-based polymers, biodegradable, or both [8]. An important clarification to make is that the adjective "bio-based" is not synonymous with the adjective "biodegradable", with the latter indicating a material that is degraded as a result of the action of certain microorganisms. As shown by Figure 2, some bio-based materials are biodegradable, for example, polylactic acid (PLA), while others are not, such as bio-PET.

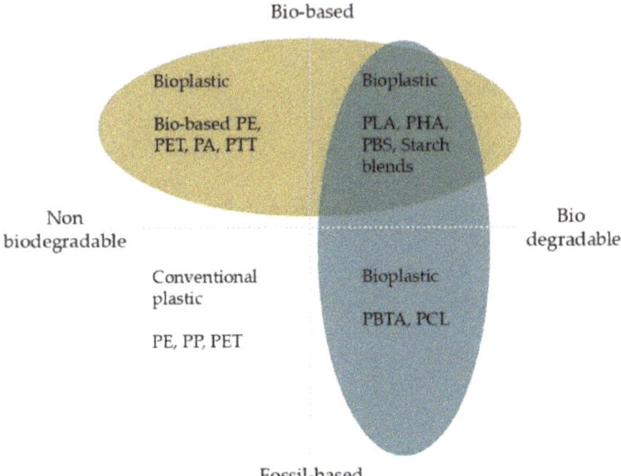

Figure 2. Classification of bioplastics according to European Bioplastics [9].

Bioplastics, therefore, are:

- Bio-based, derived from renewable natural raw materials;
- Biodegradable, of synthetic origin, but degrades quickly;
- Bio-based and biodegradable, when it possesses both properties (natural origin and degrades quickly).

2.2.1. Bio-Based Plastics

Bio-based plastics refer to those materials or products that are wholly or partially derived from organic biomasses (bio-based materials), such as corn, sugarcane, and cellulose.

2.2.2. Fossil-Based but Biodegradable Plastic

All plastics that are derived from fossil raw materials but degrade rapidly belong to this category. Actually, the composition of a fossil-based plastic does not preclude the quality of biodegradability. There are plastic polymers derived from oil that, under optimal conditions, decompose more rapidly than similar organic polymers derived from biomasses. All those oxo-biodegradable plastics are excluded, i.e., conventional plastics to which a mixture of additives such as metal salts are added during production, which accelerate the degradation process in the presence of ultraviolet light, oxygen, or heat. This exclusion, proposed by European Bioplastics, is supported by the results of a study conducted by researchers at Michigan State University [10].

2.2.3. Bio-Based and Biodegradable Plastics

Bio-based and biodegradable plastics are, as in the first case, materials or products that are wholly or partially derived from organic biomasses, but with the special feature of being biodegradable, such as polylactic acid (PLA), polyhydroxyalkanoates (PHAs), polybutylene succinate (PBS), and starch blends.

2.3. The Original Raw Materials

Fossil-based bioplastics are derived from the combination of various organic compounds produced from petroleum or hydrocarbons such as propane and butane, which are contained in natural subsurface gases or oil. In contrast, most known bioplastics are derived from carbohydrate-rich biomasses. The biomasses used can have different origins; they can come from the agricultural sector or the food industry, but there are other sources. Depending on the type of biomass used, the type of bioplastic also varies. Biomass can be divided into three macro classes [11]: those of first, second, or third generation. The first-generation feedstock includes food crops of carbohydrate-rich plants such as corn, sugarcane, beetroot, potato, tapioca, and barley. Starch and dextrose sugars are extracted from these plants. Many of the bioplastics mentioned are produced from these biomasses, as they are among the most efficient for producing bio-based plastics. The second generation of biomass includes materials with high lignocellulosic content not intended for food production. These substances represent the inedible by-products of food crops such as straw, corn stover, sugarcane waste, wood, or plant fibers, including hemp and flax. Bioplastics produced from these substances are biodegradable and compostable. Finally, further development occurs with third-generation raw materials. These are derived from the above-ground cultivation of algae, fungi, and microorganisms, which process the waste biomass. By using this family of materials, it is possible to contain the problems related to land consumption that are very pronounced in the first-generation biomass. To date, the construction industry has developed according to a logic based on oil extraction; examples include concrete, metals, and plastics. What is needed is a change of mindset that pushes toward a cultivated building materials approach. The "Cultivated Materials" concept implies the possibility and desire for a closed-loop building sector, where raw materials are derived from biomasses that can be returned to the soil after use. The idea behind this approach is that if a building can be cultivated, that is, whose materials that make up the various building components are derived from a cultivation process, it could also be composted at the end of its life and, thus, become the source of a new cultivation process [12]. A lot of cultivated/natural materials, like wood, were used in the past, nowadays substituted by other materials with improved properties. The challenge for the future is to recover the approach to cultivated materials, adapting them to the needs of current times, taking advantage of new technologies and knowledge.

2.4. End-of-Life Options

The European Commission has established a hierarchy of steps and alternatives to which products can be subjected before irreversible collection for landfill, which remains the last choice. To overcome the problem of pollution due to the accumulation of waste in the environment and to alleviate the waste load that goes to landfills every day, there are several alternatives that contribute to better disposal. Efficient waste management is of paramount importance to the European Commission, which, through Directive 2008/98/EC, has defined the five stages of waste disposal, classifying the treatments to which waste can be subjected according to its characteristics [13]. The advantage of using bioplastics is that organic recycling is used as a method of waste disposal. In fact, while the other types of end-of-life of a product can also be applied to conventional plastics, organic recycling is a prerogative of biodegradable bioplastics. Based on the model described, the priority treatment in waste management is the prevention and minimization of resource use, maximizing the functional performance of the end product, such as the number of times an object can be reused without losing its performance. Precisely in this regard, the European Commission issued Directive 904/2019, known as Single-Use Plastic (SUP), banning plastic disposable products in all EU countries from January 2022. The first end-of-life scenario for products is recycling; this practice can only be implemented through separate collection of plastic and the wet fraction. Recycling can be of three types: mechanical, chemical, and organic. The latter is practicable only for the disposal of bioplastics. When bioplastics are biodegradable, three new organic recycling options open up: home composting (for unprocessed or treated organic matter), industrial composting, and anaerobic digestion. Organic recycling includes "circular" end-of-life options for biodegradable or compostable plastic products. These are materials that biodegrade over a specific period, under particular conditions of temperature, moisture, and in the presence of microorganisms. The European standard EN 134325:2002 defines the requirements a product must satisfy to be considered biodegradable: it is defined as such if it dissolves 90% within 6 months and breaks down into simple molecules such as water, CO_2, and methane. If a material is biodegradable, however, it is not necessarily also compostable; in fact, these two terms are not synonymous. A compostable product is not only biodegradable, but as it degrades it turns into compost, which can be used in agriculture as a natural fertilizer by decomposing at least 90% within 3 months. If the previous options are not available, the best solution turns out to be energy recovery through waste-to-energy: bioplastic waste is incinerated, generating partially renewable energy through heat recovery [14]. When bio-based plastics are involved, incineration of these releases the same amount of carbon dioxide originally absorbed by plants, thus closing the loop. Because of its inherent environmental risks, landfilling is, according to the waste hierarchy, the least viable end-of-life option. Although these materials can be effectively recycled, in many countries, landfilling is still the most common disposal solution.

3. Applications in Construction

3.1. Biopolymers as Admixtures in Concrete and Mortar

One of the main applications of bioplastics in the construction industry concerns their use in concrete mixtures and dry premixed mortars, as additives that optimize these products. In several applications, bio-based aggregates compete on par with those of synthetic origin; their market is, therefore, likely to expand, especially with the increasing advances achieved by technology. The advantage of adding additives in building materials has been known since ancient times. If the main components of the mixture cannot provide the required properties, then it is necessary to implement these by introducing new components. Depending on the properties required, specific additives are added into the mixtures. The main applications of bio-based admixtures are concrete mixtures and mortars, although they are also used in paints, exterior and interior coatings, gypsum board, stucco, and joint mortars. Concrete is among the most commonly used building materials by humans, so it constitutes an important market for admixtures: about 15 percent of the total volume of concrete contains them [15]. Organic admixtures that are used in concrete mixtures include lignosulfonates, protein hydrolysates, and Welan rubber. Lignosulfonates constitute, by volume, the largest category of admixtures used in the construction industry. These can be used either for plasticization of concrete by improving its fluidity and workability, or water reduction; thus, high mechanical properties and durability can be obtained. In precast concrete, lignosulfonates are used to achieve greater mechanical strength. Other admixtures used include protein hydrolysates, which are used to lower the surface tension of water significantly in the preparation of expanded concrete. Protein hydrolysates generate spherical foam bubbles, which provide about 20 percent more compressive strength than hexagonal bubbles produced by synthetic foams [15]. Consequently, protein hydrolysates are preferred where the need arises to foam concrete to achieve low specific gravity while also maintaining compressive strength. A significant example is also found with the use of Welan rubber in concrete preparations. It is used as a viscosity modifier to increase the performance of highly fluid concretes, such as self-compacting concretes, which have a tendency to disintegrate, indicated by the formation of a water layer at the surface, or by the settling of large aggregates. Its main advantage is to stabilize without nullifying the fluidity of the mixture [15]. Additives are also used to enhance the performance of mortars, mainly used for plasters, tile adhesives, and self-levelling underlayments. Both masonry plasters and tile adhesives require a water-retention agent to prevent water loss into the wall: among bio additives, cellulose ethers are the most widely used for this purpose. Other uses include gypsum board, which requires a dispersant to thin the gypsum slurry. In this field, ammonium lignosulfonate is the most widely used. For grouts, on the other hand, the addition of xanthan gum allows fluidity to achieve better penetration of the product into the crack. A more recent study has furthered research in this area, identifying additional biopolymers that can be used as admixtures in the production of sustainable concrete. Tests carried out by Karandikar using potato starch showed increased mechanical strength of lime mortars containing this natural polymer [16]. Again, Govin et al. documented that guar gum increases the water retention capacity of cement mortars, limiting water loss [16]. These same properties have also been identified in a viscous biopolymer derived from Opuntia Ficus Indica extract. It has been shown that the use of this natural polymer results in a significant improvement in durability (about 5–7 times) in lime mortars [16]. The biopolymer has been used as an additive in a hydraulic lime mortar, with positive effects on mechanical properties due to the adhesive nature of this material. Among the natural polymers mentioned above, cactus extract is an interesting alternative because of its ability to both improve the mechanical strength and durability of lime mortar. However, to fully understand the potential of its use, it is necessary to analyze in more detail the interactions created between the viscous biopolymer and the cement mixture. Based on the results obtained, it is possible to state that these admixtures improve the water retention of concrete, preventing premature drying of the mix, thus reducing the possibility of cracking. Nevertheless, the possible negative effects must also be considered, especially on the setting and hardening of concrete.

3.2. Cement Replacement: Starch and Alginate as Adhesive Components

A study conducted by the University of Delft [17] seeks sustainable alternatives to replace cement as a binder in building materials. The subject of the research is the possibility of using certain biopolymers, particularly cornstarch and alginate, no longer just as additives, but as binders. In fact, the concrete mix is composed of a binder, cement, aggregates that vary in grain size, additives, and water. Cement, however, has a significant impact on the environment, so the exploration of new bio-based building materials as a binder is a step toward more sustainable construction. The analysis conducted in [17] is based on experiments showing that mixing cornstarch with sand and water and heating (to a temperature not exceeding 200 °C) the mixture produces a hardened solid material, called CoRncrete, with compressive strengths with values up to 20 megapascals. During heating, the starch molecules partially dissolve and form a gel that glues the sand grains, hardening when dried. Alginate is another biopolymer that can be used as a binder. It is extracted from a wide variety of algae that inhabit the oceans. The advantage of using alginate, as opposed to cornstarch, is that the raw material from which it is extracted grows in the sea and does not require land area to grow. The quick drying and low weight of CoRncrete along with its derivation from renewable sources make it an attractive material. CoRncrete can be used as a block structural material, such as bricks, in dry environments placed in arid areas and inside buildings. Mortar to seal the joints of the blocks could be replaced using fresh CoRncrete heated in place. Alternatively, the blocks could be molded through molds into interlocking shapes. Given the low temperatures required for heating, CoRncrete is an attractive material for 3D printing buildings. Although the temperatures required to heat and dry this material are much lower than in traditional concrete production or clay brick making, the heating procedure cannot currently be applied to the scale of a building [18]. As for high degradability, this has both positive and negative effects: although it cannot be used outdoors in wetlands, "biocement" can find application in temporary structures. In fact, alginate or cornstarch biopolymers are resistant to compression as long as they are dry; however, once exposed to water, they weaken easily. Further development is needed to improve the durability of such materials so that their applications in construction can be consolidated and expanded. The use of additives or hydrophobic coatings are viable options to improve their durability, if these do not compromise degradability or recyclability factors, which are necessary to counter the waste problem. The eco-cost of CoRncrete, calculated through a cradle-to-cradle life cycle assessment (LCA), appears to be higher than traditional concrete [17]. This is mainly due to the use of fertilizers for growing corn, which results in higher ecotoxicity. If these aspects were eliminated, CoRncrete could offer potential reductions in environmental impact. Based on these results, it can be concluded that CoRncrete is a promising building material due to its light weight, good compressive strength, biodegradability, and its derivation from renewable sources. However, factors such as durability and sustainability pose challenges for its use.

3.3. Application of Biopolymers as Aggregates in Concrete Mixtures

Lightweight concrete is a mix containing lightweight aggregates whose density is lower than the aggregates used in traditional concrete. It is made by replacing all or part of the stone aggregates (gravel, crushed stone) with lighter aggregates that have a honeycomb structure. Such aggregates can be either artificial (naturally derived but obtained with an industrial process), such as expanded clay, pumice, and perlite, or synthetic, such as expanded polystyrene (EPS). The main disadvantage of using artificial aggregates is the exploitation of limited, non-renewable resources, resulting in the depletion of the planet. In fact, once incorporated into the concrete mix, all aggregates undergo irreversible transformation. On the other hand, the use of synthetic aggregates, such as EPS, derived from petroleum products, involves the use of non-renewable raw materials and a serious environmental impact due to the extremely polluting extraction and processing processes: the emission of pentane during the production of EPS is another fact that has considerable effects [19]. In addition, expanded polystyrene is non-biodegradable and resistant to photolysis. All these reasons, added to the rising price of crude oil and its derivatives, are causing a growing interest in the development of more environmentally friendly materials. The study conducted by the University of Auckland [19] investigated the possibility of using expanded polylactic acid (EPLA) as an aggregate for lightweight concrete to replace EPS. Polylactic acid is a biopolymer derived from cornstarch and sugarcane, expanded using carbon dioxide so as to drastically reduce its density. Experimental research shows positive and negative aspects: EPLA is a biodegradable biopolymer that is less polluting than EPS, but when used in concrete, it reduces its compressive strength and elastic modulus. The data cannot be considered conclusive, and further studies are needed to ascertain the validity of this alternative.

3.4. Reinforcing Biopolymers for Earthen Building Construction

Biopolymers can be used in earthen constructions in order to improve certain properties. Earthen constructions have been a type of housing used since ancient times due to the easy availability of the raw material. Today, they still constitute housing for about one-third of the world's population [20]. However, these buildings have several limitations in terms of mechanical strength and durability. They are particularly susceptible to intense weather phenomena such as rainfall and flooding, the damage caused by which poses a potential risk to humans [21]. The use of binders or reinforcing substances is, therefore, necessary to increase the stability of earthen constructions. The main binder used by humans in construction is cement; however, the process for its production has been reported to be the source of nearly 5 percent of global greenhouse gas emissions [20]. Even when used in a limited dosage, cement can significantly increase the carbon footprint of earthen construction. In particular, a study by Van Damme and Houben [22] showed that "the ratio of carbon dioxide emissions to compressive strength is greater in cement-stabilized unfired earth than in other building materials such as concrete or unstabilized unfired earth." A further disadvantage to the use of cement is major regional differences in the global distribution of its price: it is particularly expensive in developing countries, making it even more difficult to use in those geographic areas. For these reasons, numerous studies have been conducted since the early 2010 s to investigate the stabilizing effects of biopolymers in earth-based materials to identify more sustainable alternatives. In 2012, Chang and Cho [23] studied the effect of glucan biopolymer on the compressive strength of soil. They showed that adding a small amount (less than 5 g/kg) of the biopolymer under study results in a threefold increase in compressive strength of soil, while adding 10 percent cement only leads to a doubling. Gel-type biopolymers, particularly gellan gum and xanthan gum, have also been used as stabilizers in earth-based constructions. Because of its stability, even under high temperature and low pH conditions, gellan gum is used as an additive to improve strength. Similar performance is also obtained with the use of xanthan gum. The results obtained meet the strength criteria for these two biopolymers to be used as a binder in the production of rammed earth bricks. Xanthan gum is also used in combination with natural fibers to make an even more efficient building material. Earth remains a brittle mineral compound, which possesses low tensile strength and is subject to shrinkage during drying, resulting in fractures; therefore, the addition of natural fibers such as hemp, flax, and date palm is a solution to cope with these problems. The results collected [24] show that the addition of fibers alone decreases the compressive strength of the soil compared to the sample that does not contain them. In contrast, the combined addition of xanthan gum and natural fibers creates a synergistic effect that significantly improves the properties of the soil sample, giving it higher compressive and tensile strength and better ductile properties. To increase the poor water resistance of earthen architecture, a study conducted by Perrot, et al. [25] investigated the feasibility of using chitosan as an additive or material to produce an exterior coating to improve the resistance of rammed earth and adobe blocks to water-induced degradation. One of the most attractive aspects of chitosan concerns its low cost since it is commonly extracted from crustacean waste from the food industry. In addition, it has proven properties such as biodegradability, antibacterial activity, and nontoxicity. In the aforementioned study, the soil material samples, both treated with the biopolymer and untreated, were subjected to a number of tests to evaluate their compressive strength, tensile strength, and flexural behavior, as well as sensitivity to water-induced degradation. The outcome of the tests showed a positive influence of the presence of chitosan both as a surface coating and as an additive added to the soil mixture. Low concentrations are sufficient to make a water-repellent outer coating while, when incorporated into the manufacture of the mixture, the results show that at least 3 percent is necessary to provide some hydrophobicity. In summary, the analysis shows that even low concentrations of chitosan can greatly improve performance with regard to water-induced degradation and mechanical properties. However, it is important to continue studies concerning the effectiveness of chitosan about long-term durability.

3.5. Biopolymers as a Material for 3D Printing

3D printing, also known as additive manufacturing (AM), consists of a series of processes for fabricating components by adding material one layer after another, each corresponding to a cross section belonging to the 3D model. This technology enables the production of complex and irregular structures, thanks to the ability to design and customize the digital model that will later be printed. Although the most commonly used materials for 3D printing are plastics, metal, and ceramics, a number of examples demonstrate the possibility of using bioplastics as a material for 3D printing, also in the field of buildings and construction [26]. Substances such as lignocellulose, starch, algae, and chitosan-based biopolymers are possible naturally derived alternatives for the application of additive printing to the construction sector through an environmentally friendly approach. Polylactic acid is used as a bioplastic filament for 3D printing, proving suitable due to its low melting temperature. In addition to PLA, numerous other bioplastics designed and patented specifically for specific projects have been applied over the past decade. Berlin-based start-up Made of Air has developed a bioplastic composed of forest and agricultural waste that is combined with a binder derived from sugarcane to create thermoplastic granules that can be melted and molded. This material absorbs carbon and can be used for different applications from furniture to building facades. Since the material stores more CO_2 from the atmosphere than it emits during its life cycle, Made of Air is a carbon-negative material. This product was first installed on a building in April 2021, to clad an Audi dealership in Munich with hexagonal panels called HexChar. Earthen constructions are traditional structures for many countries around the world, despite for which development is still limited due to the long time required for labor and curing of the material [19]. For these reasons, recently, some methods have been analyzed to make the construction process easier and faster, to achieve a construction process of traditional earthen architectures compatible with modern standards in terms of productivity and cost. As a result, 3D printing was also applied in the production of earthen buildings, in an attempt to combine state-of-the-art techniques with the most common and ancient material in the history of construction. To achieve its goals, alginate biopolymer was added to the earth to increase the setting speed of the material and make its extrusion more efficient.

3.6. Cultivating Building Elements: A Bioplastic That Is Grown

Particularly promising for the future are those bioplastics produced from third-generation raw materials, waste biomass from food processes, which do not involve land consumption. Producing a new sustainable material from waste products makes it possible to minimize environmental impact by avoiding the burning of agricultural waste, which is a major source of pollution, especially in developing countries [27]. A number of studies, first carried out on packaging production, demonstrated the ability of organisms such as algae, fungi, and microorganisms to process waste biomass by transforming it into a bioplastic, capable of fostering a zero-waste industry. Over the years, research has focused on the potential of the fungal mycelium, the vegetative apparatus of fungi, which is able to grow thanks to the lignocellulosic content present in agricultural waste material [27]. The cellulose contained in the biomass acts as both food and structure for the growth of the mycelium which, solidifying in about a week, goes on to unite all the smaller portions of the tissue forming a larger whole. The fungal tissue acts as a binder, and as it solidifies, it will harden into whatever form chosen, constituting a light and strong material. The cellular structure of the mycelium consists of chitin polymers, a feature that distinguishes fungi from vegetative matter and gives the material plastic behavior once processed. For this reason, in tests carried out [12], materials grown from fungal mycelium compare favorably with some plastics and other composites made from fossil-based resins and glues. There are numerous interesting features of mycelium-based bioplastic; in fact, it is 100 percent compostable and is obtained through a process that requires no energy and releases no carbon emissions. In addition, mycelium-based bioplastic can be made from different types of agricultural waste, depending on the availability of the location, making it possible to produce it in any region of the globe from local biomass remains. For example, Pelletier et al. made mycelium-based bioplastics from waste products such as grass, rice straw, sorghum, flax, and hemp, while Appels et al. used agricultural wastes such as straw, sawdust, and cotton [27]. Empirically, some properties of the new material have been observed that make it particularly promising for use in the construction sector, especially in the production of sound-absorbing, heat-insulating and bio-brick panels. Through incidental tests, in fact, Hari Dharan [12] has shown that the bioplastic analyzed has high fire resistance, a good index of both thermal and acoustic insulation, and is free of toxic volatile organic compounds. In addition, the compressive strength and flexural characteristics are comparable to those of fossil-based compounds and those of engineered wood. These findings inspired the cultivation of some bricks, composed of sawdust and mycelium (Figure 3), later assembled in Philip Ross's design for a tea house exhibited in Germany.

Figure 3. Mycelium-based composite (Source: Wikimedia Commons).

The spread of knowledge about this material led to several experiments: in the New Children's Museum in San Diego, a room composed of 1200 shaped mycelium bio-bricks was set up for an interactive exhibition. These blocks were created to be used in hands-on games, as if they were Lego bricks. When the initiative ended in 2014, the museum earmarked all the bricks for San Diego's municipal composting program, leaving no waste behind. The use of mycelium is not only about the materials themselves, but also involves a rethinking of standard production methods. The manufacture of each individual component no longer involves mass industrial production. The material is cultivated and needs its own development time to which it forces us to adapt. Mycelium-based plastics can be used for a variety of purposes: they can be grown to make a monolithic element, through the use of large-scale formwork; but they can also be made into individual building elements, later assembled mechanically as in David Benjamin's Hi-fi project. These individual elements can be used both in the structural field, through the production of bio-bricks, and for the production of panels for thermal and acoustic insulation. At the same time, research on these materials is producing knowledge that allows the materials themselves to inspire and transform architectural design, changing its aesthetics, but more importantly, the thinking behind it. Today, interdisciplinary teams are working in this emerging field to integrate fungal bioplastics into the construction industry [28]. As a prototype, the Mushroom Tiny House, a small mobile home that exceeds just 5 m^2 of usable area and possesses high energy performance, was made by Ecovative. The insulation, interposed to two layers of pine wood (Figure 4), grows in a few days, generating a light and homogeneous compound, comparable to classic polystyrene [29]. Other interesting uses of mycelium in architecture can be found in projects such as Breeding Space by Maria Mello [30] or Grown Structures by Aleksi Vesaluoma in collaboration with Astudio [31].

Figure 4. Mycelium-based composite. (Source: Courtesy of Ecovative).

4. Conclusions

Bioplastics provide an attractive sustainable alternative to some construction materials, contributing to solve the main problems caused by the use of plastics, such as the consumption of non-renewable raw materials and the production of non-biodegradable waste. Despite the many advantages of bioplastics, such as being bio-based, biodegradable, or both, not all aspects are positive and, therefore, their disadvantages must also be considered. However, there is a need to further study and test these new materials for application in the construction sector in order to make the most of their many potentials. To bring about a transformation, however, it is not enough to replace traditional plastics with bioplastics: a substantial change in the economic and production model is required. It is urgent to complete the transition from a linear to a circular economy model, designed to be regenerative.

Author Contributions: Conceptualization, I.O.; investigation, A.P.; resources, A.P.; writing—original draft preparation, A.P.; writing—review and editing, I.O. and A.P. All authors have read and agreed to the published version of the manuscript.

Funding: This research received no external funding.

Institutional Review Board Statement: Not applicable.

Informed Consent Statement: Not applicable.

Data Availability Statement: Not applicable.

Conflicts of Interest: The authors declare no conflict of interest.

References

1. World Economic Forum. Ellen MacArthur Foundation and McKinsey & Company, The New Plastics Economy—Rethinking the Future of Plastics 2016. Available online: http://www.ellenmacarthurfoundation.org/publications (accessed on 5 June 2022).
2. Istituto Oikos. Life Beyond Plastic. Available online: https://www.istituto-oikos.org/mareinclasse/cosa-dice-la-scienza (accessed on 5 June 2022).
3. Napper, I.E.; Thompson, R.C. Environmental Deterioration of Biodegradable, Oxo-biodegradable, Compostable, and Conventional Plastic Carrier Bags in the Sea, Soil, and Open-Air Over a 3-Year Period. *Environ. Sci. Technol.* **2019**, *53*, 4775–4783. [CrossRef] [PubMed]
4. Pilla, S. Engineering Applications of Bioplastics and Biocomposites—An Overview. In *Handbook of Bioplastics and Biocomposites Engineering Applications*, 1st ed.; Pilla, S., Ed.; Publisher: Salem, MA, USA, 2011; pp. 1–16.
5. European Bioplastics. Fact and Figures. EUPB 2021. Available online: https://docs.european-bioplastics.org/publications/EUBP_Facts_and_figures.pdf (accessed on 14 July 2022).
6. Altroconsumo. Available online: https://www.altroconsumo.it/casa-energia/pulizie/news/plastica-biodegradabile-facciamo-chiarezza/ (accessed on 5 June 2022).
7. Kissinger, G.; Herold, M.; De Sy, V. *Drivers of Deforestation and Forest Degradation: A Synthesis Report for REDD+ Policymarkets*, 1st ed.; Lexeme Consulting: Vancouver, BC, Canada, 2012; pp. 4–13.
8. Ozdamar, E.G.; Ates, M. Architectural Vantage Point to Bioplastics in the Circular Economy. *J. Archit. Res. Dev.* **2018**, *2*. [CrossRef]
9. European Bioplastics. What are Bioplastics? EUPB 2016. Available online: https://docs.european-bioplastics.org/2016/publications/fs/EUBP_fs_what_are_bioplastics.pdf (accessed on 5 June 2022).
10. European Bioplastic. Oxo-Biodegradable Plastics and Other Plastics with Additives for Degradation. EUPB 2015. Available online: https://docs.european-bioplastics.org/publications/bp/EUBP_BP_Additive-mediated_plastics.pdf (accessed on 5 June 2022).
11. European Bioplastic. Renewable Resources. EUPB 2016. Available online: https://docs.european-bioplastics.org/publications/fs/EuBP_FS_Renewable_resouces.pdf (accessed on 5 June 2022).
12. Hebel, D.E.; Heisel, F. *Cultivated Building Materials*, 1st ed.; Birkhauser: Basilea, Switzerland, 2017.
13. European Bioplastic. End of Life Options for Bioplastics. EUPB 2017. Available online: https://docs.european-bioplastics.org/publications/fs/EuBP_BP_End-of-life.pdf (accessed on 5 June 2022).
14. European Bioplastic. Energy Recovery. EUPB 2015. Available online: https://docs.european-bioplastics.org/publications/bp/EuBP_BP_Energy_recovery.pdf (accessed on 5 June 2022).
15. Plank, J. Applications of Biopolymers and Other Biotechnological Products in Building Materials. *Appl. Microbiol. Biotechnol.* **2004**, *66*, 1–9. [CrossRef] [PubMed]
16. Shanmungavel, D.; Selvaraj, T.; Ramadoss, R.; Raneri, S. Interaction of a viscous biopolymer from cactus extract with cement paste to produce sustainable concrete. *Constr. Build. Mater.* **2020**, *257*, 119585. [CrossRef]
17. Kulshreshtha, Y.; Schlangen, E.; Jonkers, H.; Vardon, P.; van Paassen, L. CoRncrete: A corn starch based building material. *Constr. Build. Mater.* **2017**, *154*, 411–423. [CrossRef]
18. Paassen, L.V.; Kulshreshtha, Y. Biopolymers: Cement Replacment. In *Cultivated Building Materials*, 1st ed.; Hebel, D.E., Heisel, F., Eds.; Birkhäuser: Basel, Switzerland, 2017; pp. 116–123.
19. Sayadi, A.; Neitzert, T.R.; Clifton, G.C. Influence of poly-lactic acid on the properties of perlite concrete. *Constr. Build. Mater.* **2018**, *189*, 660–675. [CrossRef]
20. Chang, I.; Jeon, M.; Cho, G.-C. Application of Microbial Biopolymers as an Alternative Construction Binder for Earth Buildings in Underdevelopment Countries. *Int. J. Polym. Sci.* **2015**, *2015*, 326745. [CrossRef] [PubMed]
21. Aguilar, R.; Nakamatsu, J.; Ramírez, E.; Elgegren, M.; Ayarza, J.; Kim, S.; Pando, M.A.; Ortega-San-Martin, L. The potential use of chitosan as a biopolymer additive for enhanced mechanical properties and water resistance of earthen construction. *Constr. Build. Mater.* **2016**, *114*, 625–637. [CrossRef]
22. Van Damme, H.; Houben, H. Earth Concrete. Stabilization Revisited. *Cem. Concr. Res.* **2017**, *114*, 90–102. [CrossRef]
23. Chang, I.; Cho, G.C. Strengthening of Korean residual soil with b-1,3/1,6 glucan biopolymer. *Constr. Build. Mater.* **2012**, *30*, 30–35. [CrossRef]
24. Benzerara, M.; Guiheneuf, S.; Belouettar, R.; Perrot, A. Combined and synergic effect of Algerian natural fibers and biopolymers on the reinforcement of extruded raw earth. *Constr. Build. Mater.* **2021**, *289*, 123211. [CrossRef]
25. Perrot, A.; Rangerard, D.; Courteille, E. 3D Printing of Earth-Based Materials: Processing Aspects. In *Construction and Building Materials*; Elsevier: Amsterdam, The Netherlands, 2018; Volume 172, pp. 670–676.
26. Liu, J.; Sun, L.; Xu, W.; Wang, Q.; Yu, S.; Sun, J. Current advances and future perspectives of 3D printing natural-derived biopolymers. *Carbohydr. Polym.* **2019**, *207*, 297–316. [CrossRef] [PubMed]
27. Joshi, K.; Meher, M.K.; Poluri, K.M. Fabrication and Characterization of Bioblocks from Agricultural Waste Using Fungal Mycelium for Renewable and Sustainable Applications. *ACS Appl. Bio Mater.* **2020**, *3*, 1884–1892. [CrossRef] [PubMed]
28. Lekka, D.A.; Pfeiffer, S.; Schmidts, C.; Seo, S. A review in architecture with fungal biomaterials: The desired and the feasible. *Fungal Biol. Biotechnol.* **2021**, *8*, 17. [CrossRef] [PubMed]
29. Mushroom Tiny House. Available online: https://mushroomtinyhouse.com/ (accessed on 5 June 2022).

30. Breeding Space. Available online: https://www.plataformaarquitectura.cl/cl/946524/ (accessed on 16 July 2022).
31. Grown Structures. Available online: https://www.dezeen.com/2017/06/20/aleksi-vesaluoma-mushroom-mycelium-structure-shows-potential-zero-waste-architecture/ (accessed on 16 July 2022).

Review

Iranian Household Electricity Use Compared to Selected Countries

Dorsa Fatourehchi [1],*, Masa Noguchi [1] and Hemanta Doloi [2]

[1] ZEMCH EXD Lab, Faculty of Architecture, Building and Planning, The University of Melbourne, Parkville, VIC 3010, Australia

[2] Smart Villages Lab, Faculty of Architecture, Building and Planning, The University of Melbourne, Parkville, VIC 3010, Australia

* Correspondence: dfatourehchi@student.unimelb.edu.au

Abstract: Buildings account for nearly 40% of energy use in global contexts and climatic conditions tend to contribute to consumption. Human activities are also influential in energy consumption and carbon dioxide (CO_2) emissions that lead to global warming. Residential buildings are responsible for a considerable share. There are countries aggravating this situation by heavily relying on fossil fuels. Oil-rich countries are allocating an energy subsidy to the public, making energy cheaper for their consumers. This may result in negative consequences, including households' inefficient energy use behaviours in countries such as Iran. Beyond the impact of energy subsidy allocation, this study aims to explore the climatic and non-climatic factors that affect the increase in domestic electricity use, particularly in Iran. For this purpose, this study begins with a comparative analysis between countries with and without the energy subsidy to examine the trends in domestic electricity use. Afterwards, the tendency of households' electricity use in Iran will be analysed in consideration of climatic and non-climatic factors among several provinces in Iran. This study exploited published statistical data for the analysis. The results indicate the tendency of increased domestic electricity use due to the country's generous subsidy offered to the public as well as climatic and non-climatic factors in Iran. These results may provide an opportunity for future studies regarding building occupants' inefficient energy use behaviours for policy enactment in Iran and other oil-rich countries.

Keywords: behavioural factors; developing and developed countries; domestic sector; electricity use per capita; energy subsidy; Iran

Citation: Fatourehchi, D.; Noguchi, M.; Doloi, H. Iranian Household Electricity Use Compared to Selected Countries. *Encyclopedia* **2022**, *2*, 1637–1665. https://doi.org/10.3390/encyclopedia2040112

Academic Editors: Massimiliano Lo Faro and Raffaele Barretta

Received: 12 July 2022
Accepted: 19 September 2022
Published: 23 September 2022

Publisher's Note: MDPI stays neutral with regard to jurisdictional claims in published maps and institutional affiliations.

1. Introduction

The significant impact of human activities on climate change and global warming has been highlighted by the Intergovernmental Panel on Climate Change in various reports [1,2]. Although several resources are known to contribute to this issue through CO_2 emissions, energy has a large share of 73% of the overall emissions [3]. The International Energy Agency (IEA) analysis showed that the most critical sectors responsible for CO_2 from energy consumption are industry, transportation, and buildings, amongst which buildings account for 17.5% of total emissions. However, with 19% of contributions, residential buildings have a considerable share among other types of buildings [3,4] (Figure 1).

In the special report on global warming (SR15) [5], IPCC demonstrated that there is a possibility to achieve the Paris target of 1.5 °C if CO_2 reductions reach 45% by 2030 and a net-zero emission by 2050. However, issues have made net-zero emissions harder to achieve in the future. One of these issues is the energy subsidy allocated to consumers in oil-rich countries to reduce their energy costs. This has negatively encouraged the overconsumption of fossil fuels [6–8] while reducing the incentives for further investments in renewable energies [9,10]. According to the IEA [11], some countries allocate an energy subsidy to reduce costs for energy consumers. However, unfortunately, such an energy subsidy is mainly given to fossil fuels [12,13], in which non-OECD countries form a large portion of the subsidy in the world [14].

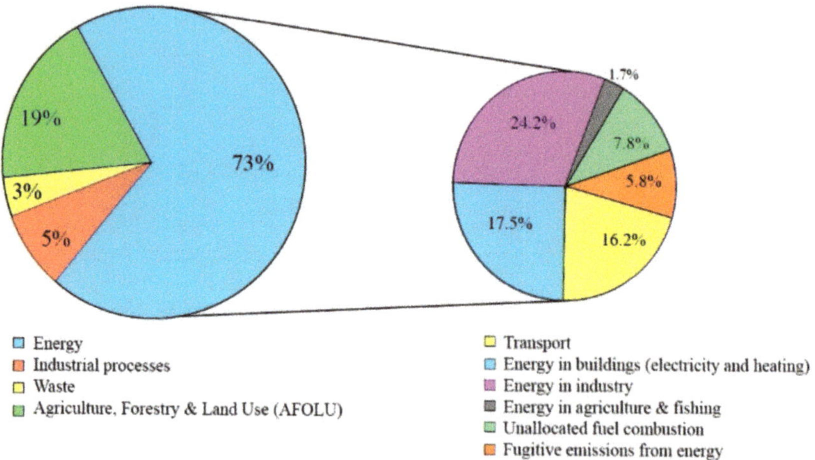

Figure 1. The share of global greenhouse gas emissions (%) from buildings consuming energy worldwide.

As with other oil-rich countries negatively affected by the energy subsidy, Iran is struggling with challenges imposed by the subsidy; through heavily relying on fossil fuels to generate energy for buildings, Iran is facing critical issues in meeting sustainable development goals. Moreover, as a country with a low energy price, the tendency of households towards inefficient energy consumption in the domestic sector is considerably increasing [8]. Therefore, there is a need for oil-rich countries with low energy prices to manage and reduce domestic users' energy demands and habits, which is mainly associated with the energy subsidy policy [15]. The increasing demand for electricity in oil-rich countries has raised concerns regarding blackouts mainly due to this energy subsidy [16].

Several previous studies have highlighted the negative consequences of the energy subsidy on sustainable development goals to be achieved by oil-rich countries. Moreover, these studies pointed out the issue of energy demand growth of occupants in different countries where the energy subsidy was in their energy policies. For instance, Al-Marri (2018) [15] investigated the current energy use of occupants in the domestic sector in Qatar to understand the challenges of consumer behaviour in terms of energy efficiency. They found that occupants were not inclined to alter their energy use behaviour due to the energy subsidy. Moreover, several researchers have pointed out the negative implication of the fossil-fuel subsidy for energy efficiency [17–19]. For instance, Sun (2015) and Dube (2003) [20,21] highlighted that higher-income households benefit more than low-income households from energy subsidies, which has further affected their energy use behaviour. Studies conducted by Ouyang and Lin (2014) and Adom and Adams [10,22] pointed out the excessive subsidy for fossil fuels in China, which has led to wasteful energy consumption. Fattouh and El-Katiri (2013) [23] highlighted the inefficiency of domestic energy consumption in the Middle East and North African regions due to a high energy subsidy. According to Oryani et al. (2022) [24], Iran has ranked as the largest consumer of energy among other Middle East and North African countries.

Although there are studies on the energy subsidy and the impact on energy consumption for different countries [25–28], little research has been conducted to fully understand the effect of a fossil fuel subsidy and prices on the energy consumption trend in the international compared to Iranian context. For instance, in the studies mentioned above, the energy consumption was investigated from the economic perspective rather than climatic or social aspects in the oil-rich countries. Moreover, the important aspect of energy consumption in countries with or without the energy subsidy and its significant effect on the way that occupants consume energy has not been studied comparatively, especially in residential buildings. It is worth mentioning that in the studies which have considered Iran as their case study, most of the researchers have solely highlighted the challenges of removing the energy subsidy and how low-income occupants can be negatively affected. Therefore, there is a need to reconsider the solutions to the subsidy by understanding important factors (climatic and non-climatic) influencing occupants' energy consumption in Iran. Therefore, instead of proposing different solutions regarding the removal of subsidies, occupants' behaviour change might have the potential to be an alternative solution to the removal of the subsidy. This necessitates understanding the effective factors in occupants' energy consumption. Therefore, further investigations are needed to explore the negative consequences of fossil fuel subsidies on households' energy consumption patterns in a global and local context.

Moreover, the tendency of oil-rich countries and the negative consequences of the energy subsidy on occupants' energy use is not clearly understood or well defined in the domestic sector in oil-rich countries, especially Iran, compared to the global context. It is worth mentioning that removing the fossil fuel subsidy alone may not address the issues with which the oil-rich countries are struggling [8]. As indicated by Gangopadhyay et al. [29], energy subsidy reduction in developing countries necessitates further support through other policies to alleviate the adverse effects. However, since the energy subsidy acts as a barrier to not only efficient energy use behaviour but also additional investments in renewable energies [15], raising the awareness of the negative consequences of this issue can globally and locally encourage future studies regarding the consideration of human behaviour for policy enactment as well as possible solutions to be proposed for the reduction of fossil fuel consumption in generating electricity in the domestic sector in Iran and other oil-rich countries.

The research objective is first to investigate the driving factors affecting fossil fuel reliance of oil-rich countries compared to other developed countries without an energy subsidy. To achieve this objective, different countries were compared in terms of their reliance on fossil fuels and its impact on their increasing/decreasing trend in energy consumption to generate electricity. Afterwards, Iran, Australia, and Germany were compared and analysed in terms of electricity costs and demands per capita to understand further consequences of the energy subsidy on households' behaviour in electricity use. The data for this purpose were collected from different statistical data centres worldwide. The data for the global context were obtained from the IEA [3,30–33], the World Bank [34,35], World Data [36], Our world in data [37,38], the World Economic Outlook Database [39], Global Carbon Project [40], Australian Bureau of Statistics [41], Australia Institute of Family Studies [42], Statistisches Bundesamt [43], and World Resources Institute [4]. In addition, the tendency of electricity use in the domestic sector was analysed in detail by comparing different provinces in Iran to have insights into the consumption behaviour of provinces known for their high amount of electricity use in Iran. For instance, to understand the significant share of the Tehran province in electricity use in Iran, this study gathered raw data from different resources. For the Iranian context, this research exploited data from the National Statistics Centre of Iran [44,45], the Plan and Budget Organisation [46], the Ministry of Energy, and the Tavanir Company [47].

This paper is divided into four parts. The first part focuses on comparing developed and oil-rich countries in terms of their status regarding the fossil fuel subsidy to explore the effect of energy price on households' electricity demand in oil-rich countries compared to developed countries. Then, the crucial factors affecting the increasing or decreasing trend in domestic electricity demand were explored through a comparative analysis between countries with and without energy subsidies. The second part implements a local comparison of domestic electricity use between different provinces of Iran to further discuss and delve into the impacts of behavioural-related factors on the electricity use of Iranian households.

The third part highlights the solutions proposed by previous studies to reduce energy use and emphasises the importance of behaviour as an effective solution to be included in the policy enactment process.

Moreover, the limitations of current energy policies in Iran were investigated for the domestic sector. The sequence of the study is illustrated in Figure 2 below. The conclusions are presented in the final section. One of the contributions of the study is to explore crucial factors affecting the behavioural aspect of domestic electricity demand in oil-rich countries by comparing them with developed countries. Another contribution is examining and comparing one of the highest electricity-consuming provinces, Tehran, with other cities to inform energy practitioners, designer, and policymakers about the critical issues related to behavioural aspects of domestic electricity use in Iran.

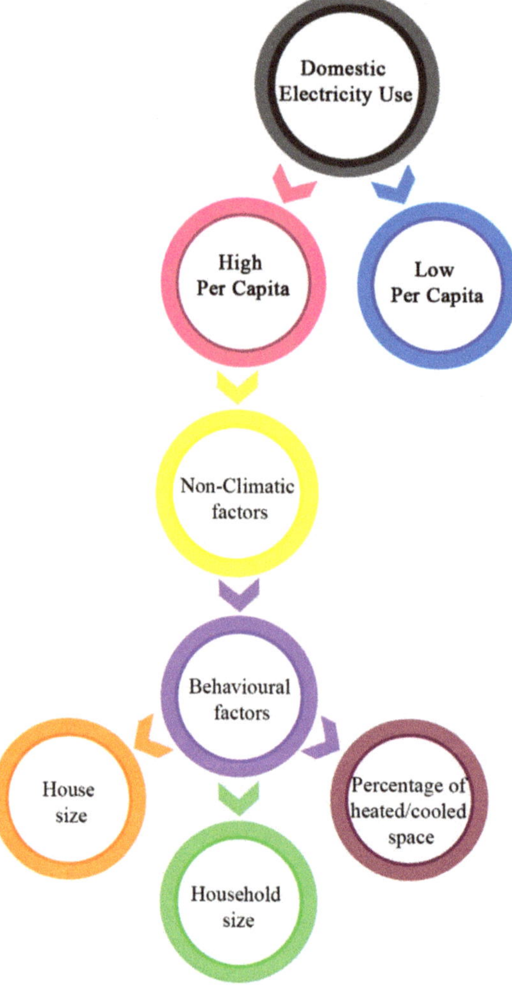

Figure 2. Factors affecting high per capita domestic electricity use in provinces of Iran.

2. Comparison between Iran and Other Countries

2.1. Comparison of the Fossil Fuel Subsidy in Oil-Rich Countries and Iran

Oil-rich countries mostly consider the fossil fuel or energy subsidy to decrease energy costs while enhancing consumers' living standards [48]. However, reducing energy costs has further caused the overuse of fossil fuels to generate energy and electricity for buildings. Different countries have different portions of the subsidy allocated to fossil fuels, such as oil, electricity, gas, and coal (Figure 3). Iran provides a significant amount of subsidy for electricity, making it one of the lowest prices among countries' electricity (Figure 4). This has resulted in domestic electricity demand growth in Iran over the years [23]. According to Fattouh and El-Katiri [23], energy use in the Middle East and North African regions has more than quadrupled, while electricity use increased over six times over the past 30 years.

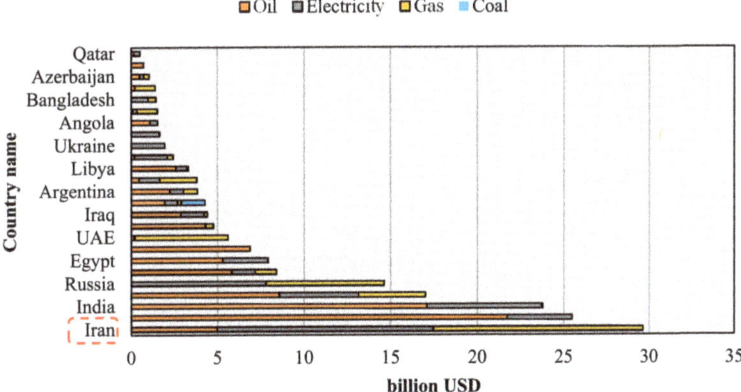

Figure 3. The energy subsidy of oil-rich countries and Iran by fuel type in 2020 (Data source: [30]).

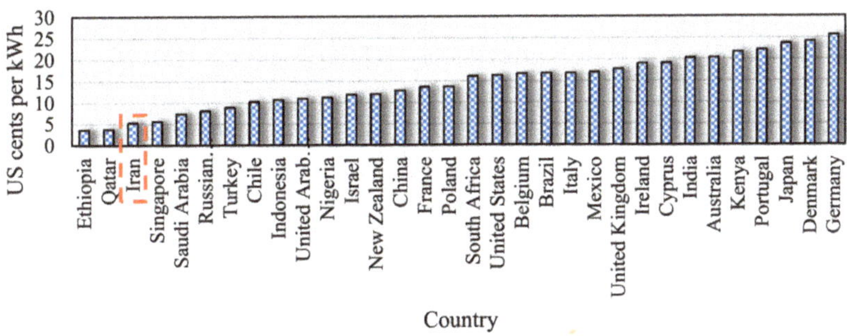

Figure 4. Electricity prices in different countries and Iran in 2019 (Data source: [35]).

In general, Iran's primary energy resources are crude oil, natural gas, coal, nuclear, and renewable resources [49]; however, as illustrated in Figures 2 and 3, the most heavily subsidised resource in Iran is natural gas, which has also led to cheap electricity prices in this country. This has led to the growing trend in CO_2 emissions from natural gas compared to other primary energies (Figure 5). This issue was an expected result in Iran since power plants' leading resource for providing buildings energy for heating and cooking as well as electricity for lighting and cooling is natural gas (Figure 6). This demonstrates the significant contribution of electricity in buildings to energy consumption across the world and within Iran.

2.2. Reliance on Fossil Fuels in Countries with High and Low Electricity Prices

As indicated previously, the energy subsidy has reduced the electricity costs in oil-rich countries, which has caused a heavy reliance on fossil fuels for electricity generation (Figure 7). This study gathered data regarding different countries' electricity generation from fossil fuels [37]. The lower the electricity price in a country, the more its dependence on fossil fuels. Therefore, with cheap electricity, Iran has continued its reliance on fossil fuels, resulting in a rapid rise in the tendency to use fossil fuels to generate electricity. It is worth mentioning that, as In Iran, most countries, especially those without energy subsidies, had previously relied on fossil fuels to generate electricity, especially before 2007. However, the most important point in those countries is their decreasing trend in relying on fossil fuels over the years.

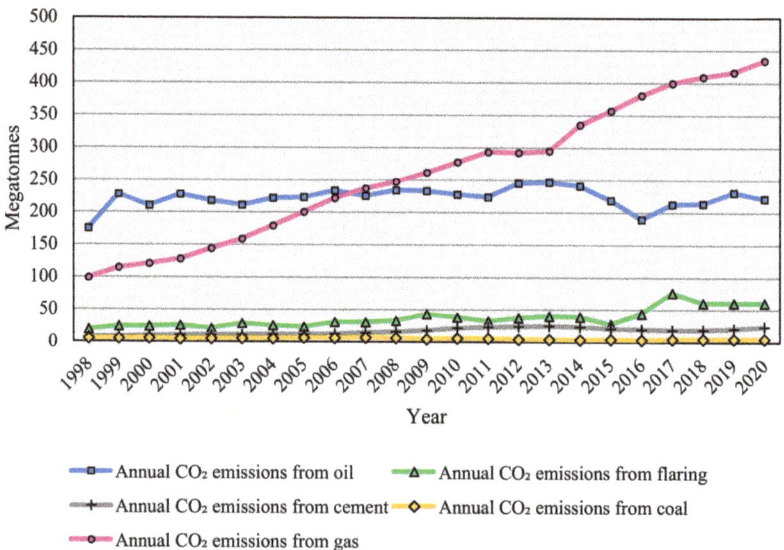

Figure 5. CO_2 emissions from various resources in Iran [40].

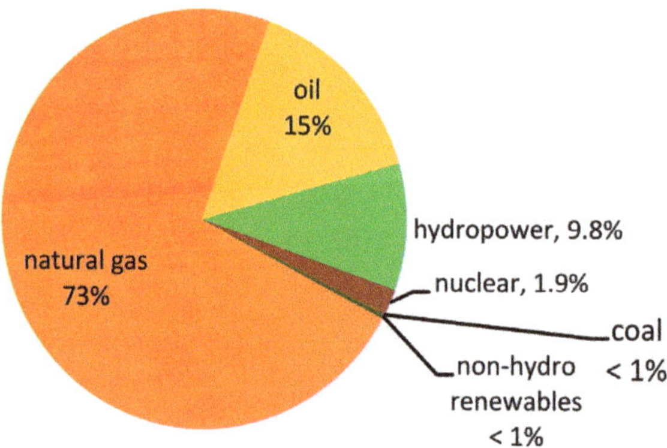

Figure 6. Primary energy consumption share in Iran, 2019 (Data source: [49]).

This study collected data from [41,43,44,49] to further explore some of the developed countries. After analysing the data, this study found that, like other oil-rich countries, Germany and Australia's fossil fuel consumption was previously significant before 2015 but suddenly changed in recent years. Therefore, Australia and Germany were selected to compare with Iran, which has recently increased its energy consumption from fossil fuels. Regarding Germany, the high electricity use from fossil fuels can be explained due to the nearly similar population to Iran. However, Australia has a lower population than Iran, which may raise a question regarding Australia's large portion of fossil fuel consumption compared to Iran before 2011. With about 22.03 million people in 2010, Australia used to generate more electricity from fossil fuels than Iran, with a population of 73.76 million [34]. However, the comparative analysis between Iran, Australia, and Germany revealed that the electricity use per capita for Australia and Germany was always larger than for Iran (Figure 8). This does not necessarily relate to inefficient use of electricity since the electricity use per capita can be different based on a variety of reasons, such as climate, housing types or other influencing factors. Moreover, it is worth mentioning that the electricity generated from fossil fuels illustrated in Figure 6 belongs to various sectors and is not specifically related to the domestic sector.

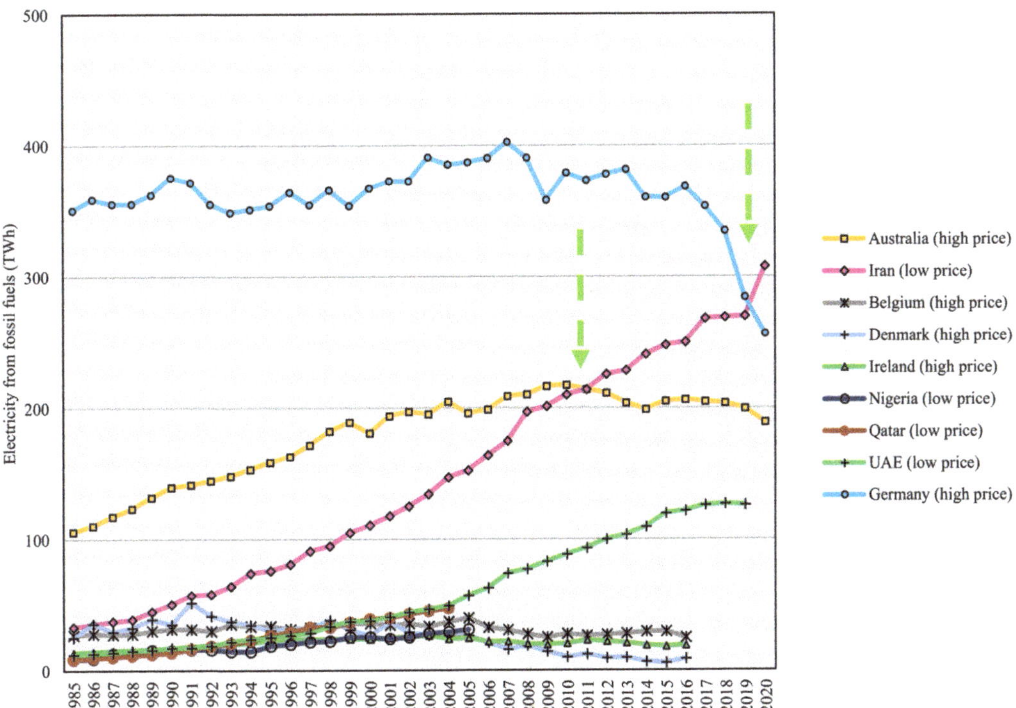

Figure 7. Comparing electricity generated from fossil fuels between countries with low and high electricity prices.

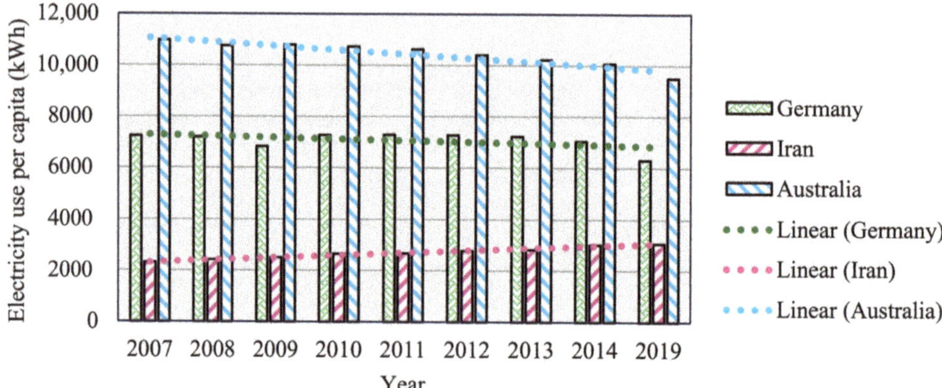

Figure 8. Electricity use per capita in Australia, Germany, and Iran (kWh).

The most crucial point in the above figure is that despite the high electricity usage per capita in those countries, after 2011, there was a rapid decrease in fossil fuel usage, demonstrating both countries' tendency towards consuming cleaner energy resources in recent years. Therefore, unlike Iran, Australia and Germany have gradually started decreasing the electricity usage per capita by increasing their reliance on renewable energy resources over time. Moreover, Australia and Germany's per capita consumption has fallen in recent years due to higher electricity prices and lower reliance on fossil fuels due to significant investments in renewable energies (Figure 9). It is worth noting that although Germany still has nearly the same population as Iran, surprisingly, electricity use per household has been decreasing in recent years (Figure 10). The reason was revealed after comparing the electricity price per capita in most oil-rich countries and Iran (Figure 11), which was collected from [50]. Therefore, this study found that the lower cost of electricity in Iran has significantly affected the increase in electricity use per capita. In contrast, Germany's electricity usage per household has a decreasing trend in recent years.

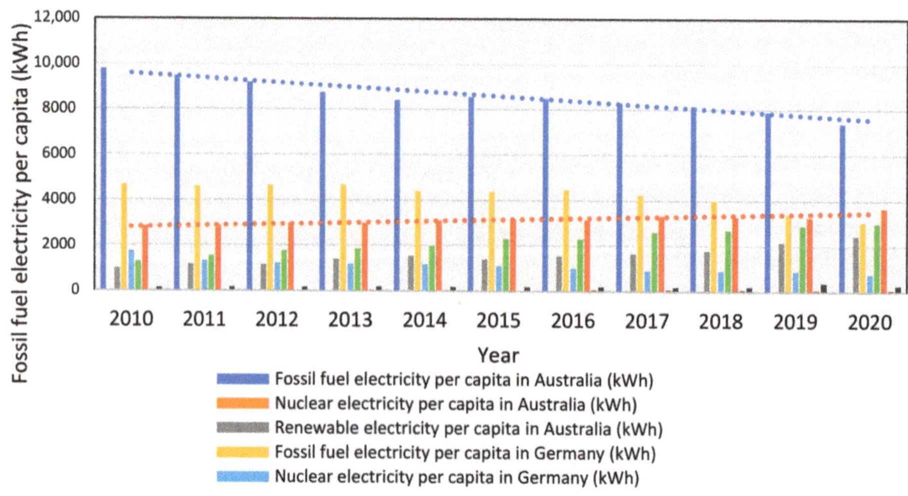

Figure 9. Fossil fuel electricity per capita (kWh) of Iran, Australia, and Germany.

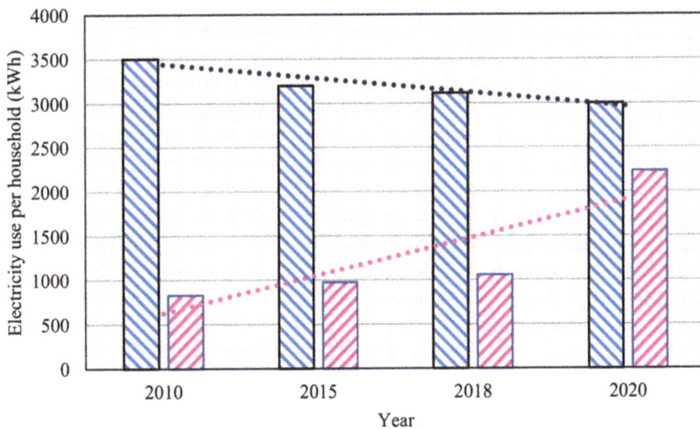

Figure 10. Electricity use per household in Germany versus Iran.

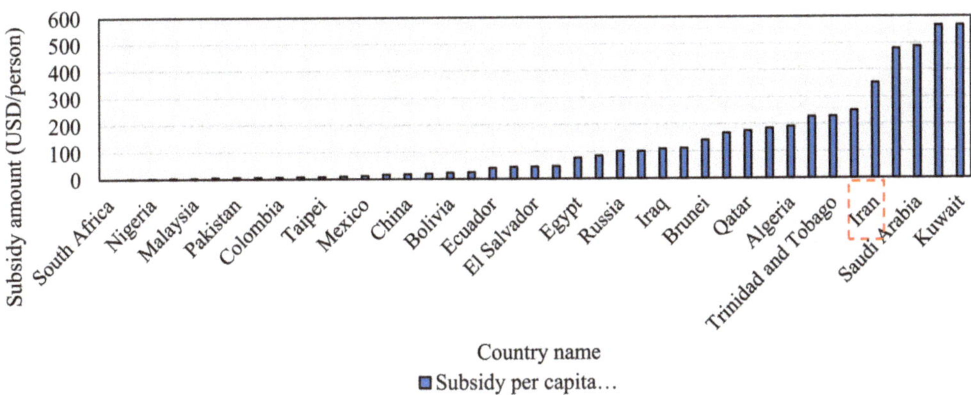

Figure 11. Subsidy of other countries and Iran per capita in 2020.

In the comparative analysis of other countries with higher electricity prices, such as Belgium and Denmark, this study observed the same results as Australia and Germany (Figure 12). The analysis revealed a decreasing trend of reliance on fossil fuels over the years, as the investments in nuclear and renewable energies increased. However, in other countries with lower electricity prices, such as Qatar and United Arab Emirates (UAE), the use of fossil fuels for electricity generation has increased over the years (Figure 13).

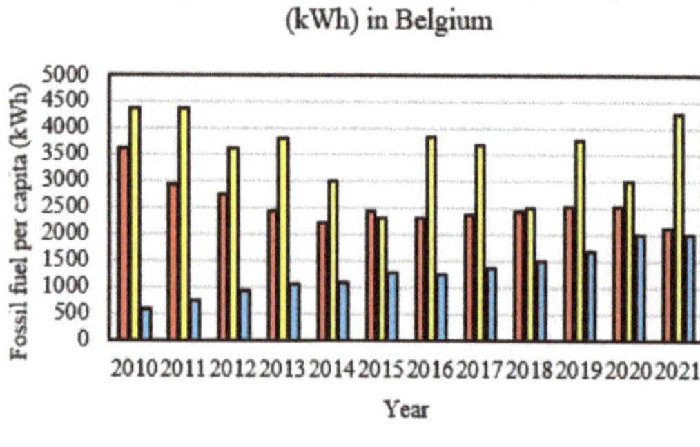
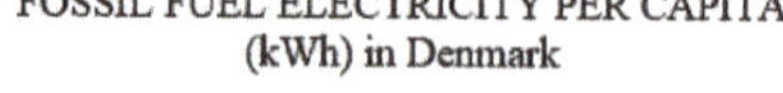
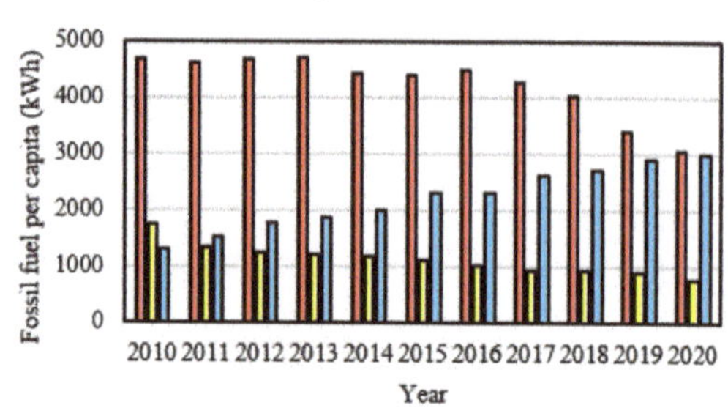

Figure 12. Fossil fuel electricity per capita (kWh) in Belgium and Denmark.

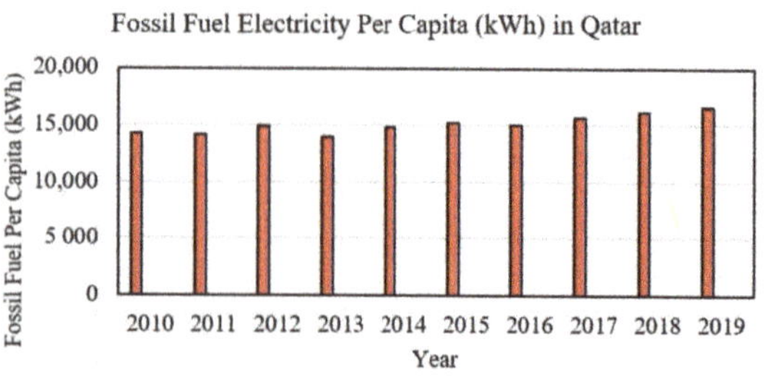
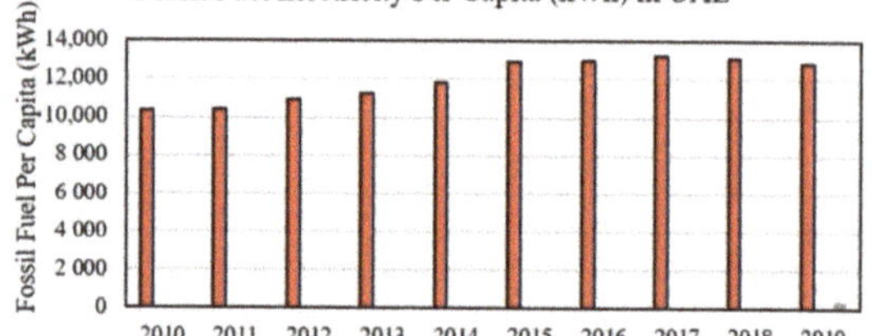

Figure 13. Fossil fuel electricity per capita (kWh) in Qatar and UAE.

Comparing countries revealed that Iran has an increasing trend towards fossil fuel reliance with decreasing renewable energy use due to low energy prices, slow rate of increase in renewable energy share, and other limitations. This not only had a negative effect on the climate but significantly encouraged households' behaviour to increase their electricity consumption. Therefore, the energy subsidy has dramatically affected their consumption behaviour.

The Effect of Fossil Fuel Reliance on the Rapid Rise of Domestic Electricity Use in Iran

It is worth mentioning that the sector which is mostly contributing to CO_2 emissions due to a large amount of natural gas consumption in Iran is the domestic sector, according to [45]. After comparing the amount of gas and electricity consumption in the residential sector for the past ten years [45], this study found that the increase in electricity use was significantly higher than the increase in natural gas consumption (Figure 14). Over the last ten years, there was a 55% increase in domestic electricity consumption but only an 18% rise in natural gas consumption, demonstrating the importance of electricity consumption in the residential sector of Iran.

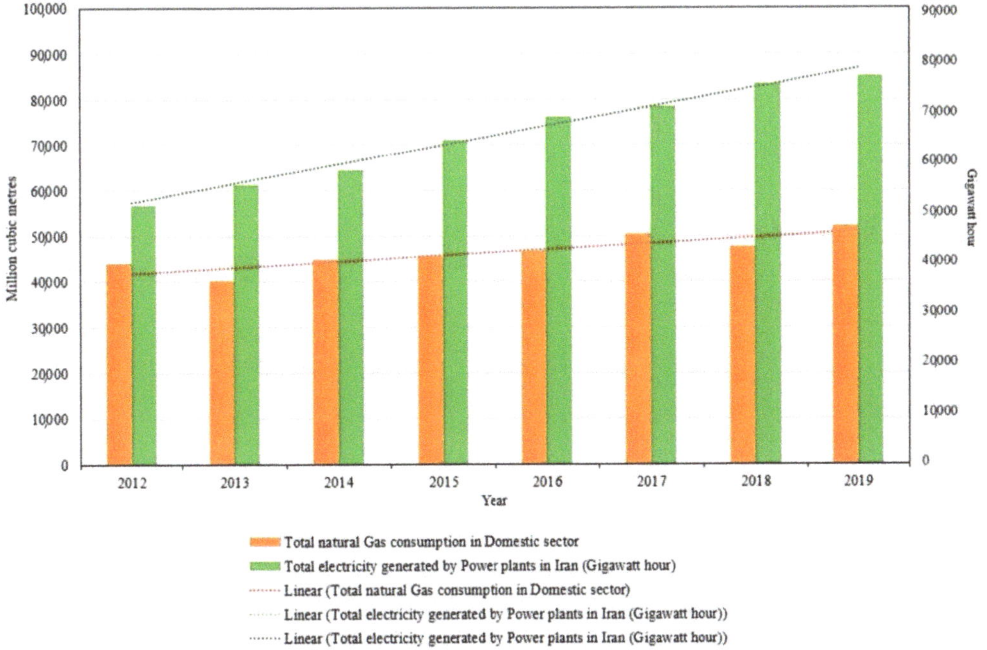

Figure 14. The rise in end-use electricity and end-use natural gas consumption of occupants in the domestic sector in Iran from 2012 to 2019.

The same trend can be inferred from comparing the residential sector with other sectors after gathering data from [51]. For instance, when comparing the residential sector with the industry sector, the same amount of electricity consumption is mostly seen in both sectors over the years, except in 2015, when the residential sector surpassed the industry sector (Figure 15). In general, the electricity consumption in the residential sector is almost the same as the industry sector and sometimes even higher. The electricity consumption of the residential sector has always shown an increasing trend over the years, making it a crucial sector to investigate in terms of factors affecting domestic electricity use in Iran.

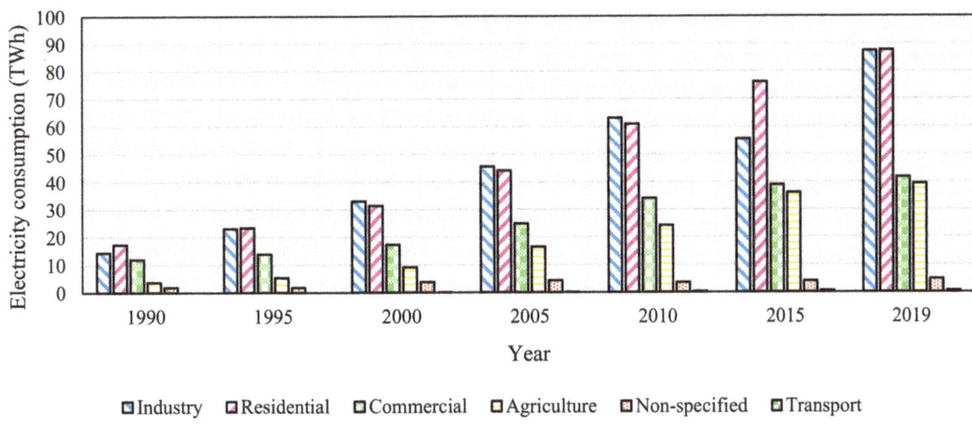

Figure 15. Electricity consumption by different sectors in Iran.

Overall, as previously explained (Figure 11), the reason for the high electricity demand in residential buildings in Iran can be explained by the subsidy considered not only for energy but for the electricity usage per capita. According to IEA [33], the subsidy allocation was about $US350 per person in Iran, making it a cheap resource for consumption. The inexpensive electricity of oil-rich countries such as Iran may raise a question regarding the effect of low and high electricity prices in countries on their consumption patterns in the domestic sector. Therefore, comparing Iran with other countries with higher electricity prices may provide essential insights into households' tendency towards electricity consumption not only for Iran but for other compared countries.

2.3. The Effect of Electricity Price on the Electricity Demand in the Domestic Sector of Iran Compared to Other Countries

After comparing different countries with high/low electricity prices (Figures 4 and 11), a general ascending trend was seen in electricity consumption for mostly oil-rich countries (i.e., Iran and UAE) and descending trend for countries with higher electricity prices (i.e., Germany and Australia) (Figure 16). The data were collected from the IEA [52] and compared in this study. According to the comparison in this study, it can be concluded that this not only applies to the whole domestic electricity consumption but also to the electricity consumption per capita. After collecting data from [53], this study found that Iran is above the world's average domestic electricity consumption per capita (Figure 17). As a result, this dramatic increase in electricity consumption in Iran, as opposed to other countries, can be attributed to the high subsidy given to domestic electricity consumption per capita.

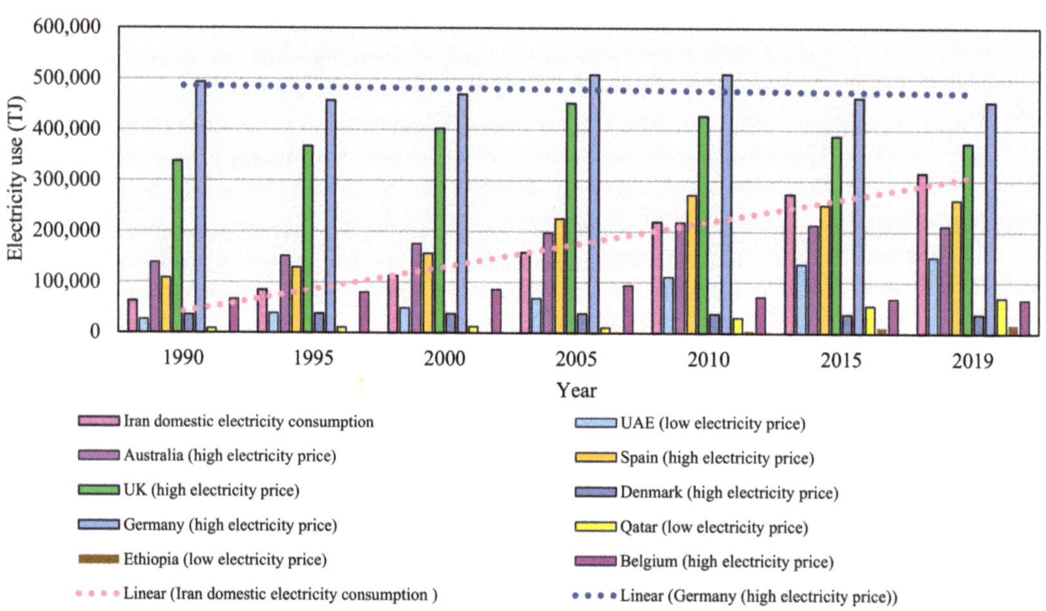

Figure 16. The trend in domestic electricity consumption of low and high electricity price countries.

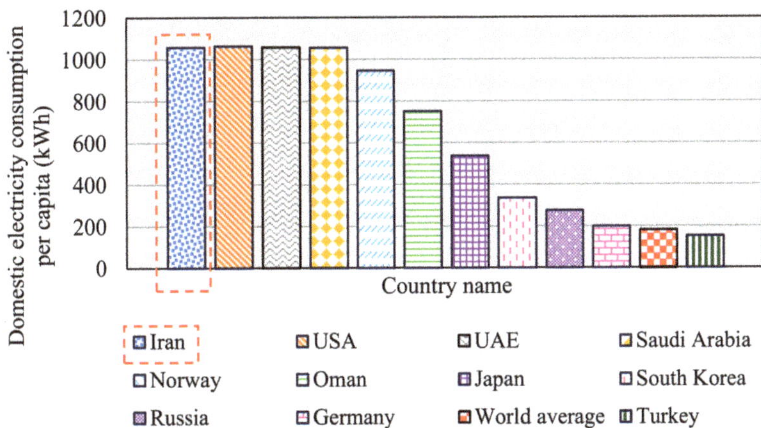

Figure 17. Domestic electricity consumption per capita in different countries and Iran.

For countries especially with high electricity costs, the price seems to be correlated with the decreasing demand for domestic electricity use. The electricity cost data were collected in this study from [54] to explore further the current and previous status of electricity prices in Iran. After analysing the data, it is worth mentioning that Iran started implementing a subsidy reform to increase electricity prices after 2010 (Figure 18). Unfortunately, it has not affected the increasing demand for domestic electricity use in Iran. A significant gap can still be observed when a price comparison between Iran and Denmark is implemented. This may explain the growing trend in electricity demand in Iran. Accordingly, although there was a sharp increase in the electricity price in Iran, a negligible amount of households' income belongs to energy costs (Figure 19). According to the data from the National Statistics Centre of Iran, the highest and lowest income of Iranian households allocate 2.5% and 6.8% of their income to energy costs, respectively.

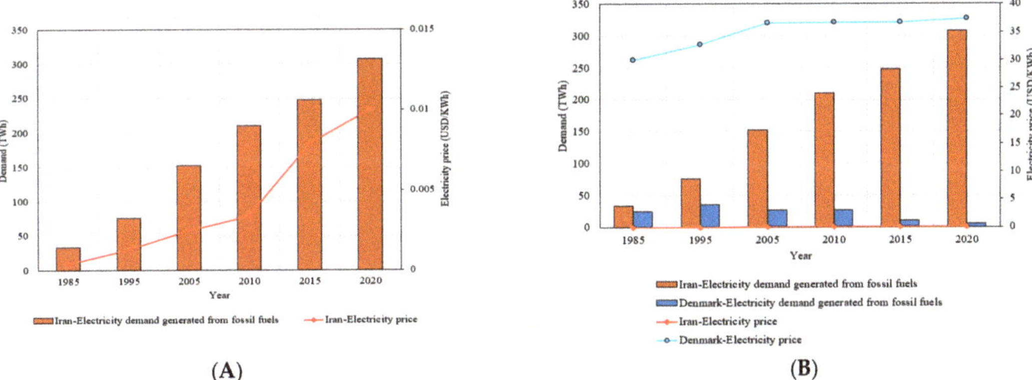

Figure 18. (A) Electricity demand vs. electricity price in Iran over the years (B) Comparison of electricity demand and price countries with Denmark.

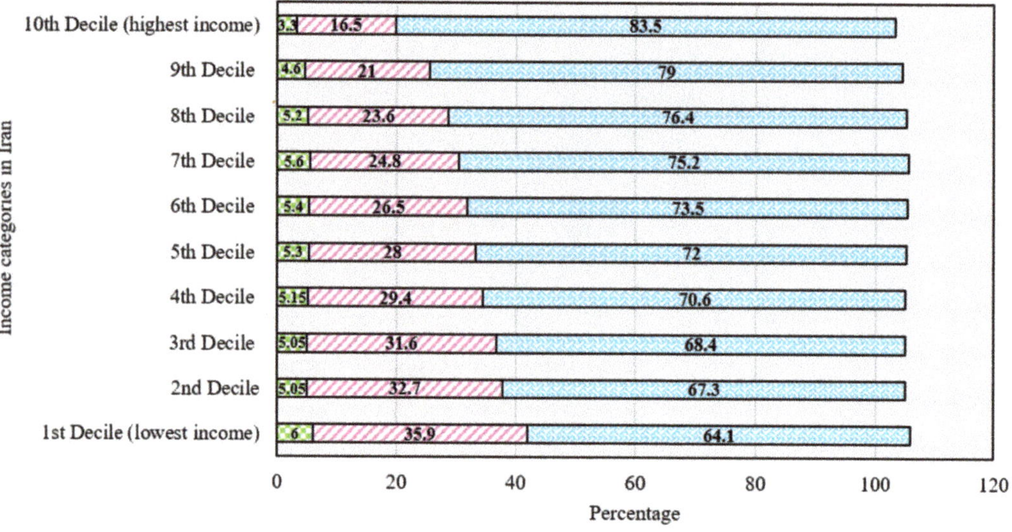

Figure 19. The portion of energy in living costs of income categories in Iran in 2016.

According to the growing trend in electricity demand due to the lower price of electricity in Iran, it seems that there is a necessity for a mental shift in households to have a conscious way of living to reduce electricity consumption, since it appears that there are important factors that are currently influencing electricity consumption in Iran related to their behaviour. Therefore, this study found that the electricity price is not the only solution to address the increasing or decreasing trend in domestic electricity demand in Iran. For further understanding of additional behavioural-related factors, deeper comparisons between Iran and other countries with reduced domestic electricity use (i.e., Australia and Germany) can be beneficial to reveal other factors correlated with domestic electricity use, which are explained in the following sections.

2.4. Correlation between Increasing/Decreasing Trend in Electricity Consumption and Different Factors in Iran

To further explore other factors affecting electricity consumption in countries with high and low electricity prices, this study gathered data to compare Iran, Australia, and Germany's electricity use per capita and their household size over the years [41–44].

After analysing the average household size over the years (Figure 20), a correlation was found between the decrease in household size with the reduction in electricity use in Australia and Germany. This result could be expected since in the previous section, a decreasing trend in the overall domestic electricity demand was seen in Australia and Germany. Therefore, the analysis in this study showed that as their average household size decreased, the electricity use demand decreased in these countries. However, this correlation was not seen in Iran. In contrast to the decreasing trend of household size in Iran, domestic electricity consumption always increased from 2011 to 2020.

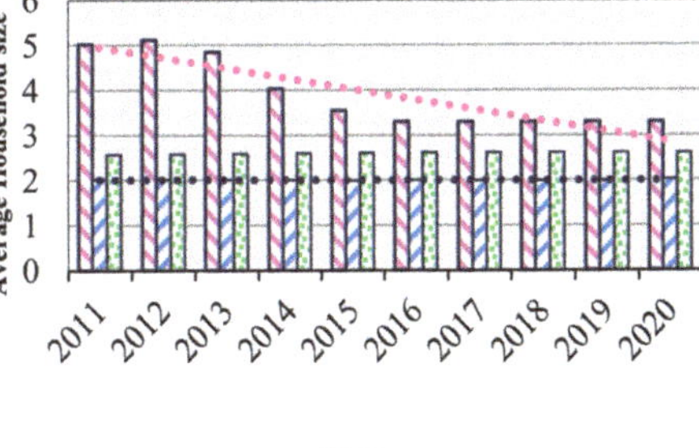 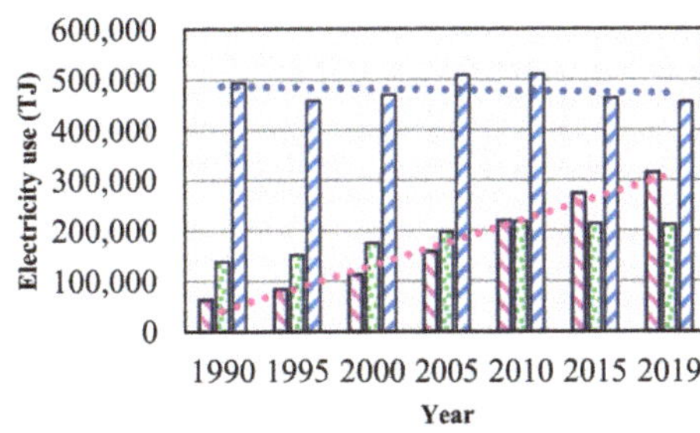

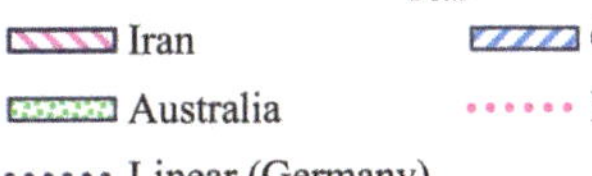

Figure 20. Average household size vs. domestic electricity uses in Iran, Germany, and Australia.

The comparison between the electricity usage per capita and household size of developed countries and oil-rich countries showed that Iran seems to be not very efficient in terms of domestic electricity consumption. Therefore, when considering domestic electricity consumption per capita, other factors may become important in addition to climatic factors, such as households' behavioural factors. Therefore, although the electricity usage per capita may be related to climatic factors in a country with higher per capita electricity usage, in another country, the behaviour of the households can also be affected by different reasons, such as their policies regarding energy subsidy as well as electricity price. Another important factor that can affect electricity consumption is the living space per person. For instance, according to statistics, the average per capita residential area in Germany and Australia is estimated to be 44.6 and 87 square metres, respectively [55,56]. This can also explain the reason for their high electricity consumption as compared to Iran. However, it is worth mentioning that the living space per person has been increasing in recent years in Iran, thus, contributing to the rise in the electricity consumption. Therefore, regarding the current living space situation, the per capita residential area is also variable in different provinces of Iran. On average, the per capita residential floor space in Iranian provinces can be between 20 and 50 square meters. According to the statistical centre in Iran, three dimensions for living area per person can be observed based on the average density of provinces. In low, middle, and high densities, the maximum living space per person is 59.2, 41.7, and 37 square meters, respectively [57].

The above comparative analysis of domestic electricity uses in developed countries, and Iran pointed out the significant impact of Iranian households' behaviour on domestic electricity demand. Therefore, further exploration of behavioural-related issues is crucial to reducing the domestic electricity use inefficiency in Iran.

3. Comparison of Domestic Electricity Uses between Provinces of Iran to Distinguish between Behavioural-Related Factors and Other Factors

One of the most crucial behavioural-related issues affecting electricity use is house size, which seems to be related to Iranian households' tendency to live in a larger unit area (Figure 21). Compared to 2012, smaller household sizes (one to two persons) tended to choose larger houses in 2018 despite their small population. Interestingly, the data showed a change in accommodation of one- to two-person households from one- to two-bedroom unit apartments to more than two bedrooms in 2018. The same result was seen for two- and three-person households, who prefer to live in two- and three-bedroom units. In general, the demand for one-bedroom units has decreased, while the larger unit areas have gained popularity among households. It is worth mentioning that the four- and five-bedroom units are reduced due to an increase in the number of people living in apartments rather than single-family houses. Living in a larger unit area can significantly affect households' electricity consumption.

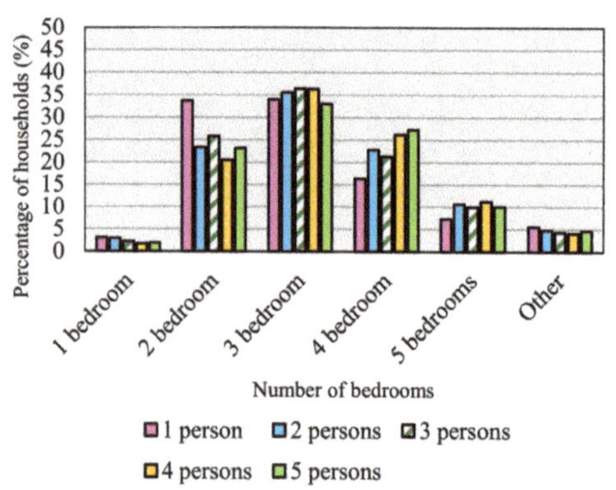

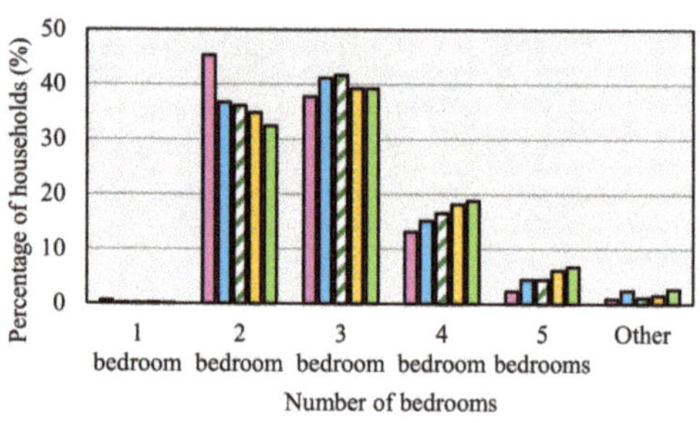

Figure 21. Percentage of household size with type of unit apartments in 2012 vs. 2018 in Iran.

For further understanding of other factors, this study compared the provinces of Iran to find out whether there is a difference in their electricity demand in the following section.

3.1. Electricity Use in Provinces of Iran and Important Factors Affecting Demand

One of the factors which can affect electricity demand is household subscribers. Therefore, among the provinces of Iran, such as Khuzestan, Isfahan, Kerman, Yazd etc., Tehran is playing a pivotal role in decreasing the growth in total electricity demand of the country due to a high number of subscribers. According to Figure 22, with a significant share of electricity use, Tehran accounts for 13.6% of the total percentage, while the remaining provinces are below 11.3%. The findings support the view of previous studies that pointed out the importance of population on energy consumption [58–60].

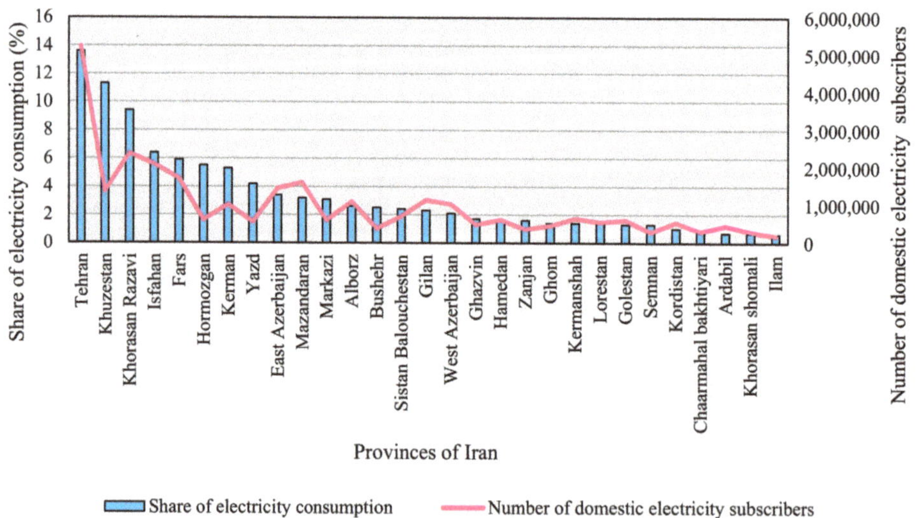

Figure 22. The comparison between the share of electricity use of provinces in the total consumption of the country vs. the number of household electricity subscribers in 2020.

Although in cities like Tehran, the number of subscribers can significantly affect the overall electricity use in a province, other factors can be as crucial as the subscribers. For instance, in Khuzestan, electricity use is extremely higher than its portion of subscribers, drawing attention to other important factors affecting electricity consumption.

3.1.1. Climatic Factors as One of the Reasons behind the High per Capita Electricity Use

According to Batliwala and Reddy (1993) [61], energy demand depends on per capita energy use. Therefore, to further understand other influencing factors of domestic electricity demand, Tehran and Khuzestan were compared in terms of overall characteristics and electricity use per capita. As indicated previously, the household subscribers were not wholly associated with the electricity use in Khuzestan due to the lower number of electricity subscribers but significantly higher electricity use than in Tehran. A question may arise regarding the inefficient use of electricity in Khuzestan. To understand whether households in Khuzestan use electricity inefficiently, the electricity uses per capita of Tehran and Khuzestan were compared. After comparing the climate of Tehran and Khuzestan (Figure 23), this study found the reason behind the large share of electricity use in Khuzestan to be the harsh climate, which in turn has affected the per capita use of electricity. Accordingly, unlike Tehran, the temperature in Khuzestan can be extremely high, based on which the per-capita electricity use would inevitably increase. Therefore, this study found that the high electricity usage per capita in Khuzestan does not necessarily imply their inefficient use. In other words, the climate drives the occupants to use more electricity to alleviate discomfort levels experienced in a harsh climate.

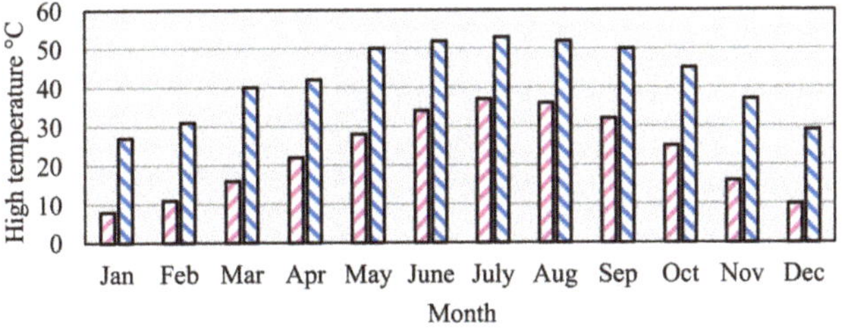

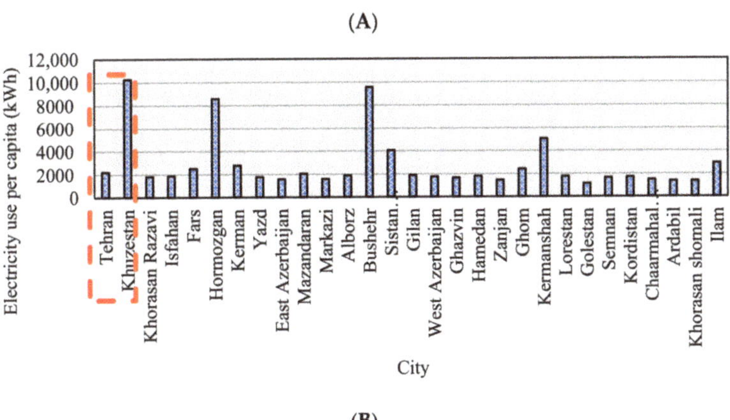

Figure 23. Comparison between Tehran and Khuzestan. (**A**) Monthly highest temperature; (**B**) electricity use per capita.

3.1.2. Non-Climatic Factors as One of the Reasons behind the High per Capita Electricity Use

While climatic factors can be crucial in electricity use per capita, non-climatic factors can also be influential. As discussed previously, an increase in Khuzestan's electricity usage per capita may not necessarily equate to the inefficient electricity use of households due to climatic factors. This can also be seen through the electricity price allocated for each province based on the climate conditions of that specific province. For instance, the electricity price for Khuzestan province is much lower per kWh electricity consumption of the occupants than in Tehran province. However, the climate is not the only factor on which the efficient use of electricity per capita is based. Therefore, additional aspects need to be investigated. For this purpose, this study explored and compared the average unit areas being heated or cooled in Tehran and Khuzestan to determine the non-climatic aspects which can lead to the inefficient use of electricity per capita (Figures 24 and 25). This study found that although Khuzestan's electricity use per capita is higher than in Tehran, the average and percentage of cooled unit areas in Khuzestan are significantly lower than in Tehran. Surprisingly, despite the harsh climate in Khuzestan, the portion of areas being cooled is lower than in Tehran. Interestingly, this study revealed that with a milder temperature in Tehran, occupants still heat and cool a more significant portion of their unit areas, nearly 96.1%. This comparison revealed that non-climatic factors have led to a higher percentage of unit areas being cooled despite the milder climate in Tehran.

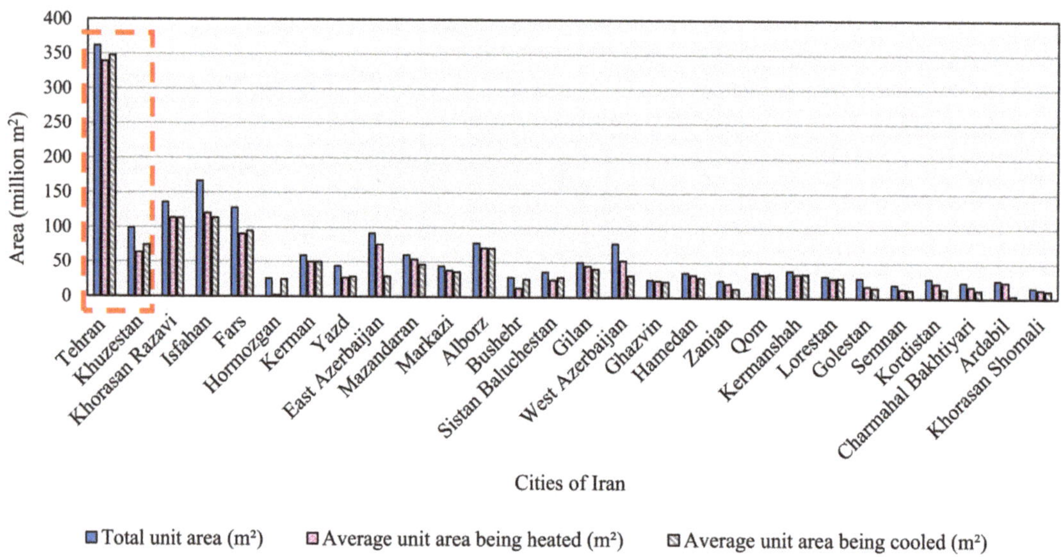

Figure 24. The average area of domestic units being heated or cooled in provinces of Iran in 2018.

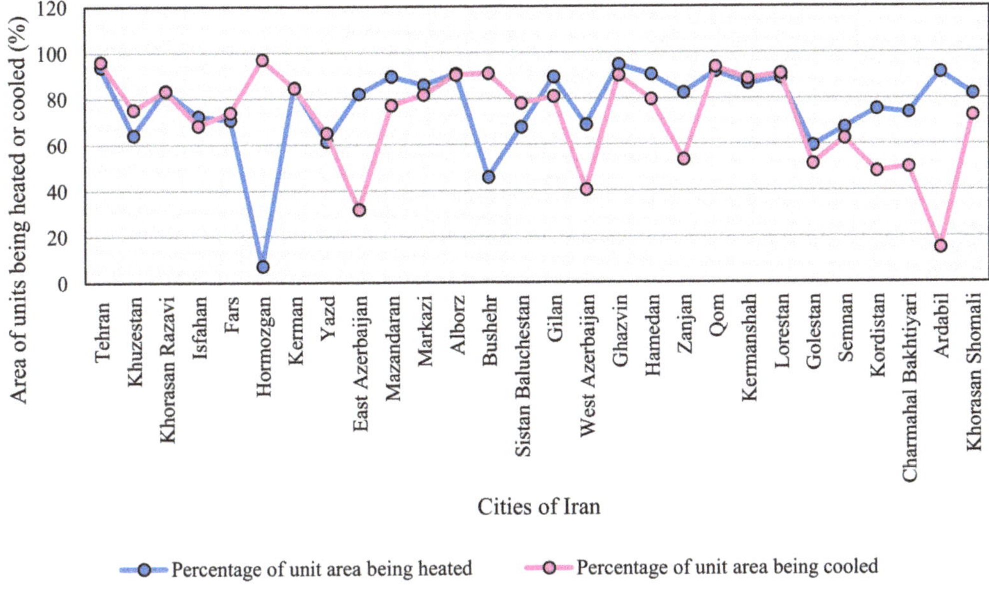

Figure 25. Percentage of units being heated or cooled in provinces of Iran.

This study found that in situations where the cities' climate differs, a harsh climate could explain a province's high per capita electricity use. The comparison also showed that there could be provinces with milder temperatures but higher electricity usage per capita than other provinces with a harsh climate. To further explore the non-climatic factors affecting electricity demand, this study selected a province with nearly the same climate as Tehran. Figure 26 depicts the high and low temperatures in different months of Tehran and Isfahan. Although Isfahan experiences higher and lower temperatures during summer and winter, the amount can be negligible as no provinces can have the same climate. Therefore, considering the temperature as a control variable, a comparative analysis of the percentage of unit areas being heated and cooled in Tehran and Isfahan was conducted.

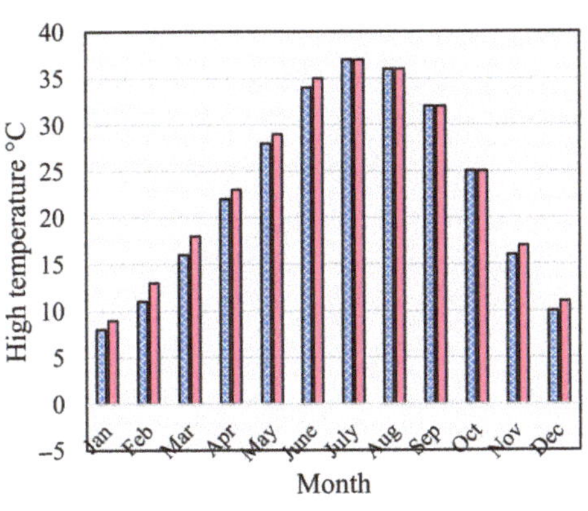

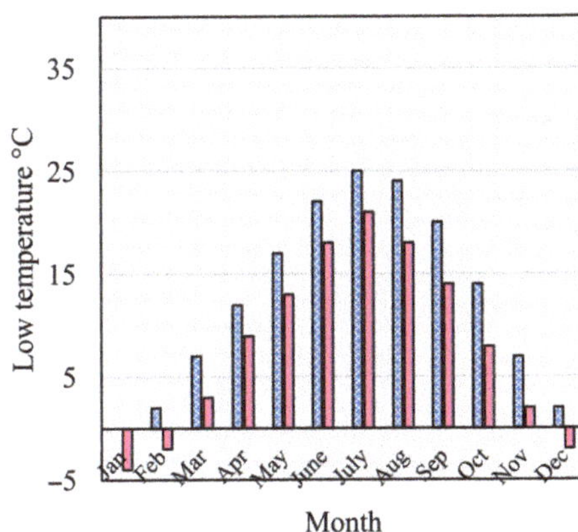

Figure 26. Temperature comparison between Tehran and Isfahan.

Interestingly, the analysed results revealed that even with the same temperature, Tehran includes a higher percentage of their spaces for heating and cooling than Isfahan. Therefore, it can be concluded that the non-climatic factor can be an essential reason for increased electricity use per capita in Tehran, primarily related to the households' behavioural issues. For instance, although Isfahan experiences higher temperatures during summer or lower temperatures during winter, the households employed a lower percentage of space areas to cool and heat their living areas than Tehran. It is worth mentioning that the higher percentage of space areas for cooling and heating purposes may also depend on the higher number of occupants in a living space. The dependency of energy use on household size was also indicated by other studies that investigated the energy use behaviour of occupants in different countries and Iran [62–65]. However, as this study compared the household size of Tehran and Isfahan over the years, a decreasing trend was observed in the household size of both provinces. However, unlike Isfahan, as the household size decreased, the electricity consumption pattern increased in Tehran (Figure 27).

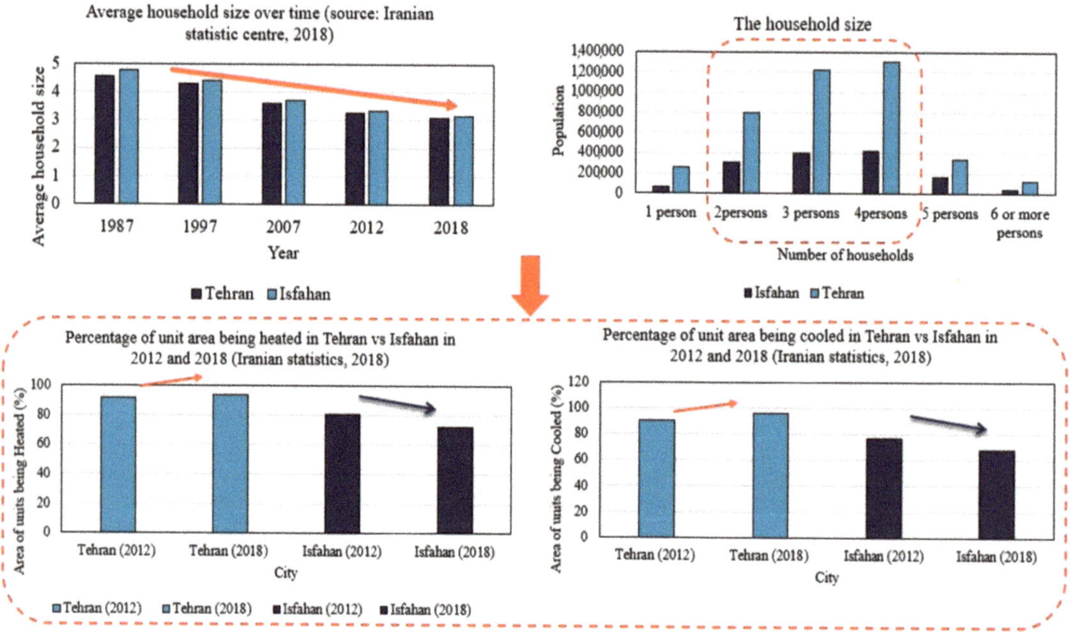

Figure 27. The effect of household size on electricity consumption in Tehran compared to Isfahan.

The comparative analysis between Tehran and other provinces revealed that in addition to climatic reasons, behavioural factors, namely house size, heated or cooled area percentage of space, and household size, could play a crucial role in increasing the amount of electricity used per capita.

Overall, in the global as well as local context, behavioural factors can explain most of the increasing trend in either "total" electricity demand or "per capita" electricity demand in Iran and Tehran when compared to other countries and provinces in Iran (i.e., Isfahan and Australia). Similar to Iran, a study by [66] revealed that energy use per capita in Qatar was one of the highest in the world, which is heavily subsidised for consumers [67]. It is worth mentioning that in Qatar, building practitioners have started to explore new approaches to reducing energy use by enhancing energy efficiency [68]. The Gulf Organization for Research and Development has developed the Global Sustainability Assessment System (GSAS) as their building sustainability assessment system to achieve energy efficiency while considering the local needs of Qatar. However, such a sustainability rating system has not been implemented in Iran for different types of buildings.

The above comparison in this research showed that unlike Australia in a global context and Isfahan in a local context, the reduction in household size had not affected the portion of space heating or cooling in Iran and Tehran. Based on the comparative analysis, non-climatic factors seem to be the main reason for domestic electricity use in Iran. Regarding climatic factors, the high per capita electricity use of occupants was seen to be associated with the temperature. However, the high per capita use due to non-climatic factors seems to be related to behavioural aspects of the households rather than the temperature, which have been affected by the energy subsidy of this country.

4. Behaviour as a Crucial Factor to Be Included in the Policy Enactment of Iran to Address High per Capita

In terms of energy, several regulations were developed by the Ministry of Housing and Urban Planning to achieve energy efficiency in the design and construction of buildings in Iran. The national energy efficiency regulation, known as Code No. 19, is one of the energy regulations which includes policies for external wall insulation materials, double glazing windows and their materials, and installation of building control systems based on the different climatic situations of different provinces of Iran [69]. Although Code No. 19 is mandatory for the buildings, there are barriers to the implementation. One of the barriers is the lack of specific codes for different building types, especially for residential buildings. It is also worth mentioning that, in contrast to most developed countries, there is no specific assessment system in Iran based on which the evaluation of the energy performance can be performed.

The most important barrier to the implementation of the Code No. 19 is the low energy price in this country, which has negatively affected the implementation of this regulation in a comprehensive manner. Therefore, many architects may not consider the regulations due to the high cost of implementing energy efficiency standards. Due to the cheap energy in Iran, the building industry has also faced the issue of insufficient trained architects and supervisors for the Code No. 19 implementation.

This cheap energy has also caused other critical issues regarding energy consumption in Iran. As discussed previously, the energy subsidy in Iran, as an oil-rich country, has significantly affected occupants' behaviour. This subsidy has made it difficult for these countries to achieve efficiency through occupants' energy use behaviour. Many researchers proposed revisions to subsidy policies after the issues created by the energy subsidy. For instance, studies conducted in China [20,70] pointed out that energy subsidy reform can be a solution to reduce energy use by reallocating a subsidy among different income groups in the domestic sector. Wang et al. (2009) and Liu and Li (2011) [71,72] also indicated that subsidy reform for fossil fuels could improve energy consumption while encouraging energy conservation. Other studies [23,73] highlighted the necessity for other solutions to substitute the fossil fuel subsidy with renewable energies or other effective social security systems, which could prevent significant inefficiencies in the Middle East regions' energy use patterns.

Although removing the energy subsidy is indicated to be an effective solution for energy use reduction by many researchers [74], it may have some drawbacks. For instance, Fattouh and El-Katiri (2013) [23] pointed out that removing the energy subsidy can directly and indirectly affect households. First, it can result in higher prices for households' electricity use and second, it may result in higher costs for other goods that use energy as an intermediate input. Therefore, removing the fossil fuel subsidy may bring about potential socio-economic issues [22].

In addition to subsidy removal proposed by researchers, policymakers endeavoured to reduce the increase in occupants' energy use by raising energy prices. In 2010, the Iranian subsidy reform aimed to remove the energy subsidy to reduce energy demand. Although this policy effectively reduced electricity consumption in the short term, the effect was not sustainable in the long term (Figure 28). According to data gathered from IEA (2022), although the subsidy had a decreasing trend from 2018 to 2020, domestic electricity had an increasing demand (Figure 29).

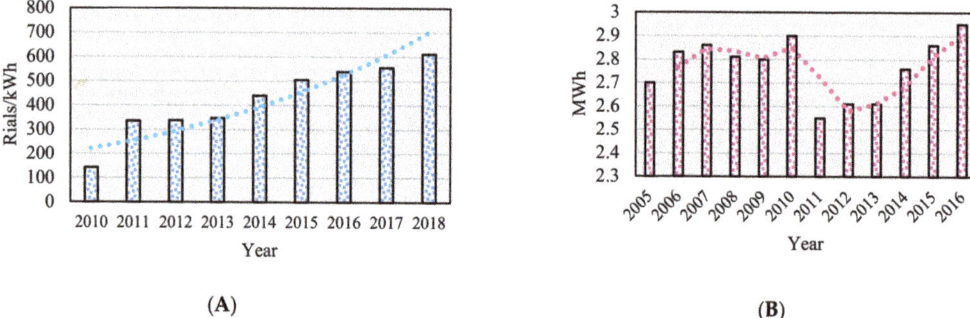

(**A**)

(**B**)

Figure 28. The increase in electricity price in Iran (**A**) and the effect on domestic electricity use per household (**B**).

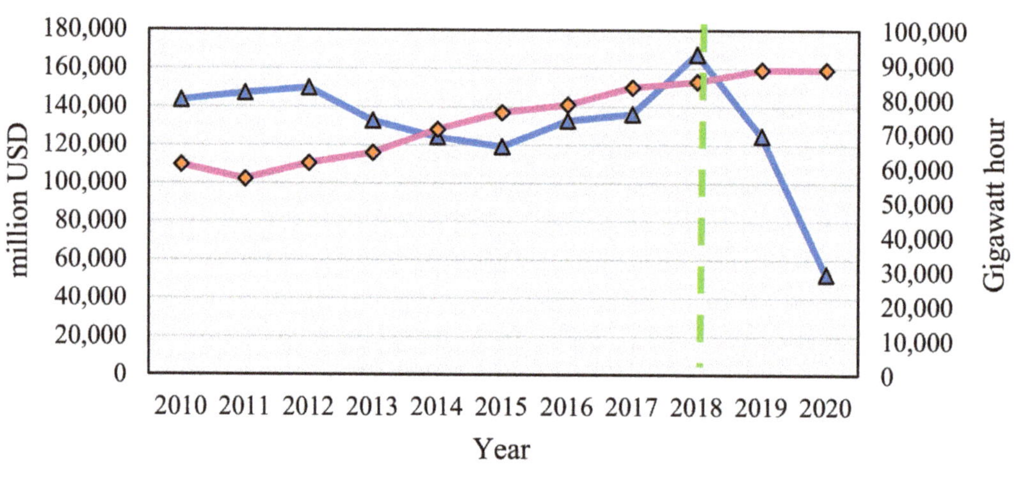

Figure 29. The electricity subsidy for Iran and the effect on domestic electricity consumption.

As the subsidy reform was not successful in decreasing the domestic electricity demand, other policies were enacted as incentives and penalties to encourage or persuade households to reduce inefficient electricity use under "Energy Consumption Pattern Reform". Based on this reform, the Ministry of Energy and the Ministry of Housing and Urban Planning of Iran defined an efficient average electricity consumption per m^2. For households exceeding the average energy use pattern, the electricity price would have a 100% increase to further allocate the money to low-income people under the subsidy reform. However, the increasing electricity demand of Iranian households showed that monetary rewards seem unsustainable in terms of changing households' behaviour in the long term. A study implemented by [75] concluded that increasing fossil fuel prices have little impact on reducing fossil fuel consumption in Iran.

Interestingly, Wang et al. [76] concluded that the energy price had a negligible effect on per capita energy use. Moreover, different studies have highlighted that energy efficiency policies may bring about a rebound effect, which should not be ignored [77]. This can be explained by the crucial role of households' behaviour on electricity demand, as concluded in this study. Therefore, when considering energy use per capita, it seems necessary for future studies to pay more attention to behavioural aspects of households to change their behaviour while reducing energy use. Unfortunately, the current energy policy gives little attention to behaviour change in households. For instance, in the "Energy balance report" published in 2018, in the section of "Efficiency in energy supply and demand" (p. 63) and subsection of "Building and domestic energy appliance" (p. 67), less attention was paid to energy use behaviour of households. For instance, under the subsection policy of "Optimisation of supply and distribution of energy and environmentally friendly technologies", it proposed solutions to prevent power plants' supply limitations. In addition, it developed proposals for overcoming the summer peak of 2018 and informed the power plants about the proposals. However, it was more related to power plant capacity increase rather than the behaviour, which is concluded to have a significant effect on electricity use in Iran. Moreover, under the policy of load management of the demand side (p. 73), the Ministry of Energy has paid most of the attention to the industrial, agricultural, and commercial sectors and less to the domestic sector.

The main action implemented by the Ministry of Energy, which was slightly related to behaviour, were the 'enhancement of efficiency of technical room, replacement of heating and cooling appliances', and 'improving materials and infrastructure of the building' under section 12 of the policy of removal of production barriers in the construction sector. Although under the policy of "Financial support" of "Building and domestic energy appliances", the Ministry of Energy is providing monetary rewards for the reduction in energy intensity which can be related to behaviour change (*e.g., shelters for evaporative coolers, replacing an old refrigerator, washing machine, iron, etc., with high efficiency*), it is mostly on the improvements in the building appliances. However, there is less broad education for households in this regard.

Despite the existing challenges regarding the removal of energy subsidies, some developing countries have proposed solutions to reduce electricity consumption instead of increasing the electricity price. For instance, several incentives for renewable energies were offered as priority policies in African and Latin American countries to decouple economic growth and carbon emissions [78]. As a result, to address the issue of high electricity demand at an affordable price, they suggested supporting the use of energy efficient equipment together with education initiatives on efficient use of energy and appliances [79]. Moreover, some countries have made a lot of progress in managing the political challenges of the subsidy reform. For instance, in Kenya, a gradual approach was employed to eliminate the subsidy over the course of nearly eight years through a combination of tariff increases and reductions in technical losses [80]. This can be a very feasible approach for Iran where an initial price spike and inflation occurred after implementing subsidy reform. In Ghana and Nigeria, a comprehensive public information campaign was created before implementing the subsidy reform. Also, there are some countries which have proposed lifeline tariffs for low-income occupants for a specific amount of energy consumption in a month, such as Uganda and Kenya. Research in the Arab world showed that transforming the reform of domestic energy prices to a set of effective mitigation measures can create opportunities for the government to lower the socio-economic cost of subsidy reforms. This can protect low-income occupants from an erosion of their incomes, while raising public acceptance of pricing reforms effectively [81]. Similarly to Iran, Gulf states are one of the highest in terms of per capita living standards. To address high electricity consumption, some Gulf states have been trying to raise electricity prices and increase income subsidies, which can be an option in reducing their growth of electricity consumption.

Another attempt to reduce electricity consumption is proposing effective solutions in terms of behaviour change interventions to motivate occupants towards energy efficiency rather than removing the subsidy. In this regard, different approaches have been implemented to reduce energy use, namely behaviour change interventions and behaviour prediction models. However, previous research concluded that further improvements are required to achieve a better solution in terms of interventions. According to Santangelo and Tondelli (2017) [62], three important solutions are necessary to reduce electricity use. In the first step, the occupants' awareness needs to be increased. A study conducted in Qatar [15] suggested user education, engagement, and awareness to be effective solutions for domestic energy conservation. In another study conducted by Oryani et al. (2022) [24], the need for a rise in public environmental awareness as well as diversifying the energy mix in Iran were highlighted as important solutions. Although under the education and awareness policy, there is a defined section for educating housewives in terms of the culture of correct energy use, it has not been practically implemented in Iran. Moreover, in terms of behaviour prediction models, inflexibility in behaviour predictions was seen [65], which necessitates further investigations regarding household behaviour variations and everyday habits. Nowadays, paying attention to consumer behaviour for energy prediction models is becoming increasingly important for researchers [65,82–84]. The second important factor is to understand their motivation towards energy conservation. Lastly, the feasibility of energy-efficient behaviour needs to be achieved. It should be noted that energy use behaviour change is a multi-dimensional factor that not only includes psychological and socio-cultural aspects to understand the way occupants think about and use energy [85,86] but also implies further consideration of physical, cognitive, emotional, personalised, and habitual aspects of the behaviour to achieve more feasible energy policies in Iran.

5. Conclusions and Prospects

This study endeavoured to explore the driving factors affecting fossil fuel reliance and its correlation to households' behaviour towards the increasing trend in total and per capita electricity use in the domestic sector on a global and local scale. In the global context, this study revealed that due to the energy subsidy, the oil-rich countries' reliance on fossil fuels increased over the years while developed countries grew their investments in renewable energies. This, in turn has significantly affected oil-rich countries, such as Iran, in terms of their domestic electricity demand, resulting in a more inefficient use of electricity, which was mostly related to households' behaviour. In the local context, the trend and tendency of electricity use in the domestic sector were analysed to understand the outcomes of the energy subsidy in Iran. Findings showed an increasing trend towards electricity generation from fossil fuels over the years, affecting households' inefficient energy use behaviour. However, in countries such as Denmark, the electricity price has contributed as an intervention in households' increasing electricity consumption behaviour, while the energy subsidy in oil-rich countries acted as an encouragement for households' domestic electricity use.

In addition, this study explored the climatic and non-climatic factors affecting domestic electricity use by comparing several provinces of Iran to explore crucial behavioural factors regarding high domestic electricity use per capita. The results revealed that not only does Tehran have higher electricity consumption due to more subscribers than other provinces in Iran, but the households' electricity use per capita was also affected by behavioural issues. After comparing Tehran with Isfahan and Khuzestan, results showed that the high electricity use per capita in Tehran mainly was due to house size, heated or cooled area percentage of space, and household size. Interestingly, although Isfahan and Khuzestan have higher temperatures during summer, their electricity use behaviour seems more efficient than in Tehran.

It is worth mentioning that the aim of this study is to investigate the important factors which have affected energy consumption of residential buildings in Iran and to compare Iran with other developed countries with high energy costs. Although there are studies and available data regarding the high energy use in Iran, there is a lack of research to compare the main factors contributing to Iran and other countries' electricity consumption. Therefore, to answer this important question, the authors needed to explore the important differences between countries with low electricity use and countries with high electricity use and whether this high electricity use can be related to the per capita usage or not. After studying different countries' electricity consumption, this study found that the energy subsidy can play a crucial role in oil-rich countries' high electricity use in residential buildings. However, this was not enough to conclude that the removal of the energy subsidy is the only way to overcome this situation. This has also been highlighted by previous researchers in the field of economy. For instance, various oil-rich countries have started to develop their own sustainability assessment systems to rate their residential buildings based on their efficiency. Therefore, removing the fossil fuel subsidy alone may not address the issues with which the oil-rich countries are struggling, due to lower income occupants. Moreover, as indicated previously, the removal of the energy subsidy can only be effective in the short term, due to the increasing electricity usage per capita. This electricity usage per capita can be explained by occupants' behaviour. Therefore, a question may arise regarding new solutions proposed based on the investigation of aspects affecting human behaviour. This can create opportunities for policy makers to include occupants' energy behaviour in their policy enactment, to reduce their electricity consumption in the domestic sector. To further understand the effect of energy subsidies in other oil-rich countries, future studies might be beneficial to investigate the solutions proposed by some of the oil-rich countries to reduce electricity use.

Therefore, after studying the electricity usage per capita in countries with or without an energy subsidy, it was revealed that high per capita electricity usage in most countries is due to occupants' behaviour and there are several important factors affecting this behaviour in various ways. The high electricity usage per capita can be explained by climatic factors and non-climatic factors. The former may not provide effective solutions because the occupants have minimal expectation from their living environments due to comfort. However, non-climatic factors can shed light on one of the important solutions to be considered in future policies (e.g., different electricity prices based on different house or household sizes).

The results of the study could shed light on the issues that oil-rich countries have faced due to energy subsidy allocation to electricity generation from fossil fuels. Raising the awareness of energy practitioners, policymakers, and designers regarding the important causes of the increase in the domestic electricity demand issues faced by oil-rich countries and Iran could encourage future studies to consider behavioural factors in energy policies. This can also lead to more practical solutions for reducing fossil fuel consumption while reducing the domestic electricity use per capita in Iran and other oil-rich countries.

Author Contributions: D.F. led overall research activities, conceptualisation, methodology, literature review, formal analysis, research outcomes and manuscript drafting; M.N. contributed to principal research supervision and refinement of manuscript development; and H.D. contributed to partial research supervision. All authors have read and agreed to the published version of the manuscript.

Funding: This research received no external funding.

Institutional Review Board Statement: Not applicable.

Informed Consent Statement: Not applicable.

Acknowledgments: The authors would like to express their sincere gratitude to the Melbourne School of Design, Faculty of Architecture, Building and Planning, The University of Melbourne, for providing access to the facilities required for this research activity and a full PhD scholarship given to the first author of this paper.

Conflicts of Interest: The authors declare no conflict of interest.

References

1. Intergovernmental Panel on Climate Change. *IPCC First Assessment Report*; WMO: Geneva, Switzerland, 1990.
2. IPCC. *Climate Change Fifth Assessment Report*; WMO: Geneva, Switzerland, 2014; p. 4.
3. Bouckaert, S.; Pales, A.F.; McGlade, C.; Remme, U.; Wanner, B.; Varro, L.; D'Ambrosio, D.; Spencer, T. *Net Zero by 2050: A Roadmap for the Global Energy Sector*; IEA: Paris, France, 2021.
4. World Resources Institute. 2020. Available online: https://www.wri.org/data/climate-watch-cait-country-greenhouse-gas-emissions-data (accessed on 10 March 2022).
5. IPCC. *Global Warming of 1.5 °C: An IPCC Special Report on the Impacts of Global Warming*; Intergovernmental Panel on Climate Change: Geneva, Switzerland, 2018.
6. Aryanpur, V.; Ghahremani, M.; Mamipour, S.; Fattahi, M.; Gallachóir, B.Ó.; Bazilian, M.D.; Glynn, J. Ex-post analysis of energy subsidy removal through integrated energy systems modelling. *Renew. Sustain. Energy Rev.* **2022**, *158*, 112116. [CrossRef]
7. Solaymani, S.; Kardooni, R.; Kari, F.; Yusoff, S.B. Economic and environmental impacts of energy subsidy reform and oil price shock on the Malaysian transport sector. *Travel Behav. Soc.* **2015**, *2*, 65–77. [CrossRef]
8. Solaymani, S. Energy subsidy reform evaluation research—Reviews in Iran. *Greenh. Gases Sci. Technol.* **2021**, *11*, 520–538. [CrossRef]
9. Solaymani, S. A Review on Energy and Renewable Energy Policies in Iran. *Sustainability* **2021**, *13*, 7328. [CrossRef]
10. Ouyang, X.; Lin, B. Impacts of increasing renewable energy subsidies and phasing out fossil fuel subsidies in China. *Renew. Sustain. Energy Rev.* **2014**, *37*, 933–942. [CrossRef]
11. IEA. *World Energy Outlook 2017 Executive Summary*; Energy Policy; IEA: Paris, France, 2017; p. 90024-4.
12. OECD. *Improving the Environment through Reducing Subsidies*; Organisation for Economic Cooperation and Development: Paris, France, 1999; Volume 1–3.
13. OECD. *Reforming Energy and Transport Subsidies: Environmental and Economic Implications*; Organisation for Economic Cooperation and Development: Paris, France, 1997.
14. Burniaux, J.-M.; Martin, J.P.; Oliveira-Martins, J. *The Effect of Existing Distortions in Energy Markets on the Costs of Policies to Reduce CO_2 Emissions: Evidence from GREEN*; the Economic Costs of Reducing CO_2 Emissions (OECD Economic Studies No 19); Organisation for Economic Cooperation and Development: Paris, France, 1992.
15. Al-Marri, W.; Al-Habaibeh, A.; Watkins, M. An investigation into domestic energy consumption behaviour and public awareness of renewable energy in Qatar. *Sustain. Cities Soc.* **2018**, *41*, 639–646. [CrossRef]
16. El-Katiri, L.; Fattouh, B. A Brief Political Economy of Energy Subsidies in the Middle East and North Africa, International Development Policy. *Rev. Int. Polit. Développement* **2017**, *7*, 58–87.
17. Verme, P. Subsidy Reforms in the Middle East and North Africa region: A review. In *The Quest for Subsidy Reforms in the Middle East and North Africa Region*; Springer: Cham, Switzerland, 2017; pp. 3–31.
18. Clements, M.B.J.; Coady, M.D.; Fabrizio, M.S.; Gupta, M.S.; Alleyne, M.T.S.C.; Sdralevich, M.C.A. *Energy Subsidy Reform: Lessons and Implications*; International Monetary Fund: Washington, DC, USA, 2013.
19. Moerenhout, T.; Irschlinger, T. *Exploring the Trade Impacts of Fossil Fuel Subsidies*; International Institute for Sustainable Development: Winnipeg, MB, Canada, 2020.
20. Sun, C. An empirical case study about the reform of tiered pricing for household electricity in China. *Appl. Energy* **2015**, *160*, 383–389. [CrossRef]
21. Dube, I. Impact of energy subsidies on energy consumption and supply in Zimbabwe. Do the urban poor really benefit? *Energy Policy* **2003**, *31*, 1635–1645. [CrossRef]
22. Adom, P.K.; Adams, S. Energy savings in Nigeria. Is there a way of escape from energy inefficiency? *Renew. Sustain. Energy Rev.* **2018**, *81*, 2421–2430. [CrossRef]
23. Fattouh, B.; El-Katiri, L. Energy subsidies in the Middle East and North Africa. *Energy Strategy Rev.* **2013**, *2*, 108–115. [CrossRef]
24. Oryani, B.; Kamyab, H.; Moridian, A.; Azizi, Z.; Rezania, S.; Chelliapan, S. Does structural change boost the energy demand in a fossil fuel-driven economy? New evidence from Iran. *Energy* **2022**, *254*, 124391. [CrossRef]
25. Charap, M.J.; da Silva, M.A.R.; Rodriguez, M.P.C. *Energy Subsidies and Energy Consumption: A Cross-Country Analysis*; International Monetary Fund: Washington, DC, USA, 2013.
26. Al Iriani, M.A.; Trabelsi, M. The economic impact of phasing out energy consumption subsidies in GCC countries. *J. Econ. Bus.* **2016**, *87*, 35–49. [CrossRef]
27. Mousavi, B.; Lopez, N.S.A.; Biona, J.B.M.; Chiu, A.S.; Blesl, M. Driving forces of Iran's CO2 emissions from energy consumption: An LMDI decomposition approach. *Appl. Energy* **2017**, *206*, 804–814. [CrossRef]
28. Gazder, U. Energy Consumption Trends in Energy Scarce and Rich Countries: Comparative Study for Pakistan and Saudi Arabia. *E3S Web Conf.* **2017**, *23*, 7002. [CrossRef]
29. Gangopadhyay, S.; Ramaswami, B.; Wadhwa, W. Reducing subsidies on household fuels in India: How will it affect the poor? *Energy Policy* **2005**, *33*, 2326–2336. [CrossRef]

30. IEA. Fossil Fuel Subsidies Database. 2021. Available online: https://www.iea.org/data-and-statistics/data-product/fossil-fuel-subsidies-database (accessed on 10 March 2022).
31. IEA. World Energy Balances. 2021. Available online: https://www.iea.org/reports/world-energy-balances-overview (accessed on 8 March 2022).
32. IEA. Data and Statistics. 2021. Available online: https://www.iea.org/data-and-statistics/data-browser/?country=WORLD&fuel=Electricity%20and%20heat&indicator=ElecIndex (accessed on 10 March 2022).
33. IEA. Value of Fossil-Fuel Subsidies. 2019. Available online: https://www.iea.org/data-and-statistics/charts/value-of-fossil-fuel-subsidies-by-fuel-in-the-top-25-countries-2019 (accessed on 11 March 2022).
34. The World Bank. Fossil Fuel Energy Consumption. 2010. Available online: https://data.worldbank.org/indicator/EG.USE.COMM.FO.ZS (accessed on 16 March 2022).
35. The World Bank. *WBG—Doing Business*; World Bank Publications: Washington, DC, USA, 2019.
36. World Data. Climate. 2021. Available online: https://www.worlddata.info/asia/iran/ (accessed on 5 March 2022).
37. Our World in Data. Per Capita Electricity Generation from Fossil Fuels. 2021. Available online: https://ourworldindata.org/grapher/fossil-electricity-per-capita. (accessed on 5 March 2022).
38. Ritchie, H.; Roser, M.; Rosado, P.; CO_2 and Greenhouse Gas Emissions. Our World Data. Available online: https://ourworldindata.org/co2-emissions?utm_source=coast%20reporter&utm_campaign=coast%20reporter%3A%20outbound&utm_medium=referral (accessed on 10 March 2022).
39. World Economic Outlook Database. 2021. Available online: https://www.imf.org/en/Publications/WEO/weo-database/2021/October (accessed on 4 March 2022).
40. Global Carbon Project. *Supplemental Data of Global Carbon Project 2021(1.0) [Data Set]*; Global Carbon Project: Canberra, Australia, 2021; pp. 1–191. [CrossRef]
41. Australian Bureau of Statistics. *Census of Population and Housing 2011 and 2016*; Australian Bureau of Statistics: Canberra, Australia, 2017.
42. Qu, L. *Families Then & Now: Households and Families*; Australian Institute of Family Studies: Melbourne, Australia, 2020.
43. Statistisches Bundesamt (Destatis). Household Projections in Germany. 2020. Available online: https://www.destatis.de/EN/Themes/Society-Environment/Population/Households-Families/Tables/projection-household.html (accessed on 10 March 2022).
44. National Statistics Centre of Iran. 2021. Available online: https://www.amar.org.ir/%D8%AF%D8%A7%D8%AF%D9%87%D9%87%D8%A7-%D9%88-%D8%A7%D8%B7%D9%84%D8%A7%D8%B9%D8%A7%D8%AA-%D8%A2%D9%85%D8%A7%D8%B1%DB%8C (accessed on 10 March 2022).
45. National Statistics Centre of Iran. 2020. Available online: https://www.amar.org.ir/%D9%BE%D8%A7%DB%8C%DA%AF%D8%A7%D9%87-%D9%87%D8%A7-%D9%88-%D8%B3%D8%A7%D9%85%D8%A7%D9%86%D9%87-%D9%87%D8%A7/%D8%B3%D8%B1%DB%8C%D9%87%D8%A7%DB%8C-%D8%B2%D9%85%D8%A7%D9%86%DB%8C/agentType/ViewType/PropertyTypeID/1 (accessed on 4 March 2022).
46. Plan and Budget Organisation. *National Centres for Environmental Information*; National Statistics Centre of Iran: Tehran, Iran, 2022.
47. Tavanir Company. *Statistics of Iran's Electricity Industry, Power Distribution Sector in 2020*; Deputy of Research and Human Resources; Information and Communication Technology and Statistics Office, Tavanir Company: Tehran, Iran, 2020.
48. Kaygusuz, K. Energy for Sustainable Development: A case of Developing Countries. *Renew. Sustain. Energy Rev.* **2012**, *16*, 1116–1126. [CrossRef]
49. Looney, B. *BP Statistical Review of World Energy 2021*; BP: London, UK, 2020.
50. IEA. *Organisation for Economic Co-Operation and Development and International Monetary Fund via United Nations Global SDG Database*; IEA: Paris, France, 2020.
51. Guerrero-Lemus, R.; Shephard, L.E. Current Energy Context in Africa and Latin America. In *Low-Carbon Energy in Africa and Latin America*; Springer: Cham, Switzerland, 2017; pp. 39–73.
52. IEA. 2020. Available online: https://www.iea.org/data-and-statistics/data-product/world-energy-statistics-and-balances (accessed on 7 March 2022).
53. IEA. *IEA Energy and Carbon Tracker 2020*; IEA, Ed.; IEA: Paris, France, 2021.
54. Ministry of Energy. Energy Balance. 2020. Available online: https://isn.moe.gov.ir/%D8%A8%D8%AE%D8%B4-%D8%A8%D8%B1%D9%82-%D9%88-%D8%A7%D9%86%D8%B1%DA%98%D9%8A/%D8%B9%D9%85%D9%84%D9%83%D8%B1%D8%AF/%D8%AA%D8%B1%D8%A7%D8%B2%D9%86%D8%A7%D9%87-%D8%A7%D9%86%D8%B1%DA%98%D9%8A (accessed on 10 March 2022).
55. Ministry of Energy. Statistics and Information Network. 2021. Available online: https://moe.gov.ir/%D8%A7%D8%B7%D9%84%D8%A7%D8%B9%D8%A7%D8%AA-%D9%88-%D8%A7%D9%85%D8%A7%D8%B1 (accessed on 9 March 2022).
56. Bierwirth, A.; Thomas, S. Almost Best Friends: Sufficiency and Efficiency; Can Sufficiency Maximise Efficiency Gains in Buildings? In Proceedings of the First Fuel Now: ECEEE 2015 Summer Study, Toulon/Hyères, France, 1–6 June 2015.

57. Stephan, A.; Crawford, R.H. The Relationship between House Size and Life Cycle Energy Demand: Implications for Energy Efficiency Regulations for Buildings. *Energy* **2016**, *116*, 1158–1171. [CrossRef]
58. Azizi, M.M. An Analysis on the Relation Between Per Capita Land Uses and City Size in Iran Urban Comprehensive Plans. *Honar-Ha-Ye-Ziba* **2014**, *18*, 25–36.
59. Shaari, M.S.; Rahim, H.A.; Rashid, I. Relationship among population, energy consumption and economic growth in Malaysia. *Int. J. Soc. Sci.* **2013**, *13*, 39–45.
60. Lee, C.-C.; Chang, C.-P.; Chen, P.-F. Energy-Income Causality in OECD Countries Revisited: The Key Role of Capital Stock. *Energy Econ.* **2008**, *30*, 2359–2373. [CrossRef]
61. Bartleet, M.; Gounder, R. Energy consumption and economic growth in New Zealand: Results of trivariate and multivariate models. *Energy Policy* **2010**, *38*, 3508–3517. [CrossRef]
62. Batliwala, S.; Reddy, A.K. Energy Consumption and Population. In *Population: The Complex Reality*; The Royal Society: London, UK, 1994.
63. Santangelo, A.; Tondelli, S. Occupant behaviour and building renovation of the social housing stock: Current and future challenges. *Energy Build.* **2017**, *145*, 276–283. [CrossRef]
64. Tavakoli, E.; Nikkhah, A.; Zomorodian, Z.S.; Tahsildoost, M.; Hoonejani, M.R. Estimating the Impact of Occupants' Behaviour on Energy Consumption by Pls-SEM: A Case Study of Pakdel Residential Complex in Isfahan, Iran. *Front. Sustain. Cities* **2022**, *4*, 700090. [CrossRef]
65. Soltani, M.; Rahmani, O.; Ghasimi, D.S.; Ghaderpour, Y.; Pour, A.B.; Misnan, S.H.; Ngah, I. Impact of household demographic characteristics on energy conservation and carbon dioxide emission: Case from Mahabad city, Iran. *Energy* **2020**, *194*, 116916. [CrossRef]
66. Sepehr, M.; Eghtedaei, R.; Toolabimoghadam, A.; Noorollahi, Y.; Mohammadi, M. Modeling the electrical energy consumption profile for residential buildings in Iran. *Sustain. Cities Soc.* **2018**, *41*, 481–489. [CrossRef]
67. El-Katiri, L.; Husain, M. *Prospects for Renewable Energy in GCC States—Opportunities and the Need for Reform*; Oxford Institute for Energy Studies: Oxford, UK, 2014.
68. Sdralevich, M.C.A.; Sab, M.; Zouhar, M.Y.; Albertin, M.G. *Subsidy Reform in the Middle East and North Africa: Recent Progress and Challenges Ahead*; International Monetary Fund: Washington, DC, USA, 2014.
69. Qader, M.R. Electricity Consumption and GHG Emissions in GCC Countries. *Energies* **2009**, *2*, 1201–1213. [CrossRef]
70. Riazi, M.; Hosseyni, S. Overview of current energy policy and standards in the building sector in Iran. *Sustain. Dev. Plan. V* **2011**, *150*, 189–200.
71. Zhang, M.M.; Zhou, D.Q.; Zhou, P.; Chen, H.T. Optimal design of subsidy to stimulate renewable energy investments: The case of China. *Renew. Sustain. Energy Rev.* **2017**, *71*, 873–883. [CrossRef]
72. Liu, W.; Li, H. Improving energy consumption structure: A comprehensive assessment of fossil energy subsidies reform in China. *Energy Policy* **2011**, *39*, 4134–4143. [CrossRef]
73. Wang, Q.; Qiu, H.-N.; Kuang, Y. Market-driven energy pricing necessary to ensure China's power supply. *Energy Policy* **2009**, *37*, 2498–2504. [CrossRef]
74. Kalkuhl, M.; Edenhofer, O.; Lessmann, K. Renewable energy subsidies: Second-best policy or fatal aberration for mitigation? *Resour. Energy Econ.* **2013**, *35*, 217–234. [CrossRef]
75. Farajzadeh, Z.; Bakhshoodeh, M. Economic and environmental analyses of Iranian energy subsidy reform using Computable General Equilibrium (CGE) model. *Energy Sustain. Dev.* **2015**, *27*, 147–154. [CrossRef]
76. Taghvaee, V.M.; Arani, A.A.; Soretz, S.; Agheli, L. Comparing energy efficiency and price policy from a sustainable development perspective: Using fossil fuel demand elasticities in Iran. *MRS Energy Sustain.* **2022**, 1–14. [CrossRef]
77. Wang, Q.; Su, M.; Li, R.; Ponce, P. The effects of energy prices, urbanization and economic growth on energy consumption per capita in 186 countries. *J. Clean. Prod.* **2019**, *225*, 1017–1032. [CrossRef]
78. Steren, A.; Rubin, O.D.; Rosenzweig, S. Energy-efficiency policies targeting consumers may not save energy in the long run: A rebound effect that cannot be ignored. *Energy Res. Soc. Sci.* **2022**, *90*, 102600. [CrossRef]
79. Butera, F.; Adhikari, R.S.; Caputo, P.; Facchini, A. The Challenge of Energy in Informal Settlements. A Review of the Literature for Latin America and Africa. In *Analysis of Energy Consumption and Energy Efficiency in Informal Settlements of Developing Countries*; Politecnico di Milano: Milan, Italy; Enel Foundation: Rome, Italy, 2015; pp. 1–32.
80. Alleyne, M.T.S.C.; Hussain, M.M. *Energy Subsidy Reform in Sub-Saharan Africa: Experiences and Lessons*; International Monetary Fund: Washington, DC, USA, 2013.
81. Fattouh, B.; El-Katiri, L. *Energy Subsidies in the Arab World*; United Nations Development Programme: New York, NY, USA, 2012.
82. Luh, S.; Kannan, R.; Schmidt, T.J.; Kober, T. Behavior matters: A systematic review of representing consumer mobility choices in energy models. *Energy Res. Soc. Sci.* **2022**, *90*, 102596. [CrossRef]
83. Krumm, A.; Süsser, D.; Blechinger, P. Modelling social aspects of the energy transition: What is the current representation of social factors in energy models? *Energy* **2022**, *239*, 121706. [CrossRef]
84. Schafer, A. *Introducing Behavioral Change in Transportation into Energy/Economy/Environment Models*; World Bank Policy, Research Working Paper No. 6234; World Bank Publications: Washington, DC, USA, 2012.

85. He, H.A.; Greenberg, S. *Motivating Sustainable Energy Consumption in the Home*; University of Calgary: Calgary, AB, Canada, 2008.
86. Harputlugil, T.; de Wilde, P. The interaction between humans and buildings for energy efficiency: A critical review. *Energy Res. Soc. Sci.* **2021**, *71*, 101828. [CrossRef]

Entry

Reducing CO$_2$ in Passivhaus-Adapted Affordable Tropical Homes

Karl Wagner

The President's Office, Technical University of Applied Sciences Rosenheim, D-83043 Rosenheim, Germany;
karl.wagner@th-rosenheim.de

Definition: On average, houses including those in the tropics are responsible for almost 39% of the global carbon emission caused by non-renewables, first and foremost by fuel. Looking at the worldwide map of residential buildings' contribution compared with commercial, the worldwide national maximum of 33.5% CO$_2$ of housing is caused by residential buildings in Uzbekistan. In an overwhelming number of most countries, their values are significantly lower, due to comparably lower energy demand than commercial buildings and because affordable homes increasingly use small PV to cater for their own basic needs. However, with the rising temperature and a likewise growing imperative to cool homes from about 30 °C onwards basically by split-unit air conditioners, the residential houses' portion of CO$_2$-emission might dramatically increase to survive such more common hot periods in the future. In combination with air conditioners needing some airtightness, the first purpose of this entry is to show that bv 2050 in tropical regions, there will be no alternative to relatively airtight houses if the temperatures rise at the present speed. This is one alternative to an uncontrollable and life-threatening migration of millions of people to cooler but still livable regions in 2050. To trigger necessary changes toward homes that can better avert the heat, using the method of qualitative comparative content analysis, passive houses (PH) have emerged as adaptations to the tropical climate. Therefore, the second purpose of this in-depth study with the perspective of social science, is to reveal a comparative closer qualitative look at the tropicalized PH-approach. It is probably the most civilized building energy-saving strategy on the planet and can systematically keep the threatening increasing heat outside. However, before utilizing the concept, herein need to investigate why PH-technology as a whole concept with all its modules discussed earlier has been very slow to "go South" into the tropical region (the original PH will be referred to as "PH1"). The reason is that some qualitative differences of the more affordable and more simplistic tropicalized "PH2" make it easier and more realistic to penetrate the market, without letting go meaningful R&D-insights of PH1. As a probably facilitating future solution, the result is the triple-tabled option to utilise more synergies between the usually closed PH1 and the more open and flexibly naturally ventilated PH2. Unlike the PH-platform, ZEMCH is a related concept which tries to cater specifically to the significantly growing market for lower-income homes to go for carbonless energy. The conclusion is that scaling for residential buildings as mass products using passive house technology in combination with ZEMCH could turn out to become an important topic. It comprises the question in how far low or no carbon affordable homes based on the PH-concept in combination with ZEMCH-applications also may come into play as standard and to help mother Earth's struggle for survival.

Keywords: climate change; Passivhaus; ZEMCH; affordable housing; synergies

Citation: Wagner, K. Reducing CO$_2$ in Passivhaus-Adapted Affordable Tropical Homes. *Encyclopedia* **2023**, 3, 168–181. https://doi.org/10.3390/encyclopedia3010012

Academic Editors: Masa Noguchi, Antonio Frattari, Carlos Torres Formoso, Haşim Altan, John Odhiambo Onyango, Jun-Tae Kim, Kheira Anissa Tabet Aoul, Mehdi Amirkhani, Sara Jane Wilkinson, Shaila Bantanur and Raffaele Barretta

Received: 20 July 2022
Revised: 22 September 2022
Accepted: 3 October 2022
Published: 26 January 2023

1. Two Different Passive House Concepts (PH1 vs. PH2) and Methodology

In building science, and logically in everyday life, a building anywhere on the globe, is an "environmental separator between inside and outside" [1]. Green buildings attempt to reduce carbon against the hotter years to come. They firm up under different concepts and strategies, such as low and zero energy houses, zero energy buildings, nearly zero-emission building (NZEB), zero-emission buildings to name a few prominent ones. Some of these green building concepts are to be implemented in the EU between 2020 and 2030.

Among these strategies, worldwide, the most energy efficient concept is the Passivhaus. It will be using and tabling two slightly differing passive house concepts, PH1 and PH2, applying three comparative content analyses as methodology that is widely used in social science. In the tradition of the sociologists Emile Durkheim and Max Weber, this is a kind of a bird's-eye view perspective and method of comparing the two contents with the "idea of uncovering and discovering new ideas about them" [2].

Applying this methodology, it seems consensual that passive basically means that the indoor climate is separated by the building envelope against an unhealthy or uncomfortable outdoor climate for humans. This counts also vice versa: If the outdoor conditions are more comfortable and healthier compared with indoor, why not open windows and perhaps even doors?

However, based on the methodology of comparative content analysis, "passive house technology" generated by concepts in building science has two different meanings. From a social science perspective, the content analysis of "passive" will be used in the tradition of relationism and phenomenology as the "discipline of phenomenology that may be defined initially as the entry of structures of experience, or consciousness" [3]. Literally, phenomenology is the entry of everyday life "phenomena," which states that different actors even like architects or building scientists might have different perceptions. This will depend on how science teaches to experience passive as part of the own social community and how different perspectives can work together. Accordingly, phenomenology describes and analyses the appearances of things, as they appear in experience, "the study of phenomena (= things that exist and can be seen, felt, tasted, etc.) and how experience them [4].

Hence, based on qualitative phenomenological studies, our first look will be at PH from five different perspectives; out of four principal ways how to build homes, looking at mainstream and three passive strategies, and within two concepts for passive houses:

1. **Conventional house**. Nothing is optimized. Buildings are just cheap and fast focusing only on fast ROIs. These houses stack the heat. Its type A works with relatively energy saving fans (50%), type B with A/C, at least in the master sleeping room. Type B compared to type A is characterized by a high carbon footprint.

2. **House with element(s) of passive technology**. Architects try to educate people to use more passive modules based on natural ventilation, which as a minimum requirement just means opening windows all day and night long to save CO_2: "The most important passive design strategy in the tropics is to open up houses as much as possible, even during the heat of the day, to achieve maximum cross ventilation and convective air flow" [5]. This is a common behaviour for most households in warm and hot conditions like in the tropics.

3. **PH1**, as presented below, was developed and marketed from the cold hemisphere (Central Europe, Sweden, USA/ Utah), following their credo with meticulously defined necessities of insulation and airtightness with software like PHPP which restricts the user's calculation to a clear standard saves enormously carbon.

4. **PH2** resembles PH1 but looks in depth at tropical adaptations. It is a combination of daytime closed and nighttime open windows, which is unsuitable for PH1 under tropical conditions of high temperature and absolute => relative humidity. It can lead to more RH if is it not controlled, and it may lead to higher "adaptive" temperatures of occupants. Therefore, tropically adapted Passive Houses (PH2) can accept higher temperatures. with set points up to 28 °C and any humidity. Hence, it will be seen more energy saving than the conventional PH1-standards with their maximum allowable temperature of 25 °C and 60% relative humidity integrated in the software tool PHPP.

5. In addition, the **ZEMCH** network products are not a derivation of 3 or 4 but could be closely related to one of them, making use of passive technology. ZEMCH is not restricted to its features. Its application might also be more related to nearly zero-emission building (NZEB).

Option 1 is still common but is not a not subject in this contribution, as looking towards more carbon-free solutions. Both 1 and 2 have no radar for real passive sustainability, as the house remains hot. Option 2 is just patchwork and does not lead to CO_2 saving targets like the PH 1 and 2, along with ZEMCH buildings.

Before applying phenomenology as a tool to compare the two remaining Passivhaus concepts (1) in the colder hemisphere and (2) its potential in the tropics, looking at different passive strategies to achieve more conducive thermal comfort. In the decades of approaching global warming catastrophes, this can be seen as a "higher" reason to assist to reduce climate change in the building sector. In the case of passive technology and how to achieve thermal comfort for occupants, it is about the (non-) interaction of two basically latent controversial standpoints passive technology and the renowned PH1. (Especially PH2 as a potential further development of the PH with ZEMCH which stresses on scalable affordable mass homes will follow later).

Comparing 2. Passive technology and 3. Passive Houses from the phenomenologist's perspective are like asking two building scientists who more unconsciously but inadvertently refused talking about the generic meaning of "passive" to each other. Decades to communicate and exchange ideas have passed by since the word ever occurred, probably independently, in tropical and colder climate origins.

The first one promoting passive technology will be called the "classical tropical" type approach that even at certain times maximises natural ventilation and the second one "passive" PH1 using the building envelope as a basic fortress against the heat. At certain "hot" times, air conditioning and in less hot cases cool-water-based cooling ceilings or walls are needed, protected by a closed-up building envelope. In this entry, by qualitative content analysis, the second meaning of passive related to PH1 and PH2, and, by the applied analysis will be pursued, try to reconcile both.

Since 1991, the State of the Art Passivhaus in the traces of type PH1 follows five principles. Instead of heat recovery, compulsory cooling recovery is suggested for tropical buildings and those in the colder hemisphere during the hot season. In cold regions like Iceland, back in the Middle Ages, dwellers started to build passive-like turf houses after wood became scarce [6]. It requires artificial cooling in residential buildings under hot conditions usually performed by air conditioners as clear separator between inside and outside climate was laid out around 1980 in Utah/USA by its fundamental DIYS-inventor, Amory Lovins [7]. Wolfgang Feist (formerly Institute for Housing and the Environment, /Institut Wohnen und Umwelt Darmstadt/Germany) and Bo Adamson (Lund University, Lund/Sweden) introduced and further developed the concept in Europe. The emerging "Passive House standard" originated from a conversation in May 1988 between Adamson and Feist, but was much more than a new label for a green building. Its founders agreed on strict criteria to make it the most radical "green" version of a thermally comfortable low energy house. Apart from any consumable energy, it is defined by extremely low heating and, likewise if applicable, cooling demand [8]. It results in the fact that (apart from other energy efficient features) necessary warmth or coolness can be implemented primarily by the supply of an (active) mechanical ventilation system. Adamson's idea of mechanical instead of sometimes not working natural ventilation was first rejected by Feist but was soon understood as a necessity to make PHs a reality with basic airtightness plus outside air if necessary [9]. When 90% less energy was consumed in the world's first passive residential house at Kranichstein/Darmstadt in 1991, experts said to Feist that it must be an error, but obviously it was not: Even by 2021, 30 years after the first building was set up, no decrease in terms of this 90% less energy performance could yet be reported, with still low maintenance efforts [10].

2. The Passive House Approach and an Outline for Its Dedicated Tropical Adaptation Model

The generic five elements and benchmarks of a Passivhaus (PH1) are well documented both in literature and video clips: airtightness, insulation, avoidance or thermal bridges, Passivhaus windows, and ventilation with heat recovery [11–14].

In addition, a tropically adapted holistic passive house (PH2) concept shaping up since 2016, has been elaborated on with the initiative of and collaboration of a ventilation developer and an international architecture Passive House IPHA firm from Munich/Germany. Basically, tropical buildings should adhere to a new entirety (totality) consisting of the following seven or eight revised core modules of five real passive and two or three active elements as well. The East, West, South or North "orientation" in the orange box is general, whereas the 4 elements for "passive" Insulation are indicated in the green boxes. The 3 modules for ventilation and (de-)humidification (blue boxes) point to "active" elements of a Passivhaus [15]. The modules are highlighted in the following Figure 1:

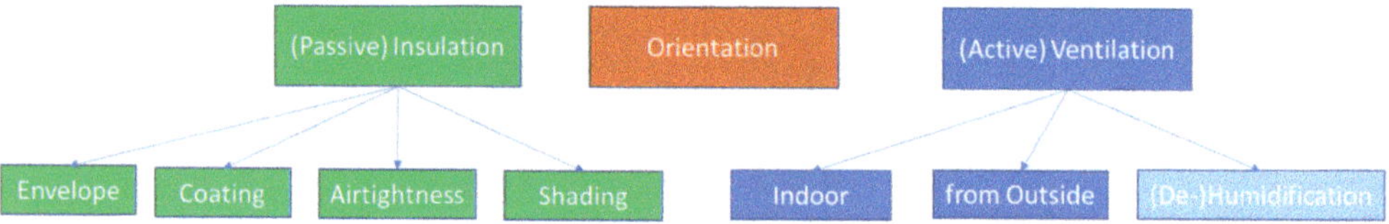

Figure 1. Output Pillars of the Passivhaus for a tropical Building.

These core modules and benchmarks of a tropical passive house are described below. They add on adapted value to the existing five of a general passive house. "Orientation" is the source for everything that happens in terms of solar radiation (the positioning is less meant for the hardly 365 days predictable natural ventilation). As shown in Figure 1, "De-"humidification carries a question mark because whether it is ultimately necessary for affordable tropical mass homes have to be discussed, which are in the focus of this entry [16]. According to discussions one month ago with an IPHA member, it was conceded that no humidification may not necessarily create mould inside the house in case of natural ventilation, when daytime hotness offsets nighttime humidity (leading to what we refer to PH 2).

In a first step of the comparison, it can be now narrowed down the scope and Table the seven or eight tropical modules as exhibited in Figure 1 above. In a second step in Table 1 below, justifying the deriving thoughts on marketability is needed. In the light of climate change with requirements toward affordability and marketability, the criteria do not necessarily coincide with the universal passive house PH1 principles [17]. Table 1 highlights the differences between the conventional Passive House (1) and the detected or derived specifications of PH2:

Table 1. Details of the Adjusted Seven to Eight Modules and Benchmarks of a Tropically Adapted Passive House (PH2).

No	Element	Description
1.	Building Orientation	Depending on the geographical location, the orientation of a planned building will be optimised by minimising or -if possible- excluding solar exposure through the windows (passive). On tropical islands or coastlines like Penang / Malaysia or Baranquilla/Colombia, the most common wind direction can determine the building's orientation. However, the wind's orientation will be exposed to seasonal fluctuations. Therefore, relying on the cool night time natural breeze by opening the windows, is not sufficient to provide indoor tropical thermal comfort.

Table 1. *Cont.*

No	Element	Description
2.	Building Envelope	Insulation of the building envlope (wall, roof, minimising thermal bridges and low-e glazing) will reduce the exterior heat infusion. At the same time, the desired ambient tropical USL temperature inside the dwelling will be pursued, as long as the occupants are inside. In the tropics, experiments have proof-cased that multiple layered glazing is not necessary, as the thermal difference between given outdoor and yielded indoor temperature is much lower than for cold countries. Tropical countries better invest into coating, less convective window frames with heat brakes and into shading (no. 5) (passive).
3.	Basic Airtightness	Like in countries of the colder hemisphere, a basic airtight tropical building envelope with minimized thermal air bridges will prevent the infiltration of "default" too hot, sometimes too humid and polluted air, whilst maintaining the desired indoor temperature. However, airtightness in shaded tropical buildings is not a radical must as in countries with seasonally huge differences between the outside and inside temperature. (Formula in the PHPP: Airtightness (AT) < 0.6 /h at a pressure difference compared inside to the outside of 50 pascal). In tropical adaptations, the concept also includes filtered mechanical exterior ventilation especially during the nighttime (passive/active).
4.	Reflective Coating	Especially in tropical buildings, outer surfaces of the building envelope like exterior walls and roofs, are coated with brighter colours which reflect solar radiation to further reduce heat gains into the building. Off-white or light yellow contribute to lower heat gains through the entire building envelope and the windows (passive). Perhaps an affordable house of the future has white walls and dark solar panels as additional sun protection in an angle of about 30° to optimise attic cross ventilation on the roof.
5.	Shading	Recently also adopted by northern hemisphere passive houses, external shading tools (fixed or movable) will stop direct solar gains inside the building or the room through the windows. Different compared to cold country passive standards which have been adapted to the tropics, reducing the heat gain by shading is not about double- or triple-glazed sun protection windows. The reason is that it is considered as not necessary for the small temperature difference between the real shaded outside and the desired inside temperature (passive with a bit of smart active for flexibly opening the shades to let natural light inside the building in case direct radiation is absent). Typical 30–32 °C outside (shaded) temperature does not justify double or even triple glazing if the set point is 26–28 °C. In buildings with a high window-wall ratio where the unshaded sun hits with 55 °C survival without intense usage of air conditioners is impossible.
6.	Dehumidification?	Allowing temporary humidity or not can turn into a very controversial question. That is why the author assigned it with a question mark for tropical countries: Considered as a must or an option for enclosed buildings or rooms, it is not applicable for the huge majority of naturally ventilated ones where neither outside temperature nor humidity is stoppable. Nocturnal typical relative humidities of 85% and above will not automatically make buildings grow mould. Basically, the issue of high humdity and its consequences is balanced out by the logical credo "what gets wet, must get dry again [18]". Hence, if wetness intrudes during the night time, it might dissipate by the lower humidity at least in the growing urban areas catalysed by the "heat island effect" [19]. And, prior to further investigation the question may be allowed: why have old highrises built in the 1970s and 1980s which were fully naturally ventilated not all become victims of mould?
7.	Efficient Outside Air Cross-Ventilation	A constant supply of fresh air from outside keeps the indoor environment healthy—if it is professionally drawn out again by an exhaust fan or at least an empty hollow [20]. This is the real natural ventilation, which is based on mechanical ventilation, which would not happen, if only the windows are open. If mechanical air supply fans are used, however, in polluted areas the outside air must be filtered. Not like in conventional Passive Houses (PH1) especially during night times, in tropical highlands the air does not have to be pre-cooled to produce also cool air before penetrating from outside. The ventilation in the tropical highlands is different to the colder hemisphere: it maximises the air stream when the temperature falls below the upper borderline of thermal comfort or cooler outside (active).

Table 1. *Cont.*

No	Element	Description
8.	Internal Air Movement	Stand-alone indoor ventilators as catalysts for 7. are not obsolete in a tropical passive house which is proposed here. On the contrary, internal air movement devices get two new meanings for residential and commercial buildings: Type 1: Different sorts of fans (stand, table or ceiling) are conceptualised of as systems, preferably in combination with 7. They permit higher set-point temperatures for tropically adapted cooling systems while providing individualised thermal comfort for occupants by blowing "personalised" air onto the occupants, hereby reducing the ambiently felt temperature. Type 2: Internal air re-produced by the A/C compressor which should be minimized, as the cooling pushes in with 11 °C. This is uncomfortable like draughts are looked upon in the colder hemisphere (active).

Non-tropicalised passive house (PH1) concepts have been adopted in estimated almost 65,000 buildings (2–5% of them certified, i.e., 95–98% are non-certified applications in all different climatic areas including 7–9 certified implementations in tropical countries). The answer depends on where the line between tropical and subtropical climate zones are drawn. Either following mainstream determining "tropical zone or not" by the latitude, or looking into climatic details as Köppen and Geiger did [21]. The following Figure 2 will scrutinize their locations.

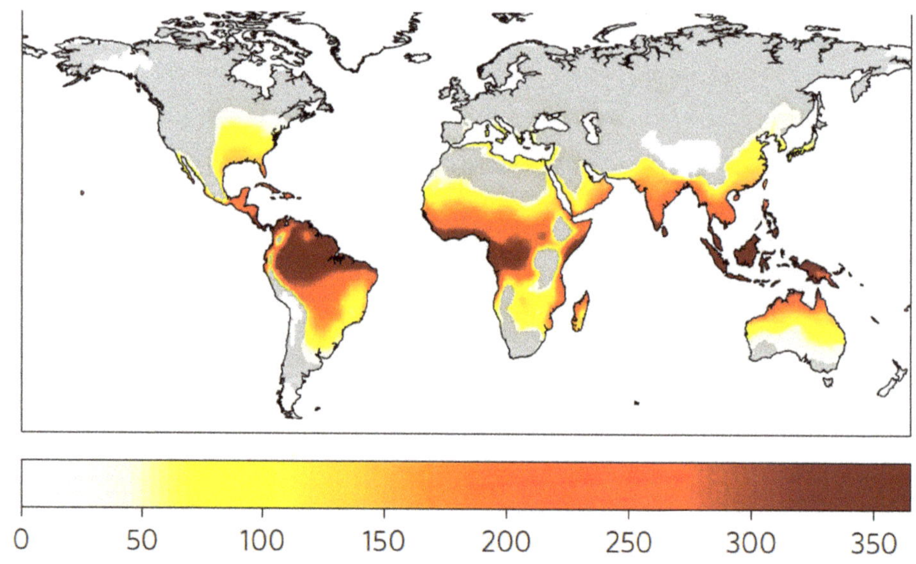

Figure 2. Global risk of deadly heat days with a Scale of annual overheated Days forecasting Climate Change in 2050 where residents need to stay in cooled areas to survive at least during the daytime. Adapted with permission from reference [22], Copyright 2017 Springer Nature.

However, talking about future protection against global warming by innovative building technology like the PH, this is where its potentially most important application of the future is found: the region where most people live. The tropical belt in 2020 was home to about 40% of the world's population, and will host more than 50% of the world population by the late 2030s or early 2040s [22]. As in the tropics it is nowhere too cold to survive, but too hot at some places, most low- and middle-income people do neither cool, but naturally use the open door and windows style, if the tropics can sustain as a place to live, which is becoming questionable [23]. This is indicated by the red and dark red regions on the globe whereby almost all tropical regions are in the danger zone of 250–365 heat-affected days in 2050:

Hence, like shown in Figure 2, the researchers around Mora described the devastating exacerbating consequences of global warming until 2050. Counting the days with temperatures of 35 °C and above, according to their global map, almost everywhere around the tropical belt the temperature is going to be deadly and would include a never-known exodus of homeless populations to colder areas [23] - unless new building codes especially for affordable houses can be found to at least (!) delay parts of the catastrophic development.

Before taking up how to go about these severe consequences, utilizing natural air is still much more carbon-free than sealed buildings mainly cooled with air cons [24]. This is practiced in the colder countries of the northern hemisphere, where in some locations carbon-triggering heating is still necessary for a couple of months or in some places almost the whole year long. Hence, most cooling is more expensive than heating, and times will change toward this direction: the hotter the climate gets, and the more people need to adopt to close them up and use air conditions, the more dreadful the emissions. This vicious circle that started long ago with worldwide sales of four AC units per minute [25] states clearly it is high time to talk about viable alternatives, if still possible.

Before shedding light on why alternative tropical passive house of the future will unlikely follow the mandatory global trend standards, Table 2 demonstrates likely differences between its cold hemisphere features and those emerging tropical ones:

Table 2. Summary: 7 Differences of PH1 (any colder hemisphere) compared with Tropical Holistic Passive Technology (PH2).

Element	Sustainable Passive Homes in Moderate Cold Hemisphere (Based on PH1-Benchmarks)	Equatorial Tropical Sustainable Homes (with Some Modifications also Counts for ZEMCH)
(1) Glazing/Façade	Triple Glazing as standard to prevent coldness at a lambda T of up to −30 °C (and colder in some Northern areas)	Ideal Types: A. No Triple Glazing. Double layer coated on window pane 2 to prevent heat transfer + Film (if sunlit) with UPVC-frame. Double layer can also be put onto the frame by any experienced carpenter B. Single coated with UPVC-frame (if fully shaded to avoid convection of 32–36 °C of common outside temperature through the windows panes): but windows do not explicitly appear C. Windows-wall ratio best not exceeding 30%
(2) Insulation	Different layers of up to above 40 cm exterior walls thickness	Lightweight concrete Walls calculated e.g., for 5° latitude sufficient as <22 cm to insulate or best using natural fibre (blow, resilient and construction board). Replacing 10 cm thick conventional walls.
(3) Green Retrofitting	Energy efficient solutions	Cladding, shutter, overhang roof, outside blinds, IAQ
(4) PV	Mono/Poly Crystalloid producing optimum yield	Higher performances by more expensive thin film technology. Parity Grid forecast e.g., for Malaysia in 2026/27 [26].
(5) Solar Thermal	warm water supply, link to PV	Warm water supply just for wealthier households with a certain market penetration compared to "red" water heaters
(6) Heat Gain e.g., by people/personal computer in small size rooms	Functional more or less during all seasons, except during hot summer days	Always dysfunctional, internal CO_2 and VOCs rising
(7) Cooling [27]	During hot summer days. e.g., terracotta tiles in conjunction with maximum nighttime opening. Geothermal—cold water as slab cooling. A few fans against the slightly increasing numbers of A/C	Smart conventional A/C (or inverter?), outside nighttime "flush" ventilation in non-coastal and certain areas in low altitude, rare cooling ceilings. Utilisation of wells with temperatures of 25 °C in 10 m depth around the equator for ceiling or wall cooling – rain or well water or heat exchanger, VRF-ventilation systems

3. Difficulties Any Passive House Concept Faces to Move into Tropical Projects

Passivhaus (which is called PH1) is a common all-climatic zones standard "from Alaska to Ghana". There is an impression, however, that the international passive house as the most energy-saving house of modern civilization does not extensively look into tropical buildings with the increasing 50% of the world population by 2030 or 2040 as mentioned in Figure 2 [28].

Over five years ago, "Passivhaus goes South" looked promising to people who understood the PH concept, some following the strict codex and some deviating as presented below in Table 3. After a brief review "PH goes south" below, a sharing of three reasons that might be responsible for why PH did not (yet) capture the heat of the moment are followed:

The first attempts going south publicly involved reaching out to Singapore as perhaps the most vibrant tropical hub for building technology (2016). Apart from very few but workable applications, the idea to include the tropical belt on a larger scale of economy was initiated and introduced to the tropics at Singapore engineered by a dedicated booth during the BEX Green Building exposition 5–7 September 2016. In December, the IPHA-expert Jürgen Schnieders in charge of tropical expansion of the Passivhaus held a speech before 20 professionals organized by the Singaporean Business Council [29]. Despite speech congratulations of most representatives who showed up instead of their senior managers, the resonance remained nil. In the database it was found even two certified passive house consultants on Singapore island, but professionally, they do not have much to do with what they were trained for. They did *patchwork* passive design, but *integrated* approaches following the entire five Passive House principles were not in the pipeline.

Subsequently, the first representation of the "Tropically adapted Passive House" as attempt to create visibility in the tropics was initiated at the International Green Building Conference Singapore, 6–10 September 2016. Four regional undisclosed Insulation, Shading and Coating partners sponsored the making of an exposition booth, issuing a brochure of the IPHA (the International Passive House Association). This first step is witnessed on the homepage "www.profwagnerkarl.com".

Apart from this self-financed and sponsored representation, the conventional PH1 as a universal approach creates viability and reliability for the construction at the same time. In 2017, an estimated number of 65,000 projects worldwide in every climatic region evolved [30]. It can be perceived as a pity to see that the runaway train called climate catastrophe gets more mileage even though Passive House related smart solution are not applied. Looking at the PH1 global map [31], the tropical region with the highest population now and even more in the future seem to be reduced to those addressed 6–7 implementations, as described depending on how we use the definition of "tropical" [32].

Table 3. Passive Homes—Content Analysis with a Synergetic Comparison of PH1 and PH2.

	Passivhaus (PH1)	PH2 (e.g. Related to ZEMCH)
Standard	Strict standards in terms of the whole building envelope.	So far no standard, instead of "zero energy", zero carbon as target.
Certification	Detailed certification based on Passivhaus software. Many houses do not apply for the costly procedure.	No certification required, justification as "zero carbon" and "mass customized homes".
Airtightness vs. natural ventilation	Airtightness is a must, no compatibility with natural ventilation. Occupants still may open the windows temporarily if they feel to do so.	Airtightness is a logical consequence, if the outside temperature is too high. Otherwise oprning windows, might depend more on the occupants' preferences and/or the seasonal necessity (e.g., hot vs. rainy vs. cool season)

Table 3. *Cont.*

	Passivhaus (PH1)	PH2 (e.g. Related to ZEMCH)
USL-(Upper Space Limit) Temperature and Humidity in tropical countries applications	USL in PHPP-software tied up to maximum 25 °C. Allowed not more than exceptional 10% excess. (>1 °C lower than tropical ASHRAE standard) The highest relative humidity international benchmark 60% cannot be exceeded in the software (absolute humidity < 20% (12 g/kg)	USL depends on user's preferences (adaptive thermal comfort, i.e., especially ambient temperature) [33] Not much effort is put on the relative humidity, because dehumdifying will cost lots of budget.
Implementations	65,000 (estimation according to Passive House publications) [34]	A few, probably less than 10, but of growing interest of ZEMCH "impact".
Material	Full flexibility, the more sustainable, the better. However, there is a preferred list in the PH-course manual.	Very flexible in terms of material selection.
Community	20,000 experts around the globe (according to Passive House officials replicated by the global map)	Board meeting virtually once a month. Around 800 followers in LinkedIn group.
Economic dimension	Payback periods compared with conventional buildings vary a lot between 5 years for low cost and 30 years for more sophisticated projects [35]	Economy/Affordability comes first for the focus of PH2

4. Four Possible Reasons Why the Passive House Concept (PH1) Has Issues When Entering the Tropical Belt

The following reasons might be responsible, and below how to propose a way out using synergies between all mentioned PH1-approaches and ZEMCH will be investigated:

(1) Numerous inquiries by government-linked companies (GLCs) and various local and international knowledge suppliers for "passive design" witness the call to identify the urgency to build in order to target awareness of the benefits compared with conventional housing with any of the PHs. However, holistic passive houses are not just a mixture of the seven or eight passive design elements. Neither are they patented or trademarked; all rules and tools are published, not pursuing or committed to a certain technology. In the colder hemisphere, basically everyone can build a holistic passive house with material and energy of their own choosing, as long as the builder can yield the astonishingly low energy consumption of 20 KW/sqm and an average heat transmission value of U < 0.15 W/m2K. Mainly, or exclusively, a PH planner/consultant with the PHPP software could handle the projects that after being successful in warm Mediterranean climates are rarely happening "further south". The software created is detailed and accurate, but it creates strict norms which also in future might be hard to achieve especially for the vast majority of affordable homes, when the upper interior temperature limit is 25 °C at a relative humidity of 60%. Hence, it is understandable that the conventional PH is restricted to just 6 or 7 tropical certified implementations, whereas the count for other climates adds up to 65,000 in a growing number [36].) Of course, the actual number of tropical installations can vary, depending on where we draw the line between "tropical" and "subtropical", as elaborated by Köppen (1894) and further detailed out by Geiger (1954), which we adapted from ref. [37].

If, by 2050, 50% of the world's population live in the tropical area without the necessity of heating, then extensive research and development needs to be undertaken to integrate the ideas of the promising Passivhaus approach.

(2) It is mandatory for the Passivhaus approach to expect their certified buildings following exactly the same strict worldwide standard. Even their Passive House Planning Package (PHPP) software, which is required to plan and after construction certification, will treat every passive house on the globe the same way what the output is concerned: energy

efficiency plus thermal comfort, today including indoor air quality. Without airtightness and a heat-recovery ventilation system, the conventional Passivhaus does not work. These are two conditions that with a few exceptions [31] PH in practice will hardly work for tropical houses. Despite the death tolls during pollution and recent heat waves in India [38] open air style is still common—not only for the underprivileged, but also for populations who are educated to dwell less upon "natural ventilation" [39]. This is a common buzzword in countries like Singapore or Brazil, failing to recognize that "ventilation" due to the lack of permanent optimum-positioned velocity is extraordinarily rare, and only smart airtight solutions often including the usage of air conditioners are hard to find.

(3) According to the previous research, the well-insulated, basically almost airtight and optimum shaded building with ventilation performs cooler in almost all cases during the rainy season and the increasing number of transition periods [40]. COVID-19 could be a driver to implement more mechanical cross ventilation not only in tropical passive houses. Without aircon and due to the lack of a moisture barrier it remained humid, especially when drawing night air into the building. However, no harm for occupants and the building envelope could be reported over a 3-year pilot run.

Therefore, passive holistic design will work best i.e., energy efficiently in a combination of (a) nighttime active usage of green cooling (i.e., cross ventilation or water-based cooling ceilings). During (b) daytime, among other related modules looked in, with the air conditioner as a substitute to generate lower temperatures, cooling is mainly based upon passive envelope features PLUS shading.

Compared to the almost 20,000 Passive House planners in Europe and the US, according to another global map of the International Passivhaus Institute, so far just about 16 consultants are certified in the tropical belt. Most of them are working as open-minded architects, but it seems difficult to convince clients to venture into complete PHs. The reason might be that airtightness is often seen as incompatible with natural ventilation, which is still the common trend for low-cost buildings. In addition, in the whole Africa (the world's second-largest, second-most populous and herein fastest growing continent accounting for already 1.3 billion people (2020) [41] with eminent problems of threatening climate change, not a single expert or project can be found. One might guess that PHs might not be the correct approach for the underprivileged. Whether a ZEMC-Home in whatever combination with the Passivhaus has the potential, which will be briefly discussed in Section 5.

Different than the PH, ZEMCH explicitly is researching and developing solutions both for the poor and the still neglected tropical area. Regions like the tropics might just be applicable for a few selected projects, and Africa simply does not provide a market place for Passive House developers who again use insulation and airtightness as necessary pillars of their concept.

(4) According to the detailed description of the global map, the Passivhaus approach focuses more on individual housing rather than on standard housing. Nonetheless, huge developments like the Bahnstadt [42] or the biggest upcoming Passivhaus settlement in China [43] are still exceptional and no mass product like former socialist residential high rises and plenty of homes built thereafter. Of course, "standardized" means saving costs as much as possible and economically cheap. Nevertheless, this label has not automatically to do with sustainability. ENDLESS unsustainable condominiums around the Globe with mid-class and high-class occupants complete the list, probably with a growing number of a few exceptions.

Apart from the costs for the less privileged, it is hard to convince simple people to follow their rules of insulation, airtightness, and mechanical ventilation to name three of the PHs better known original big five PH-elements elaborated on in Figure 1 [44].

5. Conclusions: Synergies of Passive House 1 and 2 to Make the Concept of Sustainable Homes a Reality

Our last chapter of this tabled comparative content analysis is no longer directly on the tropical adapted Passive House (PH2), which is nowhere practiced. Instead, the questions of if and how PH2 can support the implementation of more airtight well insulated Passive Houses especially in the already or soon to be overpopulated regions of tropical countries will be elaborated on. In return, how can passive house technology assist to make more affordable PH2-homes to become a reality of the hotter future to create real potential airtight and well insulated living boxes for the hot years to come [45]? In the end, PH1 and PH2 are no competitors. Scalable carbon saving is the important target point of future homes, not the intellectual origin of its developer.

In between other approaches to reduce global warming in the built environment (like the mentioned low energy houses and net zero energy buildings), finally it would be liked to point out similarities and possible differences between Passive Houses and PH2 homes. There is no automatic strong link between "passive" and "affordable homes". As a reference model, I would like to refer to the modular Living Box Home created by Antonio Frattari in his ZEMCH project (2013) [46]. Built as mobile application, it utilises the airtight Passive House technology and other adapted features.

Set up first in Bolzano, Italy, the house indicates that the investment costs of a ZEMCH-home equipped with PV (photovoltaics) are almost equal to a conventional one. Hence, if the (usual carbon-triggering heating or cooling demands) are engineered with PV, the house has very low operational costs making questions of payback superfluous. This selling argument counts already for at least a Northern part of a Mediterranean country like Italy, but as shown above, so far it has been hardly demonstrated farther south than Europe.

Summing up, in Table 3 are the main differences as they seem to be laid out in the explicit Passivhaus approach (that called PH1) and the lenient, but more consumer-oriented PH2 had been devised in this entry.

Even though the focus as laid out in Tables 1–3 is slightly different between PH1 and PH2, plenty of potential synergies can be found. PH2's openness without certification and binding software tools and the will to go for "zero carbon homes" could pose an alternative. It is explicitly meant for "affordable" and specific for "homes", regardless if they are found in Minneapolis, Bangalore, Lagos or Bhutan. It could also occur that a PH2-home can be purposely built to be certified before and after its construction.

Reducing CO_2 in Passivhaus-adapted affordable tropical Homes must be realistic for the supply of sustainable homes for millions of people. Therefore, 4. and 5. might be attractive. So far, in terms of organization and applications the worldwide accredited PH1 is the giant, whereas PH2 appears like the little dwarf doing mainly academic groundwork. However, PH2 moves into the area of impact which is covered by PH1 just by their standards. Additionally, as mentioned in Table 3, PH2 has much higher flexibilities.

The bottom line of the tropically adapted Passivhaus PH2 that needs to be discussed further is the mid-term 2030 real and potential societal impact in the wake of the 2050 tropical heating effects that are to be expected. By then, PH2 and related housings could be a kind of a lifeline for homes that otherwise would have to be abandoned causing millions of people to migrate into still all seasons long habitable regions away from the tropical corridor.

Under the impact of the recent IPCC-reports, the idea of producing sustainable buildings which look for environmental solutions to reduce the residential buildings' carbon footprint [47] PLUS decent indoor air quality is getting more common. Green certification tools have adopted sustainability for the planet and people's health. However, as their standards were predominantly developed by architects and civil engineers, apart from (a) carbon footprint reduction and (b) better and cooler indoor air quality, they have not focused on (c) economic needs as the third dimension of a triple sustainable home. Once embarking on green, it would increase the viability of the tool as it shows the user how to bring down their current cost structures. The target defined by ZEMCH adapting to matching PH-elements as far as possible will be to find and convince developers and then buyers primarily on the mass market [48]. In so doing, the tabled commonalities and differences elaborated on and presented above can help as a guideline.

Funding: This research received no external funding.

Institutional Review Board Statement: Not applicable.

Informed Consent Statement: Not applicable.

Data Availability Statement: This entry did not report any own quantifiable data, because it used the devised ideal-typological method that is derived from the Schools of phenomenology, hermeneutics and pragmatism.

Conflicts of Interest: The author declares no conflict of interest.

Nomenclature/Glossary

A/C	Air Condition
GLC	Government-linked Company
IPCC	Intergovernmental Panel on Climate Change
IPHA	International Passive House Association
NZEB	nearly zero-emission building
PH	Passivhaus
PHPP	Passivhaus Planning Project (Software)
PV	Photovoltaics
USL	Upper Space Limit (statistical upper borderline)
VRF	Variable Refrigerant Flow
ZEMCH	Zero Energy Mass Custom Homes

References

1. Berkeley Lab. Distinguished Lecturer Series: Building Science—Adventures in Building Science. 2018. Available online: https://www.youtube.com/watch?v=rkfAcWpOYAA (accessed on 10 October 2022).
2. Sociology Group. How to Do Comparative Analysis in Research (Examples). Available online: https://www.sociologygroup.com/comparative-analysis/ (accessed on 10 October 2022).
3. Smith, D.W. Phenomenology. In *The Stanford Encyclopedia of Philosophy*; Edward, N.Z. Ed.; Stanford, CA, USA. 2013. Available online: https://plato.stanford.edu/entries/phenomenology/ (accessed on 10 October 2022).
4. Cambridge University. Phenomenology. Available online: https://dictionary.cambridge.org/dictionary/english/phenomenology (accessed on 10 October 2022).
5. Genaus, H. Passive Design in tropical Zones. 2022. Available online: https://www.housingforhealth.com/housing-guide/passive-design-in-tropical-zones/#:~{}:text=The%20most%20important%20passive%20design,ventilation%20and%20convective%2 (accessed on 23 October 2022).
6. The Turf House Tradition in Iceland. Available online: https://whc.unesco.org/en/tentativelists/5589/ (accessed on 10 October 2022).
7. Cleaning up with Michael Liebreich. Ep68: Amory Lovins 'The Einstein of Energy Efficiency'. Available online: https://www.youtube.com/watch?v=0BtpbmDBGFQ (accessed on 10 October 2022).
8. IEA Statistics © OECD/IEA 2014. Available online: https://www.iea.org/t&c/termsandconditions/ (accessed on 10 October 2022).
9. Exploring Alternatives. Passive House = 90% Home Energy Reduction. Available online: https://www.youtube.com/watch?v=Hz6qomFM_dw (accessed on 10 October 2021).
10. E-Interview based on 53 pages of intellectual exchange of the author with Wolfgang Feist, Founder of Passivhaus Institute, 2021. Unpublished work.
11. Jedamzik, M. Was ist ein Passivhaus? 2021. Available online: https://www.co2online.de/modernisieren-und-bauen/sanierung-modernisierung/passivhaus/#c99185 (accessed on 10 October 2022).
12. Hopfe, C.J. *The Passivhaus Designer's Manual: A Technical Guide to Low and Zero Energy Buildings*; Hopfe, C.J., McLeod, R., Eds.; Routledge: London, UK, 2015.
13. Walshaw, E. *Understanding Passivhaus: A Simple Guide to Passivhaus Detailing and Design*; First In Architecture: London, UK, 2020.
14. Cotterell, J.; Dadeby, A. *The Passivhaus Handbook: A Practical Guide to Constructing and Retrofitting Buildings for Ultra-Low Energy Performance*; Green Books: Cambridge, UK, 2012.
15. Wagner, K. Adaption of a Tropical Passive House as holistic Approach. In Proceedings of the ZEMCH (Zero Energy Mass Customised Homes) 8th International Conference, Dubai, United Arab Emirates, 26–28 October 2021.
16. Wagner, K. Prior to a most likely controversial discussion between pragmatism and deductionism in follow up articles, findings on (de-)humification appears in no.6 of table 1, 2022. Unpublished work.

17. IPHA. Five Basic Principles Apply for the Construction of Passive Houses. Available online: https://passiv.de/en/02_informations/02_passive-house-requirements/02_passive-house-requirements.htm (accessed on 25 October 2022).

18. Lstiburek, J. *Moisture Control for New Residential Buildings*; Buildingscience.com Corporation: Westford, MA, USA, 2009; Volume 12, pp. 1–49.

19. United States Environmental Protection Agency. Reduce Urban Heat Island Effect. 2022. Available online: https://www.epa.gov/green-infrastructure/reduce-urban-heat-island-effect (accessed on 16 October 2022).

20. Ederer, C. (Technical Director of EBM Singapore) in a interview conducted by the author, 2017. Unpublished work.

21. Kottek, J.; Grieser, J.; Beck, C.; Rudolf, B.; Rubel, F. World Map of the Köppen-Geiger Climate Classification Updated. 2006. Available online: https://www.klimadiagramme.de/Frame/koeppen.html (accessed on 10 October 2022).

22. Mora, C.; Dousset, B.; Caldwell, I.R.; Powell, F.E.; Geronimo, R.C.; Bielecki, C.R.; Counsell, C.W.W.; Dietrich, B.S.; Johnston, E.T.; Louis, L.V.; et al. Global risk of deadly Heat. *Nat. Clim. Chang.* **2017**, *7*, 501–506. Available online: https://www.nature.com/articles/nclimate3322 (accessed on 16 October 2022). [CrossRef]

23. Terra X Lesch & Co. Klimakrise—Zeit zu kapitulieren? (Climatic Crisis–Time to capitulate)? 2021. Available online: https://www.youtube.com/watch?v=s2txunrkr8M (accessed on 16 October 2022).

24. Wong, N.H.; Tan, E.; Adelia, A.S. *Utilization of Natural Ventilation for Hot and Humid Singapore*; Enteria, N., Awbi, H., Santamouris, M., Eds.; Building in Hot and Humid Regions; Springer: Singapore. [CrossRef]

25. Tan, T.C. Managing Director Built Environment Research and Innovation Institute, BCA. Welcome and opening address–Transforming Cities in Hot and Humid Climates towards more Efficient and Sustainable Energy Use. Singapore, 2017. Unpublished work.

26. Millot, P. Interview on PV; Kuala Lumpur, Malaysia, 2013. Unpublished work.

27. Givoni, B. *Passive and Low Energy Cooling of Buildings*; Energy and Buildings; Van Nostrand Reinhold: New York, NY, USA, 1994; Volume 29, pp. 141–154. Available online: https://www.wiley.com/en-us/Passive+Low+Energy+Cooling+of+Buildings-p-9780471284734 (accessed on 23 October 2022).

28. WGBC (World Green Building Council). Global Status Report 2017. Available online: https://www.worldgbc.org/news-media/WorldGBC-embodied-carbon-report-published (accessed on 10 October 2022).

29. Schnieders, J. Passivhaus Moving Southward. Presentation in Singapore on a Fact Finding Mission. 16/12/2016, 2016. Unpublished work.

30. Passivhaus-Datenbank. Available online: https://passivehouse-database.org/#d_6030 (accessed on 10 October 2022).

31. International Passive House Association. Passive Houses Open Days-Passive House Production Facility in Colombo, Sri Lanka. Available online: https://www.youtube.com/watch?v=7ng1TPjN4_M (accessed on 10 October 2022).

32. International Passive House Association. Map of Certified Passive House Buildings. Available online: https://www.passivehouse-international.org/index.php?page_id=288 (accessed on 10 October 2022).

33. IPHA, Passivhaus Basics. Chapter 1: What Is a Passive House? Available online: https://passipedia.org/basics/what_is_a_passive_house (accessed on 17 October 2022).

34. Peper, S. Schrittweise Modernisierung: Luftdichtheitskonzept. 2015. Available online: https://passipedia.de/planung/sanierung_mit_passivhaus_komponenten/luftdichheit/schrittweise_modernisierung_lueftung (accessed on 17 October 2022).

35. Gugliermetti, F.; Roselli, R. Italian Research on Eco-Efficient Housing Modules. In Proceedings of the Sustainable City 2022, Rome, Italy, 10–12 October 2022. [CrossRef]

36. De Dear, R.; Brager, G. Developing an Adaptive Model of Thermal Comfort and Preference. *ASHRAE Trans.* **1998**, *104*, 145–167.

37. Mohabuth, Y. World's biggest Passivhaus under-construction in Germany. Sustainable Development. 2016. Available online: http://www.inspiraction.news/en/2016/09/15/worlds-biggest-passivhaus-under-construction-in-germany/ (accessed on 15 October 2022).

38. Avastthi, B. Green Building Cost Analysis for Existing Buildings. 2013. Available online: https://www.green.modeling.com/sustainability/green-building-cost-analysis-for-existing-buildings.html (accessed on 15 October 2022).

39. Passive House Institute. World map of certified Buildings. Available online: https://passivehouse.com/03_certification/02_certification_buildings/02_certification_buildings.htm (accessed on 10 October 2022).

40. Geiger, R. Based on Köppen's Classification into "Tropical Rainforest Climate, Tropical Monsoon Climate and Tropical Savanna Climate. In *Klassifikation der Klimate nach W. Köppen [Classification of Climates according to W. Köppen]*; Lan-dolt-Börnstein—Zahlenwerte und Funktionen aus Physik, Chemie, Astronomie, Geophysik und Technik, alte Serie; Springer: Berlin/Heidelberg, Germany, 1954; Volume 3, pp. 603–607.

41. World Meteorological Organization. Climate Change Made Heatwaves in India and Pakistan "30 Times More Likely". Available online: https://public.wmo.int/en/media/news/climate-change-made-heatwaves-india-and-pakistan-30-times-more-likely (accessed on 10 October 2022).

42. So recently the Singapore Plan 2030 "Use Fan Instead of Air-Con". Available online: https://www.greenplan.gov.sg/take-action/as-individual (accessed on 10 October 2022).

43. Wagner, K. *Tropically Adapted Green and Energy Efficient Residential Building: A Universal Trial Based on Holistic Passive Technology*; Chapter 3 Methodology; Technical University Rosenheim: Rosenheim, Germany, 2020. Available online: http://www.profwagnerkarl/ebook (accessed on 17 October 2022).

44. Statistica. Forecast of the Total Population of Africa from 2020 to 2050. 2022. Available online: https://www.statista.com/statistics/1224205/forecast-of-the-total-population-of-africa/# (accessed on 17 October 2022).

45. Heidelberg at Bahnstadt. Available online: https://www.heidelberg-bahnstadt.de/ (accessed on 10 October 2022).
46. Passive House Institute. Energy Efficiency Highly Popular in China. 2018. Available online: https://passivehouse-international.org/upload/20180310_Pressemitteilung_Tagung_2019_China_EN.pdf (accessed on 17 October 2022).
47. Chastas, P.; Theodosiou, T.; Kontoleon, K.J.; Bikas, D. Normalising and assessing carbon emissions in the building sector: A review on the embodied CO_2 emissions of residential buildings. *Build. Environ.* **2018**, *130*, 212–226. [CrossRef]
48. PASSIPEDIA. Are Passive Houses Cost-Effective? Available online: https://passipedia.org/basics/affordability/investing_in_energy_efficiency/are_passive_houses_cost-effective (accessed on 3 November 2022).

MDPI
St. Alban-Anlage 66
4052 Basel
Switzerland
Tel. +41 61 683 77 34
Fax +41 61 302 89 18
www.mdpi.com

Encyclopedia Editorial Office
E-mail: encyclopedia@mdpi.com
www.mdpi.com/journal/encyclopedia